INTERNATIONAL CENTRE FOR MECHANICAL SCIENCES

COURSES AND LECTURES - No. 318

MODELLING MACROSCOPIC PHENOMENA AT LIQUID BOUNDARIES

EDITED BY

W. KOSINSKI
IPPT-PAN, WARSAW

A.I. MURDOCH
UNIVERSITY OF STRATHCLYDE

SPRINGER-VERLAG WIEN GMBH

Le spese di stampa di questo volume sono in parte coperte da
contributi del Consiglio Nazionale delle Ricerche.

This volume contains 53 illustrations.

Originally published by Springer Verlag Wien-New York in 1991

In order to make this volume available as economically and as
rapidly as possible the authors' typescripts have been
reproduced in their original forms. This method unfortunately
has its typographical limitations but it is hoped that they in no
way distract the reader.

ISBN 978-3-211-82327-9 ISBN 978-3-7091-2612-7 (eBook)
DOI 10.1007/978-3-7091-2612-7

PREFACE

The contents of this volume were delivered as lectures at an Advanced School held at CISM from June 25 - 29, 1990, which addressed "Modelling Macroscopic Phenomena at Liquid Boundaries". The programme was designed to bring together lecturers who approach the modelling of such phenomena from a variety of viewpoints, with the intention of providing participants with a broad framework for future researches.

The School aimed to acquaint the audience with the variety of macroscopic behaviour which can occur at liquid boundaries, to indicate various approaches to relevant continuum descriptions, to demonstrate applications of continuum models to the solution of practical problems, and to convey due appreciation of experimental aspects of the subject.

In the event the School proved to be most friutful and stimulating for all concerned (not least the lecturers themselves), due to an active audience participation which reflected a diversity of viewpoints and approaches. It is to be hoped that some of the spirit of academic honesty which characterised this School will be communicated by these lecture notes.

The co-ordinators would like to express their appreciation and thanks to CISM for its support and organization of the School, and to the Publishers for enabling these lectures to be made available to a wider audience.

W. Kosinski

A. I. Murdoch

CONTENTS

INTRODUCTION

The macroscopic behaviour of material systems in which two bulk phases are in contact can only fully be understood if their mutual boundary is invested with distinct physical attributes. Such boundaries are actually thin interfacial *regions* across which bulk material quantities may undergo great change. For example, liquid-vapour interfaces are typically only $10°A (= 10^{-9}\text{m})$ thick yet support density changes from 1 to 10^3 kg m^{-3}. In the case of a fluid-fluid interface between bulk phases which consist of several molecular species, the relative concentrations of individual species within the interface usually differ from those in either bulk phase (a phenomenon termed *adsorption*). Indeed, very small bulk phase concentrations of certain species may in equilibrium be accompanied by high interfacial concentrations of these species. Such so-called *surfactants* are responsible for much of the notorious difficulty of obtaining reproducible experimental data for interfaces.

The foregoing suggests that continuum modelling of fluid interfaces may be a matter of some subtlety. In fact, equilibrium (static) behaviour is relatively straightforward to describe: a satisfactory theory has been available for over one hundred years. However, dynamic behaviour poses problems of observation, measurement, and theory which are still to be resolved. The contents of this volume represent some approaches to these problems.

The mechanical behaviour of liquid-vapour and liquid-liquid interfaces in macroscropic equilibrium is well-described in terms of *surface tension*, a concept formalised independently by Laplace and Young some two hundred years ago. This corresponds to modelling an interface as a massless membrane of zero thickness in a state of uniform tension and acted upon by pressures from the two contiguous bulk phases. By 1878 Gibbs had established a satisfactory theory of the thermostatics of multicomponent interfaces in terms of the notions of *dividing surfaces* and associated *excess* quantities. A dividing surface S is a surface chosen to lie within the interfacial region and may be regarded as approximating the location of this region. Bulk phases are considered to be homogeneous on either side of S, and discrepancies between such a description and the actual physical situation are treated in terms of fields ascribed to S and measured per unit area of S. Such fields are termed surface excess quantities corresponding to the choice S. In particular, surface tension is a surface excess quantity: for example, the surface tension σ of a planar interface of thickness h which is subject to a bulk pressure p from either side is given by $\sigma = \sigma' + ph$, where σ' is the *intrinsic* surface tension which derives entirely from matter in the interface.[1]

Non-equilibrium situations are more difficult to model. Complications arise due to the possibility of net mass exchange between an interface and its contiguous bulk phases. In particular, interfacial thickness depends upon temperature, net evaporation may occur, and surfactant material may be in the process of migrating to the interface. Further, mass transport *within* an interface is possible, driven by variations in the (intrinsic) surface tension which may arise as a consequence of non-uniformity either of interfacial temperature or surfactant concentration. Such interfacial flows can induce observable *bulk* (so-called *Marangoni*) flows.

In the articles to follow, interfacial behaviour is discussed in terms of bidimensional considerations. That is, the interfacial region is modelled, at any given instant, in terms of a (geometrical) surface to which are ascribed fields which are related to the state of interfacial matter.

Kosiński and Murdoch discuss the motivation of balance relations which govern the time evolution of mass, momentum, moment of momentum, and energy for matter within the interfacial region, taking due account of the effect of the adjacent bulk phases and three-dimensional considerations. While their results are directly comparable, the approaches adopted by these authors are entirely different. Kosiński models the interface as a thin three-dimensional continuum 'sandwiched' between two boundaries across which net mass exchange is possible. A model surface is located between these boundaries, and balance relations for the interface (regarded as a bidimensional material system) are obtained by, roughly speaking, integrating three-dimensional balance relations along directions normal to this model surface, between the interfacial boundaries. In the same manner, a statement of the second law of thermodynamics is derived from an entropy production inequality for bulk phases. Murdoch adopts a microscopic viewpoint in which molecules are modelled as interacting point masses. Mass density and velocity fields are introduced in terms of local molecular averages computed jointly in space and time. Delineation of interfacial boundaries and selection of a model surface are effected in terms of density considerations. Balance relations are obtained modulo very weak assumptions about the nature of molecular interactions and the random nature of thermal motions of molecules. All fields are identified with space-time averages of molecular quantities.

Bedeaux' viewpoint is that of Statistical Physics, based upon the formulation of nonequilibrium thermodynamics described by De Groot and Mazur. Here the balance

[1] If $ph \ll \sigma$ then σ is clearly a good approximation to σ'. For a water-air interface at NTP, $h \sim 10^{-9}$ m, $p \sim 10^4$ Nm^{-2}, and $\sigma \sim 7 \times 10^{-2}$ Nm^{-1} so $ph/\sigma \sim 10^{-4}$.

relations involve excess quantities. The (Onsager) relations which follow from an explicit expression for the interfacial entropy production are discussed. Stochastic considerations associated with macroscopic equilibrium are explored in the context of thermal fluctuations. This approach reveals an interface as an entity whose location and profile exhibit random features, essentially identifiable with capillary waves stochastically self-excited by interfacial molecular motions. The viewpoint of Murdoch corresponds, roughly speaking, to a time-averaged version of this capillary wave description.

The remaining articles concern the application of theories of interfacial behaviour to specific phenomena.

Hutter addresses the description of ice-water phase change processes. A bidimensional model of an ice-water phase interface is related to three-dimensional considerations in much the same manner as that adopted by Kosiński: this furnishes insight into physical interpretations of surface fields and deeper understanding of terms associated with mass transport between the interface and the bulk phases. Here, however, excess surface quantities are utilised: bulk fields are extrapolated into the interfacial region up to the model surface (in a non-constant fashion, unlike the Gibbs' equilibrium approach) and intrinsic surface fields are modified in an appropriate way. New theoretical results concerning freezing and melting at curved interfaces are among those presented. All results are shown to accord well with experimental observation.

Situations in which two contiguous fluids are in contact with a relatively rigid solid are considered by Slattery. The surface S_1 modelling the fluid-fluid interfacial region intersects the solid bounding surface S_2 in a curve termed a *common line*. The *contact angle* between S_1 and S_2, as casually perceived, is shown often to be misleading, due to the presence of a thin film of one fluid in contact with S_2. In dynamic situations the notions of advancing and receding common lines (together with corresponding contact angles) are examined, and continuum modelling shown to be useful. This article carefully presents the experimental data currently available and discusses its physical interpretation, prior to developing a theoretical framework in terms of which a deeper understanding of the phenomena may be obtained.

Lectures were due to have been presented at the School by L.G. Napolitano on surface and line phases. The latter model contact lines as one-dimensional continua in their own right. Unfortunately, Professor Napolitano was unable to attend and instead one of his co-workers, C. Golia, lectured comprehensively on Marangoni flows. Both theoretical and numerical studies were presented. Readers interested in these aspects of fluid interfacial modelling and phenomena should contact these researchers at the Istituto di Aerodinamica "Umberto Nobile", Universitá di Napoli, Piazzale V. Tecchio, 80, 80125

Naples, Italy, for further details.

The Editors hope the contents of this volume will be of interest both to established workers in interfacial science and to those who wish to gain some familiarity with interfacial phenomena. The references cited in the articles herein provide ample testimony to the manifold approaches adopted by researchers from many disciplines. Here an attempt has been made to inter-relate several viewpoints, in the belief that this results in a deeper appreciation of the subject for all concerned.

ON THE PHYSICAL INTERPRETATION OF FLUID INTERFACIAL CONCEPTS AND FIELDS

A. I. Murdoch
University of Strathclyde, Glasgow, UK

Abstract. A fluid-fluid interfacial region is in general very thin. In such case its macroscopic behaviour is well-modelled in terms of a bidimensional continuum which can exchange matter with its contiguous bulk phases. In particular, balance relations for mass, linear and rotational momentum, and energy can be ascribed to a surface which models the instantaneous location of the region. Delineation of the interfacial region, selection of a model surface and appropriate velocity fields, and motivation of the appropriate balance relations, are here effected on the basis of molecular considerations. A further balance relation (redundant for bulk continua), that of *moment* of mass, is also developed. This furnishes a relationship between the geometrical ('normal') velocity of the model surface, the velocity field associated with interfacial matter, and mass transport *across* the interface. All fields are identified with local averages of molecular quantities in both space and time. The analysis is based upon very weak assumptions concerning the nature of molecular interactions and the random character of molecular thermal motions. Such assumptions are made explicitly and their physical basis is discussed. The notation employed is coordinate-free: relevant explanation is summarised in an Appendix.

1. Introduction

Fluid interfacial regions are in general extremely thin and characterised by great bulk density changes. For example, liquid-vapour interfacial thicknesses are typically of order 10 Å$(= 10^{-9}m)$ away from near-critical temperatures (cf. Rusanov (1971), Brown (1974), Rowlinson & Widom (1982)) and liquid:vapour density ratios are of order $10^3 : 1$. For ("immiscible") liquid-liquid interfaces the situation is similar (in such cases each bulk density vanishes on one side of the interface). *Mechanical phenomena* associated with fluid interfacial regions *in macroscopic equilibrium* are well-described in terms of a *surface of tension* which acts as if it were a massless membrane in a state of uniform tension (cf. Young (1805), Adam (1941), Adamson (1960), Davies & Rideal (1961), Bikerman (1970)). A satisfactory description of thermostatic behaviour of *multicomponent interfaces* was established by Gibbs (1876, 1878). However, non-equilibrium situations are somewhat complex. Any comprehensive theory must accommodate the possibilities of transport phenomena both *within* and *across* the interface: bulk motions may be induced by interfacial inhomogeneity (cf. Levich & Krylov (1969)) and evaporation can occur.

Statistical mechanical considerations have provided extensions to Gibbs' approach and a basis for the understanding of interfacial molecular behaviour (cf. Buff (1960), Defay, Prigogine & Bellemans (1966), Rowlinson & Widom (1982), Bedeaux (1986)). These will be addressed in Professor Bedeaux' lectures. An alternative approach is to model interfacial regions as bidimensional continua and to *postulate* appropriate balance and constitutive relations in the manner of contemporary continuum mechanics (cf. Truesdell & Noll (1965)). It is this latter viewpoint with which we will here be concerned.

A dynamical theory of bidimensional fluids, in the absence of mass exchange with contiguous bulk phases, was presented by Scriven (1960) and extended to take account of such mass transfer by Slattery (1964). Thermodynamic balance relations for general bidimensional continua which include the effects of mass, momentum, and energy exchange with adjoining bulk continua have been postulated by Moeckel (1975) and Napolitano (1979); both developed complete theories, the former only in the event that interfacial mass is conserved (in which case the interface is described as *material*). Gurtin & Murdoch (1975) proposed a mechanical theory of elastic material interfaces, extended by Murdoch (1976) to a thermodynamical analysis.

In order to elucidate assumptions implicit in the bidimensional continuum approach, to understand more fully the physical interpretations of quantities associated therewith, and also to examine apparent inconsistencies between some of the aforementioned works, it is desirable at least to motivate (if not derive) the balance relations postulated in these works on the basis of alternative theories. Since any interfacial region is really three-dimensional, albeit usually very thin, it might be thought reasonable to regard this as a three-dimensional continuum in which macroscopic physical quantities (such as mass density) exhibit pronounced spatial variation in directions transverse to notional interfacial bounding surfaces. (This viewpoint is apparently supported by the substantial work on interfaces undertaken within the framework of statistical mechanics. See in particular the discussions of density and transverse stress component profiles in Rusanov (1971) and March & Tosi (1976)). From such a standpoint a bidimensional theory may be derived in precisely the same manner in which so-called direct theories of thin shells may be motivated from complete three-dimensional continuum descriptions (cf., e.g., Naghdi (1972)). This viewpoint has been adopted in the present context by Eliassen & Scriven (1963) Dell'Isola & Romano (1987) and Kosiński & Romano (1990) and, roughly speaking, involves the integration of three-dimensional continuum quantities across the interface.

Professor Kosiński's lectures will carefully develop this approach. In this respect we note that the physical interpretation of bulk fields ascribed to interfacial regions requires some care, since these regions are typically only a few molecular diameters thick. Bulk fields in continuum mechanics represent *gross* molecular behaviour and are usually regarded to be fluctuation-free: that is, their values are not considered to change significantly over time intervals of the scale associated with microscopic (molecular) behaviour. Accordingly, for example, volume-continuous fields (such as mass density) cannot be interpreted (within the framework of conventional continuum mechanics) as spatial averages of molecular quantities over *arbitrarily-small* regions. Evidence of a fluctuating physical quantity and an associated length scale is provided by the erratic ("Brownian") motion of a small body immersed in a liquid (cf. Reif (1965), p.251, Heer (1972), pp.417-9, Gopal (1974), p.99 and, for an historical perspective, Brush (1968)). Such behaviour can be observed in bodies of characteristic dimension as large as $10^{-6}m$ and is the result of fluctuations in the net force exerted upon these bodies by molecules of the liquid. Only stochastic modelling can mimic such behaviour. The foregoing considerations indicate that in adopting a continuum approach *within* interfacial regions attention should be paid to the viewpoint of statistical mechanics. From such standpoint balance relations may have the same *form* as those of continuum mechanics but involve fields which are stochastic in nature. In order to obtain the usual continuum interpretations of such fields, averages of the stochastic quantities both in space and time need to be implemented. (Cf. Irving & Kirkwood (1950), particularly the discussion at the end of §III.) Expressions for mean square fluctuations of some stochastic quantities about the corresponding average values associated with small regions are well-known (cf. Landau & Lifschitz (1980), §112) and depend upon the volumes of such regions. When these regions have at least one characteristic dimension of the order of interfacial thickness (in order to yield average fields which locate the interfacial region with an accuracy commensurate with its thickness, for example by mass density considerations: cf. Rusanov (1971)) the root-mean-square (r.m.s.) fluctuations in some quantities will be non-negligible. If even smaller regions are chosen, to obtain detailed information about the variation of quantities *within* the interface, the corresponding r.m.s. fluctuations may dominate the average values.

In order to avoid the complex interpretations seen above to be implicit in the three-dimensional field-theoretic approach to interfacial regions, a different viewpoint is here adopted. Starting from molecular considerations the approach is based upon the methodology developed in Murdoch (1983, 1985). In these works the basic concepts and balance relations associated with three-dimensional continua were shown to follow in a systematic manner from a small number of physically well-motivated assumptions concerning interactions between point masses (these model atoms, ions, or molecules) and the nature of corpuscular motions in both fluid and solid phases. The paucity of molecules within interfacial regions here requires that rather careful averaging procedures be implemented. In particular it proves necessary to interpret *all* continuum fields as local averages jointly in space and time.

There are essentially two problems to be considered here:

1. delineation of interfacial boundaries and the time evolution of the model surface to which all interfacial quantities are to be referred, and

2. determination of general balance relations, which represent interfacial behaviour, to be satisfied on this model surface.

Throughout emphasis will be placed on the physical interpretation of concepts and quantities. Constitutive relations and the posing of appropriate boundary/initial-value problems will *not* be addressed, nor will modification of balance relations caused by localisation of interfacial behaviour on the model surface be considered (such modification involves the notion of *excess* quantities). As few assumptions as possible will be made and the physical basis for these discussed. Accordingly the balance relations to be obtained will be very general, and in many situations not all terms will be relevant. However, it is to be hoped that what follows will aid the formulation of theories relevant to specific phenomena by providing a systematic approach to fluid interfacial modelling.

The balance relations here developed are directly comparable with those derived by Professor Kosiński. The fields in these relations are to be regarded as subject to negligible fluctuation: that is, the viewpoint is that of conventional continuum mechanics. This description of interfaces may be viewed, roughly speaking, as a time-averaged version of the capillary wave (stochastic) approach to be discussed by Professor Bedeaux.

2. Delineation of Interfacial Boundaries

2.1 Preamble

Interfacial boundaries will be identified with singular behaviour of bulk-phase densities. Accordingly density and velocity fields for bulk phases are first considered, together with the concepts of motion and "material point". In order to delineate the interfacial region with an accuracy commensurate with its thickness it proves necessary to interpret bulk density near the interface as a special type of space-time average.

2.2 Bulk Phase Considerations

Any identifiable (instant by instant) set of atoms, ions or molecules constitutes a *material system*, [1]$\mathcal{M}$. A continuum (equivalently, macroscopic) description of $\mathcal{M}$ requires that *velocity* $\mathbf{v}$ and *mass density* ρ fields should exist as local averages associated with these fundamental discrete entities. Formally, the values of $\mathbf{v}$ and ρ at a geometrical point $\mathbf{x}$ (at any given instant t, suppressed for brevity) are given by (cf. Murdoch, (1985), p.294)

$$\mathbf{v}(\mathbf{x}) := lim_\epsilon \Big\{ \sum_i{}' m_i \mathbf{v}_i / \sum_i{}' m_i \Big\} \tag{2.1}$$

and

$$\rho(\mathbf{x}) := lim_\epsilon \Big\{ \sum_i{}' m_i / V_\epsilon \Big\}. \tag{2.2}$$

Here the mass and velocity associated with a fundamental discrete entity P_i (modelled as a *point mass* and subsequently termed *particle*) are denoted by m_i and $\mathbf{v}_i$, and the sums taken over particles within "ϵ-cells centred at $\mathbf{x}$" (this is always indicated by a superposed prime attached to the summation sign).

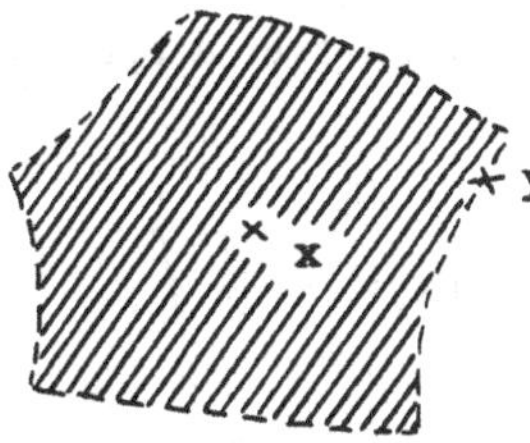

Fig 1: a typical ε-cell centred at x

Roughly speaking, an ϵ-cell centred at $\mathbf{x}$ is a simply-connected region with centroid at $\mathbf{x}$, whose boundary is piece-wise smooth (with not too many component subsurfaces and curvatures not large) and such that any boundary point $\mathbf{y}$ satisfies $|\mathbf{y} - \mathbf{x}| \sim 0(\epsilon)$; for definiteness, say

$$\epsilon/2 < |\mathbf{y} - \mathbf{x}| < 3\epsilon/2.$$

[1] It should be noted that for a given material system the set of atoms, ions or molecules may change with time. For example, the set of ions of a specific type within a reacting mixture contained in a closed vessel, or the set of $H_2 0$ molecules in the liquid phase in a glass of water.

In relations (2.1,2), V_ϵ denotes the volume of the cell and the *pseudo-limit* notation "lim_ϵ" is intended to indicate that the ratio it qualifies varies negligibly at a length scale ϵ microscopically large yet macroscopically small.[2] The mass and linear momentum associated with a *macroscopic region* R (that is, a region which can be decomposed into very many ϵ-cells of the scale which yields ρ and $\mathbf{v}$ are $\sum_{P_i \in R} m_i$ and $\sum_{P_i \in R} m_i \mathbf{v}_i$, respectively. If R is partitioned into ϵ-cells then

$$\sum_{P_i \in R} m_i = \sum_{cells} \Big(\sum_i {}' m_i / V_\epsilon \Big) V_\epsilon = \sum_{cells} \rho(\mathbf{x}) V_\epsilon .$$

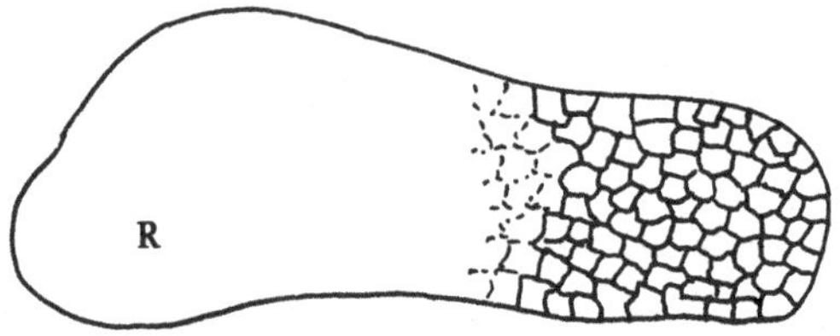

Fig. 2: a partition of R into ε-cells

Here $\sum_i {}' m_i / V_\epsilon$ denotes the sum taken over a typical cell (centre $\mathbf{x}$) divided by its volume. We make the identification

$$\sum_{cells} \rho(\mathbf{x}) V_\epsilon \longleftrightarrow \int_R \rho, \tag{2.3}$$

where a Riemann integral is intended. Similarly, the momentum in R is

$$\sum_{P_i \in R} m_i \mathbf{v}_i = \sum_{cells} \Big(\sum_i {}' m_i \mathbf{v}_i / \sum_i {}' m_i \Big) \Big(\sum_i {}' m_i / V_\epsilon \Big) V_\epsilon = \sum_{cells} \mathbf{v}(\mathbf{x}) \rho(\mathbf{x}) V_\epsilon .$$

Of course, we make the identification

$$\sum_{cells} \rho(\mathbf{x}) \mathbf{v}(\mathbf{x}) \longleftrightarrow \int_R \rho \mathbf{v}. \tag{2.4}$$

[2] Since the motion of a Brownian particle (cf. Introduction) cannot be understood on the basis of a continuum theory in which quantities fluctuate negligibly, care must be exercised in applying such a theory at length scales less than those associated with such particles. Accordingly, a "safe" – though perhaps overly-conservative – scale in this respect would seem to be $\epsilon \sim 10^{-5} m (= 10^5 \AA)$.

Remarks

1. The definitions of ρ and $\mathbf{v}$ imply that ρ and $\rho\mathbf{v}$ vary linearly at the ϵ length scale, and thus render plausible the assumptions that these fields be spatially smooth. In fact we expect very little variation in these fields at the ϵ length scale.

2. Although $\int_R \rho$ will make sense for *any* region R (modulo continuity of ρ) this number cannot be expected to represent the mass in R if R is smaller than an ϵ-cell. Further, the linearity of ρ indicated in Remark 1 implies that if R is an ϵ-cell centred at $\mathbf{x}$ this number is $\rho(\mathbf{x})V_\epsilon$, where V_ϵ is the volume of the cell. A similar observation holds for $\int_R \rho\mathbf{v}$ in respect of momentum.

The region M_t wherein both ρ and $\mathbf{v}$ are meaningful at instant t is regarded as the region "occupied" by $\mathcal{M}$ at this instant. The time evolution of $\mathcal{M}$ may be visualised in terms of a motion map defined as the solution to an initial-value problem as follows. Suppose $\hat{\mathbf{x}} \in M_0 := M_{t_0}$ (where t_0 denotes a fixed instant) and consider the solution[3] $\chi_0(\mathbf{x},.) : \mathcal{I} \to \mathcal{E}$ of

$$\dot{\chi}_0(\hat{\mathbf{x}},t) = \mathbf{v}(\chi_0(\hat{\mathbf{x}},t),t) \quad \text{where} \quad \chi_0(\hat{\mathbf{x}},t_0) = \hat{\mathbf{x}}. \tag{2.5}$$

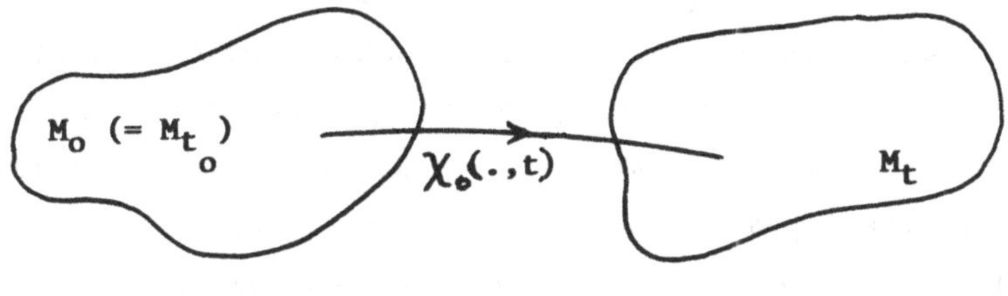

Fig. 3: the motion map

We assume that there *is* such a map for every $\hat{\mathbf{x}} \in M_0$, that $\chi_0(M_0,t) = M_t$ and $\chi_0(.,t) : M_0 \to M_t$ is injective and of class C^1 with $\mathbf{F}_0 := \nabla\chi_0$ taking values in[4] Invlin$^+$. Map χ_0 is termed the *motion* of the system (relative to the situation at instant t_0). To emphasise that χ_0 is a *derived* concept we say χ_0 is the *motion prescribed by* $\mathbf{v}$ (for system

[3] Here $\mathcal{I}$ denotes a time interval which includes instant t_0. Equations (2.5) may be regarded as posing both "initial-" and "final-" value problems. $\mathcal{E}$ denotes Euclidean space.

[4] Invlin$^+$ represents the group of all linear transformations of $\mathcal{V}$ with positive determinant, where $\mathcal{V}$ denotes the space of displacements between points of $\mathcal{E}$.

$\mathcal{M}$). We further assume that $\chi_0(\hat{\mathbf{x}},.)$ is of class C^2 for every $\hat{\mathbf{x}} \in M_0$ and that $\mathbf{F}_0(\hat{\mathbf{x}},.)$ is of class C^1. The *trajectory* $\mathcal{T}_{\mathcal{M}}$ is the common domain of ρ and $\mathbf{v}$ in space-time: that is

$$\mathcal{T}_{\mathcal{M}} := \bigcup_{t \in \mathcal{I}} M_t \times \{t\}, \tag{2.6}$$

where here $\mathcal{I}$ denotes the time interval for which χ_0 exists.

The time derivative of a field ϕ defined on $\mathcal{T}_{\mathcal{M}}$ may be computed either at fixed points in space or by moving with the material system as delineated by the motion prescribed by $\mathbf{v}$. We write

$$\partial\phi/\partial t(\mathbf{x}, t) := \frac{d}{d\tau}\{\phi(\mathbf{x}, \tau)\} \mid_{\tau=t;\mathbf{x} \text{ fixed}} \tag{2.7}$$

and

$$\dot{\phi}(\mathbf{x}, t) := \frac{d}{d\tau}\{\phi(\chi_0(\hat{\mathbf{x}}, \tau), \tau)\} \mid_{\tau=t;\hat{\mathbf{x}} \text{ fixed}} \tag{2.8}$$

where

$$\chi_0(\hat{\mathbf{x}}, t) = \mathbf{x}.$$

In the event that mass conservation holds (that is, $\dot{\rho} + \rho \operatorname{div} \mathbf{v} = 0$) $\dot{\phi}$ is termed the *material* time derivative of ϕ; $\partial\phi/\partial t$ is called the *spatial* time derivative of ϕ. In general we describe $\dot{\phi}$ as the *time derivative following the motion prescribed by* $\mathbf{v}$.

If mass conservation holds on $\mathcal{T}_{\mathcal{M}}$ we term $\mathcal{M}$ a *body*. In such case it is convenient to identify with each geometrical point $\hat{\mathbf{x}} \in M_0$ a "*material point*" whose location at instant t is $\chi_0(\hat{\mathbf{x}}, t)$. Of course, this notion is a mathematical *artefact* (albeit very useful) which under no circumstances should be regarded as a point mass. [5] If mass conservation does *not* hold (this would be the case were $\mathcal{M}$ to be a mixture constituent reacting with other constituents) it is still possible to follow the motion prescribed by the appropriate velocity field. *The foregoing considerations have been emphasised because it will turn out that* **two** *natural velocity fields can be ascribed to the interfacial model surface and mass conservation will not hold in general.*

2.3 The Concept of Interfacial Motion

In Section 3 will be discussed the family of model surfaces (parameterised by time t) $S(t)$ which will represent interfacial location. Associated therewith will be the singular surface velocity $U\mathbf{n}$, where $\mathbf{n}$ denotes a choice of (unit normal) orientation. However, there is another velocity field $\mathbf{u}_s$ which is a more natural choice (cf. (3.1)).

[5] Unfortunately, many authors use the terminology "particle" rather than material point (a label introduced by Noll precisely to avoid suggesting point mass) which can lead astray the unwary.

If t_0 is a fixed instant and $\hat{\mathbf{x}} \in S_0 := S(t_0)$ we may consider the initial-value problem for $\chi_0(\hat{\mathbf{x}}, .)$ defined by (cf. (2.5))

$$\dot{\chi}_0(\hat{\mathbf{x}}, t) = \mathbf{v}(\chi_0(\hat{\mathbf{x}}, t), t), \quad \text{where} \quad \chi_0(\hat{\mathbf{x}}, t_0) = \hat{\mathbf{x}}. \tag{2.9}$$

Here $\mathbf{v}$ denotes *either* $U\mathbf{n}$ *or* $\mathbf{u}_s$. For each choice ($\mathbf{v} = U\mathbf{n}$ and $\mathbf{v} = \mathbf{u}_s$) we assume there exists such a map for each $\hat{\mathbf{x}} \in S_0$, that $\chi_0(S_0, t) = S(t)$ and that

$$\chi_0(., t) : S_0 \rightarrow S(t)$$

is injective and of class C^1. Further, $\chi_0(\hat{\mathbf{x}}, .)$ is assumed to be of class C^2 and the values of the surface gradient [6] $\nabla_{S_0}\chi_0$ are assumed to have bidimensional ranges with $\nabla_{S_0}\chi_0(\mathbf{x}, .)$ of class C^1. The trajectory of the interfacial system $\mathcal{S}$ is (here $\mathcal{I}$ has the same sense as in (2.6))

$$\mathcal{T}_s := \bigcup_{t \in \mathcal{I}} S(t) \times \{t\}. \tag{2.10}$$

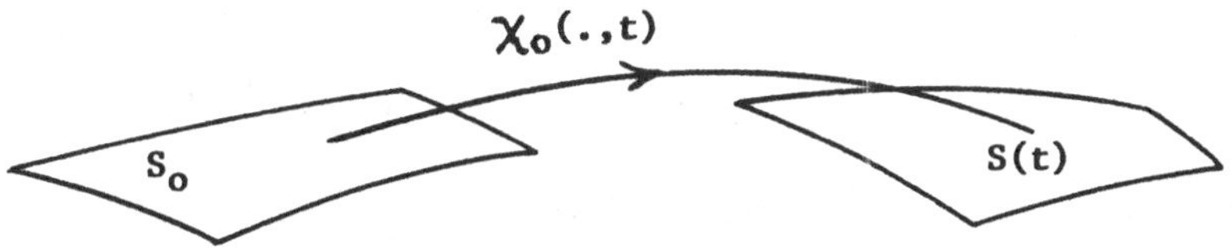

Fig. 4: the surface motion map

Map χ_0 is termed the *motion of $\mathcal{S}$ prescribed by* $\mathbf{v}$.

While $\mathcal{S}$ constitutes a material system, the lack of mass conservation in general means we do not use the terminology material point in tracing its motion. (In the special cases in which mass *is* always conserved for a material system modelled in terms of surfaces then the system is termed a *material surface* and material points *can* be ascribed thereto.)

If ϕ denotes any field defined on $\mathcal{T}_s$ then

$$\dot{\phi}(\mathbf{x}, t) := \frac{d}{dt}\{\phi(\chi_0(\hat{\mathbf{x}}, \tau), \tau)\} \mid_{\tau = t, \hat{\mathbf{x}} \text{ fixed}} \tag{2.11}$$

where

$$\chi_0(\hat{\mathbf{x}}), t) = \mathbf{x},$$

[6] Cf. Appendix: this gradient has the property A.5.

is termed the *time derivative of* ϕ *following the motion prescribed by* $\mathbf{v}$.

2.4 Delineation of the Interfacial Region

If an ϵ-cell ($\epsilon \sim 10^{-5}m$) intersects or straddles a liquid-vapour interfacial region the ratio $\sum_i{}' m_i/V_\epsilon$ will be highly shape- and size-dependent. Thus ρ, as defined by (2.2), will fail to exist in a region of thickness $\sim 2 \times 10^{-5}m$ centred on the interface; this region can be identified with the interfacial zone[7] I.

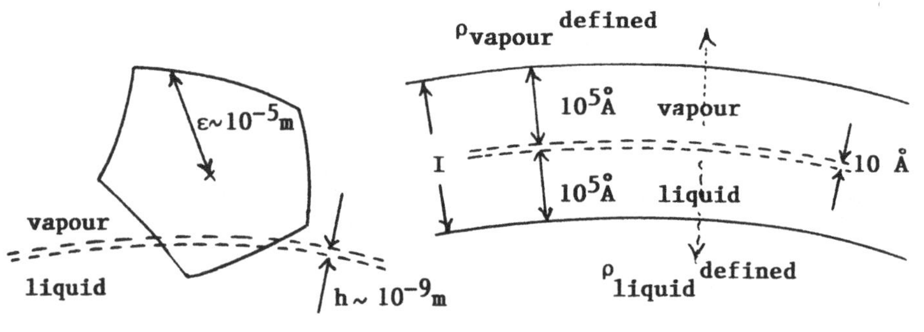

Figures 5

However, this is unsatisfactory since there are compelling reasons (related to the requirement that *excess* quantities should approximate true interfacial quantities as closely as possible) for delineating an interface with an accuracy commensurate with its thickness. Greater precision can be gained by re-interpreting ρ as a space-*time* average.

The Δ-time average of a fluctuating quantity ϕ at instant t is defined by

$$\bar{\phi}_\Delta(t) := \frac{1}{\Delta} \int_{t-\Delta}^{t} \phi(\tau) d\tau. \tag{2.12}$$

[7] A *multicomponent* interfacial zone is identified with that region in which *at least one* constituent density (defined as ρ but with sum taken only over the relevant species) fails to exist.

It may happen that over a range of Δ values which are macroscopically small, $\bar{\phi}_\Delta$ is essentially constant, with value $\bar{\phi}(t)$, say. In such case we write

$$\bar{\phi}(t) := lim\{\bar{\phi}^\Delta\}. \tag{2.13}$$

Now consider ratios $\sum_i{}' m_i/V_\epsilon$, where the sums are over ϵ-cells with $\epsilon \sim 10^{-6}m$ or less. Although such sums might fluctuate it may be possible to smooth these by time averaging. If so the smoothed value at instant t is identified with $\rho(\mathbf{x},t)$ and we write

$$\rho(\mathbf{x},t) := lim_\epsilon\left\{\overline{\sum_i{}' m_i/V_\epsilon}^\Delta\right\}\left(= lim_\epsilon\left\{\overline{\sum_i{}' m_i}^\Delta /V_\epsilon\right\}\right). \tag{2.14}$$

The boundary of the ϵ-cell is here assumed to deform with the motion prescribed by $\mathbf{v}$ and to coincide with some ϵ-cell centred at $\mathbf{x}$ at instant t. Since V_ϵ will change imperceptibly during the time-averaging period it may be regarded as the volume at instant t (this accounts for the second equality in (2.14)). *The pseudo-limit notation* $\lim_\epsilon\{\overline{\cdot}^\Delta\}$ *will consistently be used to denote a quantity which is insensitive to modest changes in* ϵ, Δ *and* ϵ*-cell shape.* Thus ρ, re-interpreted as a space-time average, may be meaningful at length-time scales (ϵ_0, Δ_0) where $\epsilon_0 < 10^{-5}m$. Intuitively we expect such averages to exist for a range of (ϵ_0, Δ_0) pairs lying on a monotonic-decreasing curve Γ in R^2 with increasing spatial accuracy (associated with ϵ_0) achieved at the expense of temporal precision (measured by Δ_0).

It is also possible to re-interpret $\mathbf{v}$ as a space-time average via

$$\mathbf{v}(\mathbf{x},t) := lim_\epsilon\left\{\overline{\sum_i{}' m_i \mathbf{v}_i/\sum_i{}' m_i}^\Delta\right\} \tag{2.15}$$

where the sums are taken over an ϵ-cell whose boundary deforms *with the motion prescribed by* $\mathbf{v}$. Such an implicit definition can be motivated as the limit of an iterative procedure.

The interfacial zone I can now be identified with that region in which ρ, as defined by (2.14) with $\mathbf{v}$ delivered by (2.15), fails to exist. It follows that the more temporal precision is sacrificed the more precisely may be delineated the interface. Further, ρ will only make sense if $\mathbf{v}$, as given by (2.15), is meaningful.

For a given temporal accuracy, delineation of I may be improved by modifying the *shape* of spatial regions over which averages are computed. To this end let S denote a smooth surface with normal curvatures very much less than 10^9m^{-1}. Consider that pair of surfaces S^+, S^- parallel to and distant $10^{-9}m$ therefrom, together with an ϵ-*subsurface* of S *centred* at $\mathbf{x} \in S$ (that is, the intersection with S of an ϵ-cell centred at **x**: cf. Murdoch (1985), p.300) where $\epsilon >> 10^{-9}m$. That region bounded by S^+, S^- and the ruled surface *("wall")* generated by normals to S through the perimeter of the ϵ-subsurface will be termed an *ϵ-wafer based on S and centred at* **x**.

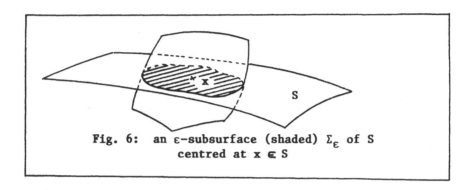

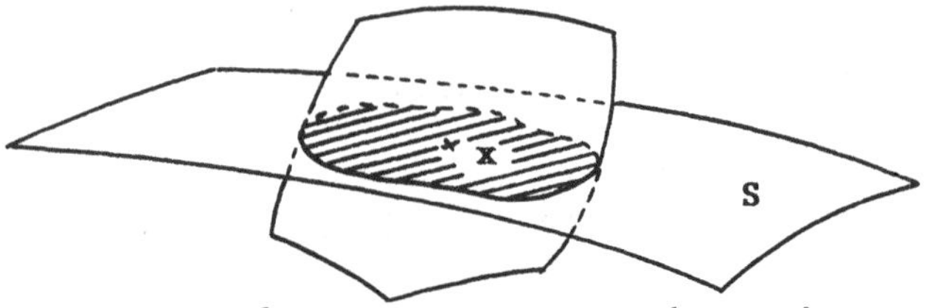

Fig. 6: an ε-subsurface (shaded) Σ_ε of S centred at x ∈ S

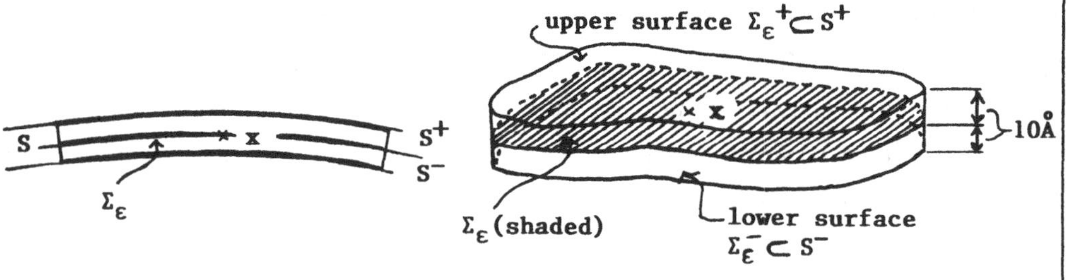

Fig. 7: an ε-wafer based on S and centred at x ∈ S

Now suppose ρ fails to exist, in the sense of (2.14) with **v** given by (2.15), in the region I bounded by distinct surfaces S_1^- and S_2^+. It may prove possible to extend fields ρ and **v** *into I* by re-interpreting these as space-time averages taken over ϵ-wafers based on families of surfaces parallel to S_1^- and S_2^+. By this we mean ρ and **v**, as given by (2.14), (2.15), make sense when the instantaneous sums are over wafer-like regions whose walls deform with the motion prescribed by **v** restricted to the surface on which the neo-wafer is based, the other bounding surfaces move with the motion prescribed by **v**, and at instant t the region is an ϵ-wafer based upon a surface parallel to S_1^- (or S_2^+) through **x**. Motivation for such an implicit interpretation of **v** is possible via iterative considerations. The interfacial zone I may now be regarded as the region in which (the new interpretation of) ρ fails to exist. We may postulate that this region has smooth bounding surfaces S_1 and S_2.

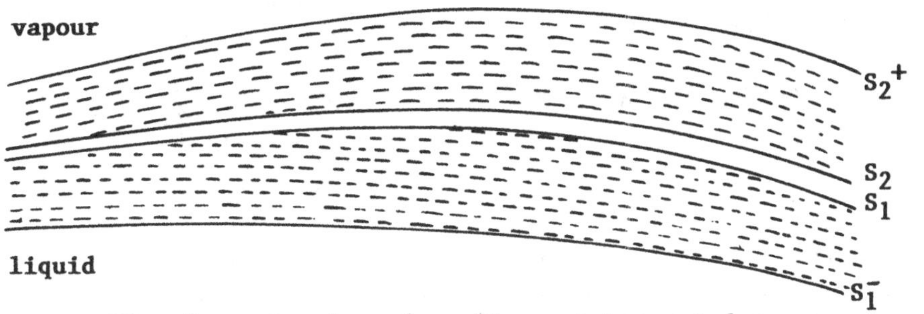

Fig. 8: extension of ρ when re-interpreted as an average involving wafer-like regions

Remarks

1. The revised manner of effecting space-time averages near the interfacial region (in terms of wafers) will enable this region to be delineated to within an accuracy[8] commensurate with its thickness, since wafers of such thickness have been selected.

2. The re-interpretation of the bulk fields ρ and $\mathbf{v}$ gives these fields increased sensitivity to change with displacement along normals to interfacial boundaries as compared with displacements parallel thereto. This is clearly compatible with large values of $\partial\rho/\partial n_\alpha$ and $\partial\mathbf{v}/\partial n_\alpha$, where derivatives are along the normal $\mathbf{n}_\alpha$ to S_α away from I ($\alpha = 1, 2$). Indeed, observed motions driven by inhomogeneous interfacial stress involve large values of $\partial\mathbf{v}_s/\partial n_\alpha$, where $\mathbf{v}_s$ denotes (roughly speaking) the tangential (to the interface) velocity component in either bulk phase.

3. Selection of a Model Surface and Interfacial Velocity Fields

The intersection with I of an ϵ-cell ($\epsilon \sim 10^{-5}m$) centred at $\mathbf{x} \in I$ is termed an *interfacial* ϵ-cell, described as *"walled"* if its boundary component between S_1 and S_2 is a ruled surface generated by normals to some smooth surface lying strictly between

[8] Of course, this depends upon the tolerance of error implicit in the definition of local space-time averages: cf. Murdoch(1985), p.294 in which, for ϵ-limits, this tolerance is quantified by the number β.

S_1 and S_2. We note that the mass centre of the set of molecules instantaneously within such a cell must be expected to fluctuate, even in conditions of macroscopic equilibrium.

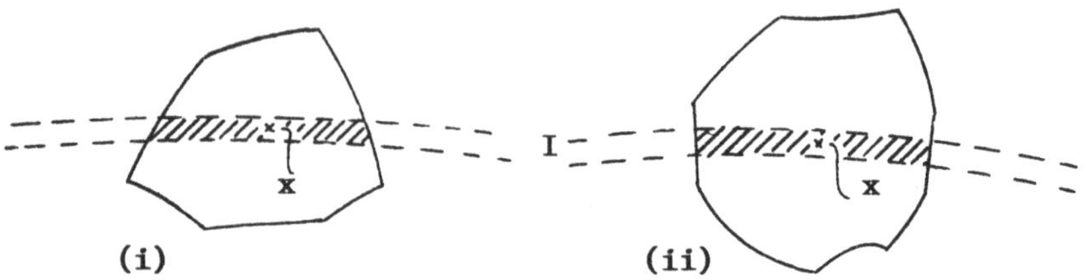

Figures 9: interfacial ε-cells (i) general (ii) walled

It proves possible to motivate, again via an iterative procedure, the following modelling assumptions:

M.A. To each instant t corresponds a smooth surface $S(t)$ which is the locus of Δ-time-averaged walled interfacial ϵ-cell mass-centre locations, where the time averaging involves wall deformations mandated by the surface motion of S prescribed by the *interfacial surface velocity*.

$$\mathbf{u}_s := \mathrm{U}\mathbf{n} + \mathbf{P}\mathbf{u}. \tag{3.1}$$

Here $\mathrm{U}\mathbf{n}$ denotes the normal velocity field on S (for definiteness we choose $\mathbf{n}$ to be that unit normal field to S directed from S_1 towards S_2) and [9] $\mathbf{Pu}$ the tangential component of the *interfacial material velocity* $\mathbf{u}$, where

$$\mathbf{u}(\mathbf{x},\mathrm{t}) := lim_\epsilon \{ \overline{\sum_i{}' m_i \mathbf{v}_i / \sum_i{}' m_i}^{\Delta} \}. \tag{3.2}$$

In (3.2) the sums are taken over a walled interfacial ϵ-cell (centred at $\mathbf{x} \in S(t)$ at instant t) whose walls deform with the motion prescribed by $\mathbf{u}_s$.

Remarks

1. The somewhat complex inter-relationships between S, $\mathbf{u}$ and $\mathbf{u}_s$ arise naturally in the iterative considerations of Murdoch & Cohen (1989).

2. Regarded as a family of surfaces parameterised by time, the trajectory of S has a well-defined normal (singular surface) velocity field $\mathrm{U}\mathbf{n}$. The interfacial material velocity $\mathbf{u}$ is defined entirely in terms of molecules within the interface and is that velocity in terms of which individual interfacial molecular thermal velocities $\tilde{\mathbf{v}}_i$ should be computed

[9] $\mathbf{P} := 1 - \mathbf{n} \otimes \mathbf{n}$ is the perpendicular projection of spatial vectors onto the relevant tangent space to S.

($\tilde{\mathbf{v}}_i := \mathbf{v}_i - \mathbf{u}(\mathbf{z}_i, t)$, where $\mathbf{z}_i$ is the closest point of $S(t)$ to the location $\mathbf{x}_i(t)$ of particle P_i at instant t). However, considerations of evaporation indicate that in general $\mathbf{u}.\mathbf{n} \neq U$ (cf. §6, Remark 2). The motion prescribed by $\mathbf{u}_s$ is that motion which allows us to remain on the surface $S(t)$, yet simultaneously move as closely as possible with the local gross interfacial molecular average velocity. The utility of $\mathbf{u}_s$ seems first to have been indicated by Ishii (1975).

3. All averages relating to interfaces, including those associated with the delineation of interfacial boundaries, have here been computed jointly in space *and* time, over regions having characteristic dimensions $\epsilon \times \epsilon \times 10^{-9} m^3$ with ϵ notionally $10^{-5} m$. The necessity of time averaging was mandated by the essentially erratic *gross* behaviour of liquids at length scales of $10^{-6} m$ and below indicated by the motions of Brownian particles (recall that Brownian motion implies purely spatial averages over volumes of order $10^{-18} m^3$ might fluctuate). Even greater chaos is to be expected in gaseous phases in view of their smaller densities. Typical molecular separations in liquid and gaseous phases are of orders $3\mathring{A}$ and $30\mathring{A}$, respectively. Thus, while spatial volumes of $(10^{-5})^3 m^3$ in liquid phases may provide sufficiently many molecules to yield averages which are essentially fluctuation-free ($\sim 4 \times 10^{13}$ molecules), volumes of $(10^{-6})^3 m^3$ do not [10] ($\sim 4 \times 10^{10}$ molecules). The ϵ-wafer averages here employed would contain $\sim 4 \times 10^9$ molecules for liquid phases and $\sim 4 \times 10^6$ molecules for gaseous phases. The fluctuations associated with instantaneous wafer sums are to be smoothed out by time averaging. For conditions of static macroscopic equilibrium the time-averaging duration may be indefinitely long, resulting, of course, in $\mathbf{u} = \mathbf{0}$. In non-equilibrium situations there is no *a priori* guarantee that average $\mathbf{u}$ should exist. Our analysis is directed towards providing interpretations of macroscopic interfacial quantities *if and when continuum modelling is valid.* It would seem that in modelling interfaces one is pushing continuum mechanics to its limit. (That continuum mechanics is not universally applicable is clear, for example, from considerations of fully-developed turbulence in fluid flows: cf. Landau & Lifschitz (1959), §31. Interestingly, in such context a time-averaged quantity, the *mean velocity*, is introduced. However, the time-averaging interval in question is not macroscopically small and the mean velocity should not be confused with the usual notion of macroscopic velocity. Of course, this mean velocity *could* be regarded as "velocity" in a continuum theory in which variables change only on length and time scales in excess of those usually appropriate to continua.)

4. The question naturally arises as to the interpretation of density profiles *within* the interfacial region. Naively it could be thought that in equilibrium situations ever-thinner

[10] Cf. Huang (1963), p.56.

wafer averages might be adopted (matched by appropriately longer time averaging periods) and that consequently ρ could exist everywhere, albeit varying by three orders of magnitude over a distance of $10\mathring{A}$. Alternatively, such profiles might be identified with expectations associated with ensemble distributions for interfaces in macroscopic equilibrium. More precisely $\rho(\mathbf{x})$ could be identified with the expectation $< \hat{\rho}(\mathbf{x}) >$ of the mass density distribution (a stochastic quantity) $\hat{\rho}(\mathbf{x})$ at an arbitrary point $\mathbf{x}$ in such case. However, unless the variance of $\hat{\rho}(\mathbf{x})$ were zero we could not in *principle* obtain experimentally reproducible values $\rho(\mathbf{x})$ at $\mathbf{x}$ for mass density. Our considerations have shown that if $\mathbf{x}$ lies within the interfacial region such reproducibility is not to be expected. Accordingly mass density (as a *continuum* quantity) does not exist, since an implicit pre-requisite of macroscopic quantities is that their values be reproducible.

Approaches to interfacial modelling based upon integration across the interface of equations which resemble formal balance relations for continua would appear to involve many implicit assumptions, and to require some care in the interpretation of quantities which appear in the pre-integrated relations.

5. Consider the temporal average of the ratio $\sum_i{}' m_i/V_\epsilon$, where the sum is over all particles in an ϵ-cell whose boundary deforms with the motion prescribed by the bulk velocity $\mathbf{v}$ (assumed to exist everywhere) and has volume V_ϵ at instant t. This average will be

$$\sum_i{}' m_i \delta_i / V_\epsilon \Delta. \tag{3.3}$$

Here the sum is taken over all particles P_i which lie within the deforming ϵ-cell in question for at least some time during the time-averaging interval: δ_i denotes the total length of time such a particle resides within the cell in this interval. (Any such particle may, of course, leave and enter the cell any number of times during the interval.) If the ratio (3.3) is insensitive to modest variations in ϵ and Δ about any point on a curve Γ as described in Section 2 (preceding (2.15)) then this expression represents the local mass density. If Γ contains a point whose ϵ value is so small that at most one particle could reside in a corresponding cell at any given instant (say $\epsilon \sim 10^{-11}m = 0.1\mathring{A}$) and all particles have mass m then (3.3) reduces to

$$\rho(\mathbf{x}, t) = m\Big(\sum_i{}' \delta_i/\Delta\Big)/V_\epsilon =: m\pi/V_\epsilon. \tag{3.4}$$

Evidently π denotes the proportion of the time interval that the cell is "occupied". In macroscopically-static situations (so that $\mathbf{v} = \mathbf{0}$) π might be expected to be independent of Δ provided only that Δ be large enough. In such case π could clearly be regarded as an operational definition of the *probability* of finding a particle in the cell at instant t. More generally, the product of expression (3.3) with V_ϵ could correspondingly be regarded as the most probable mass within the cell at instant t.

The purpose of this remark has been to draw attention to the relationship between time-averaging and statistical approaches, supporting the contention of Remark 4 that within an interface ρ be regarded as the expectation of a random variable (the instantaneous value of $\sum_i{}' m_i/V_\epsilon$ with $\epsilon << 1\mathring{A}$) which has non-negligible variance.

4. Interfacial Balance Relations: General Considerations

4.1 Preamble

We wish to establish balance relations for the interfacial region, regarded as a bidimensional continuum, by taking account of the microscopic situation. Specifically, the analysis will be based on the equations

$$\dot{m}_i = 0 \tag{4.1}$$

and

$$\sum_\ell \mathbf{f}_{i\ell} + \sum_r \mathbf{f}_{ir} + \sum_s \mathbf{f}_{is} + \mathbf{b}_i = \dot{\widehat{m_i \mathbf{v}_i}} \tag{4.2}$$

associated with any particular P_i in the interfacial region. Relation (4.1) represents mass invariance for P_i and (4.2) prescribes its motion in an (any) inertial frame. Here $\mathbf{f}_{ip}$ denotes the force exerted upon P_i by P_p; the sums are taken over *all* particles P_ℓ in the interface, *all* particles P_r in the bulk phase bordering S_1, and all particles P_s in the other bulk phase. The term $\mathbf{b}_i$ represents the net force on P_i due to the remainder of the material universe. The balance relations will result from space-time averages of sums of relations (4.1,2) and associated equations. Assumptions will be made concerning the nature of interactions and of interfacial thermal motions, and all continuum fields identified with molecular quantities suitably averaged in both space and time.

4.2 Preliminaries

Consider at instant t a connected subsurface Σ of $S(t)$ with piecewise smooth boundary $\partial\Sigma$ having outward unit normal field[11] ν. Normals to $S(t)$ through points of $\partial\Sigma$ will, at least near $S(t)$, generate a ruled surface which isolates a portion $R(t)$ of the interfacial region. Of course, the boundary $\partial R(t)$ of $R(t)$ will consist of subsets of $S_1(t)(\Sigma_1(t)$, say), $S_2(t)$ ($\Sigma_2(t)$, say) and that part of the ruled surface between $S_1(t)$ and $S_2(t)$ ($W(t)$, say). Formally,

$$\partial R(t) = \Sigma_1(t) \cup W(t) \cup \Sigma_2(t). \tag{4.3}$$

We term $R(t)$ a *walled interfacial region (with wall W(t)) based upon* $\Sigma \subset S(t)$, and $W(t)$ an *interfacial wall based upon* $\partial\Sigma$.

Suppose Σ may be decomposed into very many mutually-disjoint ϵ-subsurfaces with $\epsilon \sim 10^{-5}$m; in such case we term $R(t)$ a *macroscopic* walled interfacial region. At any instant τ we may identify that subsurface $\Sigma(\tau)$ of $S(\tau)$ which deforms with the motion of S prescribed by $\mathbf{u}_s$ (*cf*.§2.3)and coincides with Σ at instant t. The ϵ-subsurface decomposition of Σ will induce corresponding partitions of $\Sigma(\tau)$ (via the motion map) and hence partitions of $\Sigma_1(\tau), \Sigma_2(\tau)$ into ϵ-subsurfaces, and $R(\tau)$ into walled interfacial ϵ-cells: $R(\tau)$ is generated from $\Sigma(\tau)$ by erecting a wall $W(\tau)$ based upon $\partial\Sigma(\tau)$ with generators normal to $S(\tau)$ and this wall identifies the subsets $\Sigma_\alpha(\tau)$ of $S_\alpha(\tau)(\alpha = 1, 2)$. The walled interfacial ϵ-cells are similarly generated by walls based on the perimeters of the ϵ-subsurfaces of $\Sigma(\tau)$ which also induce the partitions of $\Sigma_\alpha(\tau)$.

Balance relations to be ascribed to Σ at instant t are obtained by summing relations such as (4.1,2) over all particles P_i within $R(\tau)$ and then time averaging this sum over a time interval $t - \Delta \leq \tau \leq t$, where Δ is macroscopically short (say $\Delta \sim 10^{-6}s$).

4.3 Interfacial Mass and Linear Momentum Densities

The *interfacial mass density* σ and *linear momentum density* $\mathbf{p}$ are defined by

$$\sigma(\mathbf{x}, t) := lim_\epsilon \Big\{ \overline{\sum_i {}' m_i / A_\epsilon}^{\Delta} \Big\} \Big(= lim_\epsilon \Big\{ \overline{\sum_i {}' m_i}^{\Delta} / A_\epsilon \Big\} \Big) \tag{4.4}$$

and

$$\mathbf{p}(\mathbf{x}, t) := lim_\epsilon \Big\{ \overline{\sum_i {}' m_i \mathbf{v}_i / A_\epsilon}^{\Delta} \Big\} \Big(= lim_\epsilon \Big\{ \overline{\sum_i {}' m_i \mathbf{v}_i}^{\Delta} / A_\epsilon \Big\} \Big). \tag{4.5}$$

Here the sums, at any instant τ in the averaging period $t - \Delta \leq \tau \leq t$, are over all particles P_i within a walled interfacial ϵ-cell based upon an ϵ-subsurface of $\Sigma(\tau)$ with area A_ϵ whose boundary deforms with the motion prescribed by $\mathbf{u}_s$ (so inducing the cell wall motion) and which at instant t is centred at $\mathbf{x}$. Since A_ϵ will change imperceptibly during the

[11] Thus ν is tangential to $S(t)$ but orthogonal to tangents to $\partial\Sigma$

time-averaging period it may be regarded as the area at instant t (this accounts for the second equalities in (4.4) and (4.5): cf. (2.14)). The notation "$lim_\epsilon\{\overline{\cdot}^\Delta\}$" is employed to indicate that the quantity associated therewith is insensitive to modest changes in ϵ, Δ and subsurface shape (with $\epsilon \sim 10^{-5}m$ and Δ macroscopically small).

If the total mass and linear momentum associated with particles in a *macroscopic* walled region $R(t)$ do not fluctuate they can be identified with their Δ-time averages. Thus, effecting a decomposition of the base Σ of $R(t)$ into ϵ-subsurfaces, from the deliberations of § 4.2 we have

$$\begin{aligned}\text{mass in } R(t) &= \sum_{P_i\, in R(t)} m_i \\ &= \overline{\sum_{P_i\, in R(\tau)} m_i}^{\Delta} = \sum_{subsurfaces} \left\{\overline{\sum_i{}' m_i}^{\Delta} / A_\epsilon\right\} A_\epsilon \\ &= \sum_{subsurfaces} \sigma(\mathbf{x}, t) A_\epsilon, \end{aligned} \tag{4.6}$$

where $\mathbf{x}$ denotes a subsurface centre at instant t. We identify the final sum with the Riemann integral of σ over Σ. That is

$$\text{mass in } R(t) \leftrightarrow \int_\Sigma \sigma(.,t). \tag{4.7}$$

Similarly, we make the identification

$$\text{momentum associated with } R(t) \leftrightarrow \int_\Sigma \mathbf{p}(.,t). \tag{4.8}$$

Remark

If the mass and (or) linear momentum associated with particles in $R(t)$ at instant t fluctuate(s) then the only meaningful *macroscopic* interpretation of these quantities is their Δ-time average. In such case we obtain the same integral representations (4.7) and (4.8), of course: the time averaging appears now after the first equality of relations (4.6).

We now argue that it is reasonable to identify $\mathbf{p}$ with $\sigma\mathbf{u}$ (cf. (3.2)): that is

$$identify\ lim_\epsilon\left\{\overline{\sum_i{}' m_i \mathbf{v}_i / A_\epsilon}^{\Delta}\right\} with\ lim_\epsilon\left\{\overline{\sum_i{}' m_i / A_\epsilon}^{\Delta}\right\} lim_\epsilon\left\{\overline{\sum_i{}' m_i \mathbf{v}_i / \sum_i{}' m_i}^{\Delta}\right\}. \tag{4.9}$$

Of course, instant by instant,

$$\sum_i{}' m_i \mathbf{v}_i / A_\epsilon = \Big(\sum_i{}' m_i / A_\epsilon\Big)\Big(\sum_i{}' m_i \mathbf{v}_i / \sum_i{}' m_i\Big).$$

However, (4.9) involves equality between the time average of a product and the product of two time averages. Said differently, fluctuations in the quantities

$$\phi := \sum_i{}' m_i / A_\epsilon \; and \; \psi := \sum_i{}' m_i \mathbf{v}_i / \sum_i{}' m_i$$

are not correlated over time intervals of duration Δ. Here we note[12] that if $\overline{\phi}$ and $\overline{\psi}$ denote the time averages of fluctuating quantities ϕ and ψ then we expect $\overline{\phi\psi} = \overline{\phi}\,\overline{\psi} + \overline{\phi'\psi'}$, where $\phi' := \phi - \overline{\phi}, \psi' := \psi - \overline{\psi}$ denote the fluctuations in ϕ and ψ.) As a first step we define, for any P_i in the interfacial region,

$$\tilde{\mathbf{v}}_i := \mathbf{v}_i - \mathbf{u}(\pi(\mathbf{x}_i)), \tag{4.10}$$

where time dependence has been suppressed for brevity, and $\pi(\mathbf{x}_i)$ denotes the (assumed) unique point on S at which the normal to S passes through the location $\mathbf{x}_i$ of P_i. Since $\mathbf{u}$ is not defined everywhere in the interface (but only upon S), (4.10) is the closest we can come to the concept of *interfacial thermal velocity*. Clearly, if P_i belongs to an interfacial ϵ-cell centred at $\mathbf{x} \in S$ then

A.1 $\tilde{\mathbf{v}}_i$ is well–approximated by $\mathbf{v}_i - \mathbf{u}(\mathbf{x}) - \nabla_s \mathbf{u}(\mathbf{x}) \mathbf{P}(\mathbf{x})(\mathbf{x}_i - \mathbf{x}).$ (4.11)

Here $\mathbf{P}(\mathbf{x})$ denotes the perpendicular projection of vectors upon the tangent space to S at $\mathbf{x}$ and $\nabla_s \mathbf{u}$ denotes the surface gradient (cf. Appendix) of $\mathbf{u}$. Consistent with the location of S in terms of time-averaged interfacial ϵ-cell mass centres, we shall also assume hereafter

$$\text{A.2} \quad lim_\epsilon \left\{ \overline{\sum_i{}' m_i(\mathbf{x}_i - \mathbf{x})/A_\epsilon}^{\Delta} \right\} = \mathbf{0}. \tag{4.12}$$

In (4.12) the sum is over all particles within any walled interfacial ϵ-cell centred at $\mathbf{x}$ with base area A_ϵ at the end of the averaging period ($\epsilon \sim 10^{-5} m, \Delta$ macroscopically small: cf. Murdoch (1985), p.296 for the three-dimensional analogue). Finally, we here make the first (of two) thermal motion assumption (cf. Murdoch (1985) p.298)

$$\text{T.M.1} \quad lim_\epsilon \left\{ \overline{\sum_i{}' m_i \tilde{\mathbf{v}}_i / A_\epsilon}^{\Delta} \right\} = \mathbf{0}. \tag{4.13}$$

[12] Cf. Murdoch (1985), p.295.

T.M.1, on invoking A.1, A.2 and noting that $\mathbf{u}$, $\nabla_s \mathbf{u}$ and $\mathbf{P}$ do not change perceptibly during the time-averaging period, yields[13]

$$\begin{aligned} \mathbf{0} &= lim_\epsilon \left\{ \overline{\left(\sum_i {}' m_i \mathbf{v}_i - \left(\sum_i {}' m_i \right) \mathbf{u} \right) / A_\epsilon}^{\Delta} \right\} \\ &= lim_\epsilon \left\{ \overline{\sum_i {}' m_i \mathbf{v}_i / A_\epsilon}^{\Delta} \right\} - \lim_\epsilon \left\{ \overline{\sum_i {}' m_i / A_\epsilon}^{\Delta} \right\} \mathbf{u} = \mathbf{p} - \sigma \mathbf{u}. \end{aligned}$$

That is,

$$\mathbf{p} = \sigma \mathbf{u}. \tag{4.14}$$

5. Mass Balance

Consider

$$\phi(\tau) := \sum_{P_i \, in \, R(\tau)} {}^{\tau} dm_i / d\tau = 0, \tag{5.1}$$

where the sum is taken over all particles P_i within $R(\tau)$ (cf. §4.2) at instant τ: this is the significance of the superscript τ attached to the summation sign. Calculation of $\phi_\Delta(t)$ (cf. (2.12)) yields (on noting that particles may enter and leave the deforming region, and taking full account of all such mass exchange)

$$\begin{aligned} 0 = & \frac{1}{\Delta} \left\{ \sum_{P_i \, in \, R(t)} {}^{t} m_i - \sum_{P_i \, in \, R(t-\Delta)} {}^{(t-\Delta)} m_i \right\} \\ & + \frac{1}{\Delta} \left\{ \sum_j {}^{(1)} m_j s_j{}^{(1)} + \sum_j {}^{(2)} m_j s_j^{(2)} + \sum_j {}^{w} m_j s_j^{w} \right\}. \end{aligned} \tag{5.2}$$

If the sums over P_i in $R(t)$ and $R(t-\Delta)$ do not fluctuate[14] (they are taken over macroscopic interfacial regions) they may be identified with their time averages taken over a macroscopically-short period and hence associated with (cf. (4.7)),

[13] Upon noting that since $\nabla_s \mathbf{u} \mathbf{P}$ is a macroscopic quantity it varies negligibly during a macroscopically-small time interval.

[14] Alternatively, relation (5.2) may be time averaged: cf. Remark in §4.3.

$$\int_{\Sigma(t)} \sigma(.,t) \text{ and } \int_{\Sigma(t-\Delta)} \sigma(.,t-\Delta). \tag{5.3}$$

We identify the first term involving brackets in (5.2) with

$$d/d\tau \left\{ \int_{\Sigma} \sigma(.,\tau) \right\}_{|\tau=t}. \tag{5.4}$$

The three sums in the second term in brackets in (5.2) are taken over particles P_j which cross the boundaries Σ_1, Σ_2 and W of the deforming region during the time-averaging period. Each such crossing contributes a term to the sum, with $s_j^{(1)}$, $s_j^{(2)}$ or s_j^w taking the value +1 if P_j *leaves* across Σ_1, Σ_2 or W, respectively, at some instant and the value −1 if P_j *enters* across such a surface. A decomposition of Σ into ϵ-subsurfaces induces (cf. §4.2) corresponding partitions of $\Sigma_\alpha (\alpha = 1,2)$ and we may write

$$\frac{1}{\Delta} \sum_j{}^{(\alpha)} m_j s_j^{(\alpha)} = \sum_{\epsilon-subsurfaces} \left\{ \sum_j{}'^{(\alpha)} m_j s_j^{\alpha} / A_\epsilon \Delta \right\} A_\epsilon, \tag{5.5}$$

where the superposed prime on the last summation sign indicates the sum is to be performed only over a deforming ϵ-subsurface (the area of the corresponding subsurface of Σ, with centre $\mathbf{x} \in S(t)$ at instant t, having been denoted by A_ϵ). The continuum version of (5.5) is

$$-\int_{\Sigma} s_\alpha \tag{5.6}$$

where

$$s_\alpha(\mathbf{x},t) := lim_{\epsilon,\Delta} \left\{ \sum_j{}'^{(\alpha)} m_j s_j^{(\alpha)} / A_\epsilon \Delta \right\}. \tag{5.7}$$

Here the symbol "$lim_{\epsilon,\Delta}$" indicates that the ratio it prefixes is essentially constant over ϵ-Δ length-time scales which are macroscopically small, and insensitive to modest change in ϵ-subsurface shape.

The final sum in (5.2) involves particle exchange across the deforming wall $W(\tau)$ of $R(\tau)$. In order to identify its continuum counterpart consider the intersection of an ϵ-cell centred at $\mathbf{x} \in \partial\Sigma$ with $\partial\Sigma$, so forming what we call an *ϵ-sub-boundary of $\partial\Sigma$ centred at* $\mathbf{x}$. Normals to $S(t)$ through points on this sub-boundary generate a portion of $W(t)$ we term an *interfacial ϵ-subwall centred at* $\mathbf{x}$ and *based on the given sub-boundary of $\partial\Sigma$.*

A decomposition of $\partial\Sigma$ into complementary mutually-disjoint ϵ-sub-boundaries will thus induce a partition of $W(t)$ into interfacial ϵ-subwalls. Further, such a decomposition will, via the interfacial motion map, induce sub-boundary and sub-wall decompositions of $\partial\Sigma(\tau)$ and $W(\tau)$ respectively. It follows that the final sum in (5.2) may be expressed as

$$\frac{1}{\Delta}\sum_j {}^w m_j s_j^w = \sum_{sub-boundaries\ of\ \Sigma} \Big\{\sum_j {}'^w m_j s_j^w / \ell_\epsilon \Delta\Big\}\ell_\epsilon. \tag{5.8}$$

The superposed prime denotes a sum over a single ϵ-subwall whose corresponding sub-boundary at instant t has length ℓ_ϵ. The macroscopic version of (5.8) is

$$\int_{\partial\Sigma} s \tag{5.9}$$

where

$$s(\mathbf{x},t) := lim_{\epsilon,\Delta}\Big\{\sum_j {}'^w m_j s_j^w / \ell_\epsilon \Delta\Big\}. \tag{5.10}$$

Again the pseudo–limit notation indicates insensitivity of the ratio to modest variations in ϵ and Δ scales, where these are macroscopically small. The subwall is centred at $\mathbf{x}$ at instant t.

It follows from (5.4,6,9) that (cf. Appendix, (A.22))

$$-\int_{\partial\Sigma} s + \int_{\Sigma}(s_1 + s_2) = \frac{d}{dt}\Big\{\int_{\Sigma(t)} \sigma(.,t)\Big\}. \tag{5.11}$$

Use of the appropriate transport theorem [15] yields

$$\int_{\Sigma}(\dot{\sigma} + \sigma\,\mathrm{div}_s \mathbf{u}_s - s_1 - s_2) = -\int_{\partial\Sigma} s. \tag{5.12}$$

Since the balance relation is for *macroscopic* interfacial regions it is not *a priori* clear that (5.12) should hold for *arbitrarily-small* regions Σ. However, an arbitrarily–small (curvilinear) triangular part of S may be written as the *difference between two appropriate macroscopic parts*, whence additivity of integration delivers (5.12) for all such regions. [16]

---

[15] Cf. Appendix and Murdoch (1990b): $\mathrm{div}_s \mathbf{u}_s$ denotes the *surface divergence* of $\mathbf{u}_s$.

[16] In the remainder of this work we shall localise integral balances (using this observation) without further explanation.

It follows from (5.12) that if s is *regular* [17] then there exists a tangential vector field s (the *interfacial mass flux vector*) such that

$$s = \mathbf{s}.\nu\,, \tag{5.13}$$

where ν denotes the outward unit normal to Σ. The local form of (5.12) thus becomes

$$-div_s\mathbf{s} + s_1 + s_2 = \dot{\sigma} + \sigma div_s\mathbf{u}_s \tag{5.14}$$

on using the surface divergence theorem (cf. Appendix).

Remarks

1. The time derivative which appears in (5.14) denotes that following the motion prescribed by $\mathbf{u}_s$. If another motion of S were chosen this derivative would change accordingly, and div$_s\mathbf{u}_s$ would be replaced by the surface divergence of the appropriate velocity field. Of course, in such case s_1, s_2 and s would have different interpretations.

2. Since $\mathbf{u}_s$ has a tangential component equal to the tangential component of the velocity associated with interfacial material we expect $\mathbf{s}$ to be very small. However, unlike the analogous three-dimensional situation, we feel unable to argue it to be zero since the interfacial material velocity is not here treated as existing everywhere in the interfacial region, but is ascribed to S as an average quantity. This constrained us to define $\mathbf{s}$ in terms of mass transport across a wall whose evolution is prescribed by the motion of S (and of singular surfaces (S_1, S_2), of course).

3. Terms s_1, s_2 represent the mass flux *into* the interface (across the bounding surfaces S_1, S_2, respectively) measured per unit area of S.

4. If w denotes a scalar quantity defined for every oriented curve Γ on S which depends continuously both upon position and orientation ν then w will be described as *regular*. In what follows relations of the form

$$\int_{\partial\Sigma} w = \int_{\Sigma} g \tag{5.15}$$

will often be encountered, where Σ denotes an arbitrary macroscopic subsurface of S. If w is regular then we can show the existence of a vector $\mathbf{w}$ such that

$$w = \mathbf{w}.\nu. \tag{5.16}$$

[17] Cf. Remark 4 below.

The reasoning requires several observations to be made before standard arguments (cf. 19, Gurtin & Murdoch (1975),Theorem 5.1) can be invoked, precisely because Σ in (5.15) is *macroscopic* (and so not arbitrary). If C denotes an oriented macroscopic curve segment on S with unit normal ν, and Σ_1, Σ_2 are contiguous but disjoint macroscopic subsurfaces of S with $C = \partial\Sigma_1 \cap \partial\Sigma_2$, then application of (5.15) to subsurfaces $\Sigma_1 \cup \Sigma_2$, Σ_1 and Σ_2 yields

$$\int_{|C|} w(.,\nu) + w(.,-\nu) = 0, \tag{5.17}$$

where $|\ C\ |$ denotes the set of points associated with C. Relation (5.17) will hold also for C^+, any oriented curve obtained by extending C by no matter how small an amount. The difference of (5.17) in respect of C^+ and C establishes, modulo continuity of w, the local relation

$$w(\mathbf{x},\nu) + w(\mathbf{x},-\nu) = 0 \tag{5.18}$$

for any $\mathbf{x} \in$ S. When (5.18) is taken together with the observation following (5.12), (5.16) may be deduced in standard fashion.

6. Moment of Mass Balance

Consider

$$\sum_{P_i in R(\tau)}{}^{\tau}(\mathbf{x}_i - \mathbf{x}_o)dm_i/d\tau, \tag{6.1}$$

which vanishes by (4.1). Here $\mathbf{x}_i$ denotes the location of P_i at instant τ and $\mathbf{x}_0$ an arbitrary point. This expression may be re-written as

$$\sum_{P_i in R(\tau)}{}^{\tau}(\frac{d}{d\tau}\{(\mathbf{x}_i - \mathbf{x}_0)m_i\} - m_i\mathbf{v}_i).$$

Setting (cf. (5.1))

$$\phi(\tau) := \sum_{P_i in R(\tau)}{}^{\tau} d/d\tau\{(\mathbf{x}_i - \mathbf{x})m_i\},$$

calculation of the Δ-time average of (6.1) yields (cf. (5.1,2) and (2.12))

$$\mathbf{0} = \frac{1}{\Delta}\Big\{ \sum_{P_i\,in\,R(t)}{}^{t}(\mathbf{x}_i - \mathbf{x}_0)m_i - \sum_{P_i\,in\,R(t-\Delta)}{}^{(t-\Delta)}(\mathbf{x}_i - \mathbf{x}_0)m_i \Big\}$$

$$+\frac{1}{\Delta}\Big\{\sum_j{}^{(1)}(\mathbf{x}_j - \mathbf{x}_0)m_j s_j^{(1)} + \sum_j{}^{(2)}(\mathbf{x}_j - \mathbf{x}_0)m_j s_j^{(2)} + \sum_j{}^{w}(\mathbf{x}_j - \mathbf{x}_0)m_j s_j^{w}\Big\}$$

$$-\frac{1}{\Delta}\int_{t-\Delta}^{t}\Big(\sum_{P_i\,in\,R(\tau)}{}^{\tau} m_i \mathbf{v}_i\Big)d\tau. \tag{6.2}$$

Decomposing Σ into ϵ-subsurfaces (cf. §4.2 and the discussions preceding (5.8)), writing $(\mathbf{x}_i - \mathbf{x}_0)$ as $((\mathbf{x}_i - \mathbf{x}) + (\mathbf{x} - \mathbf{x}_0))$ within interfacial ϵ-cells (and similarly for $(\mathbf{x}_j - \mathbf{x}_0)$ in respect of ϵ-subsurfaces and curves), and noting that the last term in (6.2) is

$$-\sum_{\epsilon-subsurfaces}\Big\{\Big(\overline{\sum_i{}' m_i \mathbf{v}_i}^{\Delta}/A\epsilon\Big)A_\epsilon\Big\}$$

which can be identified (cf. (4.5, 14)) with

$$-\int_\Sigma \sigma \mathbf{u},$$

relation (6.2) yields the macroscopic balance (after equating the first term with its time average and invoking A.1.)

$$d/dt\Big\{\int_{\Sigma(t)}\sigma(\mathbf{x} - \mathbf{x}_0)\Big\} + \int_\Sigma\Big\{-\mathbf{a}_1 - \mathbf{a}_2 - (s_1 + s_2)(\mathbf{x} - \mathbf{x}_0)\Big\}$$

$$+\int_{\partial\Sigma}\{\mathbf{a} + s(\mathbf{x} - \mathbf{x}_0)\} = \int_\Sigma \sigma \mathbf{u}. \tag{6.3}$$

Here (α = 1,2)

$$\mathbf{a}_\alpha(\mathbf{x}, t) := -lim_{\epsilon,\Delta}\Big\{\sum_j{}'^{(\alpha)}(\mathbf{x}_j - \mathbf{x})m_j s_j^{(\alpha)}/A_\epsilon\Delta\Big\}$$

and (6.4)

$$\mathbf{a}(\mathbf{x}, t) := lim_{\epsilon,\Delta}\Big\{\sum_j{}'^{w}(\mathbf{x}_j - \mathbf{x})m_j s_j^{w}/\ell_\epsilon\Delta\Big\}.$$

The sums and notation are precisely those of (5.7) and (5.10). Using the transport theorem, (5.13) and the surface divergence theorem, (6.3) yields

$$\int_{\partial\Sigma} \mathbf{a} = \int_{\Sigma} \Big\{ \sigma\mathbf{u} - \overbrace{\sigma(\mathbf{x}-\mathbf{x}_0)}^{\cdot} - \sigma(\mathbf{x}-\mathbf{x}_0) div_s \mathbf{u}_s + \mathbf{a}_1 + \mathbf{a}_2 + (s_1 + s_2)(\mathbf{x}-\mathbf{x}_0)$$

$$- div_s((\mathbf{x}-\mathbf{x}_0)\otimes \mathbf{s})\Big\}. \tag{6.5}$$

Hence (cf. Remark 4 of §5) if $\mathbf{a}$ is regular there exists a rank-two tensor field $\mathbf{A}$ such that

$$\mathbf{a} = \mathbf{A}\nu. \tag{6.6}$$

Balance (6.3) hence takes the form

$$-\int_{\partial\Sigma} \Big\{ \mathbf{A} + (\mathbf{x}-\mathbf{x}_0)\otimes \mathbf{s} \Big\}\nu + \int_{\Sigma} \Big\{ \mathbf{a}_1 + \mathbf{a}_2 + (\mathbf{x}-\mathbf{x}_0)(s_1 + s_2) + \sigma\mathbf{u} \Big\}$$

$$= d/dt \Big\{ \int_{\Sigma(t)} \sigma(\mathbf{x}-\mathbf{x}_0) \Big\}. \tag{6.7}$$

The local form of (6.7), upon invoking mass balance (5.14), is

$$-\mathrm{div}_s \mathbf{A} - \mathbf{s} + \mathbf{a}_1 + \mathbf{a}_2 = \sigma(\mathbf{u}_s - \mathbf{u}) = \sigma(U - u_n)\mathbf{n} \tag{6.8}$$

where

$$u_n := \mathbf{u}.\mathbf{n} \tag{6.9}$$

and we have used (3.1).

Remarks

1. While balance (6.3) is somewhat unfamiliar it will turn out that terms invoking $\mathbf{A}$ and $\mathbf{a}_1$, $\mathbf{a}_2$ will appear naturally in the balance relations to follow. For bulk continua in which mass is conserved the analogue of s vanishes as does that of $\mathbf{a}$ (cf. Murdoch (1985), §3.2, Remark 4): of course s_α and $\mathbf{a}_\alpha$ are not relevant in such context, and the analogue of the term involving the time derivative coincides with the last term since only a single velocity field is involved.

2. Notice that if the separation of S and S_α is essentially constant (h_α, say) at the ϵ-length scale then, from $(6.4)_1$ and (5.7),

$$\mathbf{a}_1.\mathbf{n} = -h_1 s_1 \text{ and } \mathbf{a}_2.\mathbf{n} = +h_2 s_2. \tag{6.10}$$

Thus the normal component of (6.8) yields (recalling **s** is tangential)

$$\sigma(U - u_n) = h_2 s_2 - h_1 s_1 - (div_s \mathbf{A}).\mathbf{n}. \tag{6.11}$$

If **a** is tangential then **A** maps tangent vectors linearly into tangent vectors and (6.11) becomes

$$\sigma(U - u_n) = h_2 s_2 - h_1 s_1 + \mathbf{A}.\mathbf{L} \tag{6.12}$$

where **L** denotes the curvature tensor [18] corresponding to orientation **n** for S. In the case of net mass transport from liquid to vapour phase, $s_1 > 0$ and $s_2 < 0$. Thus if **A**.**L** is negligible (we expect **A** always to be small; further, for plane interfaces $\mathbf{L} = \mathbf{0}$) and $s_2 = -s_1$ (steady-state evaporation) then (6.12) implies

$$\sigma(u_n - U) = s_1\, h \quad \text{where} \quad h := h_1 + h_2 \tag{6.13}$$

denotes interfacial thickness. Consideration of a plane water/water vapour interface of thickness 10$\mathring{A}$, "falling" at the rate of 3.6mm per hour due to evaporation (so that $U = -10^{-6} ms^{-1}$), and taking $\rho_1 = 10^3 kgm^{-3}, \rho_2 = 1 kgm^{-3}$, and $\sigma = (\rho_1 + \rho_2)h/2 = 5 \times 10^{-7} kgm^{-2}$, we find from $(6.13)_1$ that $u_n - U = 2 \times 10^{-6} ms^{-1}$. This indicates that $u_n - U$ should not be neglected in comparison with U.

Assuming representations

$$s_1 = \rho_1(\mathbf{v}_1.\mathbf{n} - U_1) \quad \text{and} \quad s_2 = \rho_2(U_2 - \mathbf{v}_2.\mathbf{n}), \tag{6.14}$$

where $\rho_\alpha, \mathbf{v}_\alpha$ denote limiting values of bulk density and velocity fields on S_α (which has singular surface velocity $U_\alpha \mathbf{n}$), then noting

$$\dot{h}_2 = U_2 - U \text{ and } \dot{h}_1 = U - U_1 \tag{6.15}$$

it follows from (6.12) that

$$(\sigma - \rho_1 h_1 - \rho_2 h_2)U = \sigma u_n - \rho_1 h_1 \dot{h}_1 + \rho_2 h_2 \dot{h}_2 - (\rho_1 h_1 \mathbf{v}_1 + \rho_2 h_2 \mathbf{v}_2).\mathbf{n}. \tag{6.16}$$

[18] Cf. Appendix §1; $\mathbf{A}.\mathbf{L} := tr(\mathbf{A}^T\mathbf{L})$.

7. Linear Momentum Balance

Summing relations (4.2) over all P_i in $R(\tau)$ and time averaging for $t-\Delta \leq \tau \leq t$ we obtain for the left-hand-side contributions the macroscopic representation

$$\int_{\Sigma} (\mathbf{f} + \mathbf{t}_1 + \mathbf{t}_2 + \mathbf{b}) \tag{7.1}$$

where

$$\left.\begin{aligned}
\mathbf{f}(\mathbf{x},t) &:= lim_\epsilon \left\{ \overline{\sum_i{}' \sum_\ell \mathbf{f}_{i\ell}}^{\Delta} / A_\epsilon \right\}, \\
\mathbf{t}_1(\mathbf{x},t) &:= lim_\epsilon \left\{ \overline{\sum_i{}' \sum_r \mathbf{f}_{ir}}^{\Delta} / A_\epsilon \right\}, \\
\mathbf{t}_2(\mathbf{x},t) &:= lim_\epsilon \left\{ \overline{\sum_i{}' \sum_s \mathbf{f}_{is}}^{\Delta} / A_\epsilon \right\}, \\
\mathbf{b}(\mathbf{x},t) &:= lim_\epsilon \left\{ \overline{\sum_i{}' \mathbf{b}_i}^{\Delta} / A_\epsilon \right\}.
\end{aligned}\right\} \tag{7.2}$$

At each instant in the period $t-\Delta \leq \tau \leq t$ the sums are taken over all particles P_i instantaneously within a deforming walled interfacial ϵ-cell centred at $\mathbf{x} \in S(t)$ at instant t with base area A_ϵ and all particles P_r and P_s in the bulk phases adjacent to S_1 and S_2, respectively.

The right-hand-side contributions are

$$\frac{1}{\Delta} \int_{t-\Delta}^{t} \left\{ \sum_{P_i in R(\tau)}{}^{\tau} d/d\tau(m_i \mathbf{v}_i) \right\} d\tau$$

which yield (cf. (5.2))

$$\frac{1}{\Delta}\left\{ \sum_{P_i in R(t)}{}^{t} m_i \mathbf{v}_i - \sum_{P_i in R(t-\Delta)}{}^{(t-\Delta)} m_i \mathbf{v}_i \right\} + \frac{1}{\Delta}\left\{ \sum_j{}^{(1)} m_j \mathbf{v}_j s_j^{(1)} + \sum_j{}^{(2)} m_j \mathbf{v}_j s_j^{(2)} + \sum_j{}^{w} m_j \mathbf{v}_j s_j^{w} \right\}. \tag{7.3}$$

Identifying these terms [19] with their Δ-time averages, if these do not fluctuate (the sums are over macroscopic geometrical objects), or by undertaking a second time average, the

[19] Cf.§5, following (5.4), for an explanation of the particles involved in the P_j sums. The physical interpretation of surface gradient $\nabla_s \mathbf{u}$ is given in (A.5) of the Appendix.

continuum counterpart of (7.3) is found to be

$$d/dt\Big\{\int_{\Sigma(t)} \sigma \mathbf{u}\Big\} + \int_{\Sigma}\Big\{\mathbf{d}_1 + \mathbf{d}_2 - (s_1 + s_2)\mathbf{u} - (\nabla_s \mathbf{u})\mathbf{P}(\mathbf{a}_1 + \mathbf{a}_2)\Big\}$$

$$+ \int_{\partial\Sigma}\Big\{\mathbf{d} + s\mathbf{u} + (\nabla_s \mathbf{u})\mathbf{P}\mathbf{a}\Big\}. \tag{7.4}$$

Here (α = 1,2)

$$\mathbf{d}_\alpha(\mathbf{x},t) := lim_{\epsilon,\Delta}\Big\{\sum_i{}^{\prime(\alpha)} m_j \tilde{\mathbf{v}}_j s_j^{(\alpha)}/A_\epsilon\Delta\Big\}$$

and (7.5)

$$\mathbf{d}(\mathbf{x},t) := lim_{\epsilon,\Delta}\Big\{\sum_j{}^{\prime w} m_j \tilde{\mathbf{v}}_j s_j^w/\ell_\epsilon\Delta\Big\}.$$

In deriving (7.4) use has been made of (4.14), thermal velocity approximation A.1., (5.7, 10) and (6.4).

Invoking the transport theorem and representations (5.13) and (6.6) yields a relation (on equating (7.1) with (7.4)) of form (5.15) with w generalised to a vector, here $\mathbf{d}$. Thus if $\mathbf{d}$ is regular then there exists a tensor[20] $-\mathbf{D}$ (the *interfacial diffusive stress tensor*) such that

$$\mathbf{d} = \mathbf{D}\nu. \tag{7.6}$$

Accordingly, at this stage linear momentum balance takes the form

$$\int_{\partial\Sigma} -(\mathbf{D} + \mathbf{u}\otimes\mathbf{s} + (\nabla_s\mathbf{u})\mathbf{P}\mathbf{A})\nu + \int_{\Sigma}\{\mathbf{f} + t_1 + t_2 + \mathbf{b}\}$$

$$= d/dt\Big\{\int_{\Sigma(t)} \sigma\mathbf{u}\Big\} + \int_{\Sigma}\{\mathbf{d}_1 + \mathbf{d}_2 - (s_1 + s_2)\mathbf{u} - (\nabla_s\mathbf{u})\mathbf{P}\mathbf{a}_1 + \mathbf{a}_2)\}. \tag{7.7}$$

Locally (7.7) yields

$$-div_s\mathbf{D} - (div_s\mathbf{s})\mathbf{u} - (\nabla_s\mathbf{u})\mathbf{s} - (\nabla_s\mathbf{u})\mathbf{P}div_s\mathbf{A} - \mathbf{B} : \mathbf{A}$$

---

[20] At any point $\mathbf{x} \in S(t)$ at instant t, $\mathbf{D}(\mathbf{x},t)$ maps tangent vectors into vectors.

$$+\mathbf{f}+(\mathbf{t}_1-\mathbf{d}_1)+(\mathbf{t}_2-\mathbf{d}_2)+\mathbf{b}+(s_1+s_2)\mathbf{u}+(\nabla_s\mathbf{u})\mathbf{P}(\mathbf{a}_1+\mathbf{a}_2)$$

$$= \quad (\dot{\sigma}+\sigma div_s\mathbf{u})\mathbf{u}+\sigma\dot{\mathbf{u}}, \tag{7.8}$$

where

$$\mathbf{B} := \quad (\nabla_s^2\mathbf{u})\mathbf{P}+(\nabla_s\mathbf{u})\mathbf{L}\otimes\mathbf{n} \tag{7.9}$$

and ":" denotes double composition (or, in component formulation, double contraction of indices). See (A.11) for the interpretation of $\nabla_s^2\mathbf{u}$.

Local mass and mass moment balances (5.14) and (6.8), together with the observation that $\mathbf{Pn}=\mathbf{0}$, reduce (7.8) to

$$-div_s\mathbf{D}+\mathbf{f}+(\mathbf{t}_1-\mathbf{d}_1)+(\mathbf{t}_2-\mathbf{d}_2)+\mathbf{b}=\sigma\dot{\mathbf{u}}+\mathbf{B}:\mathbf{A}. \tag{7.10}$$

Up to this point no assumptions about the nature of interactions have been made beyond accepting their existence and additivity. Consistent with what is known concerning molecular interactions we make the following very weak assumptions.

I.1 The net self-force associated with particles within any interfacial ϵ-cell,[21] or union thereof, is zero.

I.2 If P_i lies within the interface, but outside a macroscopic interfacial region $\mathcal{R}$, then the net force exerted on P_i by particles in $\mathcal{R}$ may be computed by summing interactions between P_i and particles in $\mathcal{R}$ distant less than δ therefrom, where[22] $\delta < 10^3\,\text{Å}$.

Assumption I.1 means that

$$\sum_i{}'\sum_k{}'\mathbf{f}_{ik}=\frac{1}{2}\sum_i{}'\sum_k{}'(\mathbf{f}_{ik}+\mathbf{f}_{ki})=\mathbf{0}, \tag{7.11}$$

where the sums are over all pairs of particles P_i, P_k in an interfacial ϵ-cell or union of such. Clearly from (7.11) this is true if $\mathbf{f}_{ik}+\mathbf{f}_{ki}=\mathbf{0}$ for each such pair: however, I.1 is much weaker than such a requirement.

---

[21] With $\epsilon \sim 10^{-5}m$.

[22] Molecular interactions of range $10^3\,\text{Å}$ are considered to be extremely long in the literature: cf., e.g. Hirschfelder (1967) and also Murdoch (1985), §3.1, Remark 3.

As a consequence of I.1,2 (with $\mathcal{R}$ as the interfacial complement of $R(\tau)$) the sum over P_i in $R(\tau)$ of the first term in (4.2) simplifies:

$$\sum_i \sum_\ell \mathbf{f}_{i\ell} = \sum_i \sum_k \mathbf{f}_{ik} + \sum_i \sum_p \mathbf{f}_{ip}, \tag{7.12}$$

where the "k" sum is over particles in $R(\tau)$ and the "p" sum over particles in the interface at instant τ which lie *outside* $R(\tau)$. I.1 implies the "k" sum vanishes while I.2 then reduces (7.12) to

$$\sum_i \sum_\ell \mathbf{f}_{i\ell} = \sum_i \sum_j \mathbf{f}_{ij}, \tag{7.13}$$

where the second double sum is only over particles P_i within $R(\tau)$ and particles P_j outside $R(\tau)$ with $P_iP_j < \delta$. Accordingly, particles P_i which yield non-vanishing contributions must lie within that region bounded by $S_1(\tau), S_2(\tau), W(\tau)$ (cf.§4.2) and a wall *parallel* to $W(\tau)$ lying *within* $R(\tau)$ and distant δ therefrom. For book-keeping purposes we term the intersection of such region with a walled interfacial ϵ-cell centred at $\mathbf{x} \in \partial\Sigma(\tau)$ an *interfacial wall ϵ-cell centred at* $\mathbf{x}$ *and based upon* $\partial\Sigma(\tau)$. If the aforementioned region is partitioned into interfacial wall ϵ-cells at instant t, so inducing via the motion map such a decomposition at any instant τ, then from (7.13)

$$\sum_i \sum_\ell \mathbf{f}_{i\ell} = \sum_{cells} \Big({\sum_i}' \sum_j \mathbf{f}_{ij}/\ell_\epsilon \Big) \ell_\epsilon, \tag{7.14}$$

where the "i " sum is now over P_i in an interfacial wall ϵ-cell based upon $\partial\Sigma(\tau)$ with length (measured along $\partial\Sigma(\tau)$) ℓ_ϵ at instant t. Upon time averaging (7.14) we obtain

$$\int_\Sigma \mathbf{f} = \int_{\partial\Sigma} \mathbf{t} \tag{7.15}$$

where

$$\mathbf{t}(\mathbf{x}, t) := lim_\epsilon \Big\{ \overline{{\sum_i}' \sum_j \mathbf{f}_{ij}}^{\Delta} / \ell_\epsilon \Big\}. \tag{7.16}$$

Of course, the cell associated with the sum in (7.16) is centred at $\mathbf{x}$ at instant t.

It follows that if $\mathbf{t}$ is regular then there exists a tensor field [23] $\mathbf{T}^-$ on S such that

$$\mathbf{t} = \mathbf{T}^- \nu. \tag{7.17}$$

[23] Having the same character as $\mathbf{D}$ introduced in (7.6).

The surface divergence theorem implies from (7.15) that

$$\int_\Sigma \mathbf{f} = \int_\Sigma div_s \mathbf{T}^-$$

and we accordingly

$$\textit{identify } \mathbf{f} \textit{ with } div_s\, \mathbf{T}^-. \tag{7.18}$$

Linear momentum balance (7.7) may now be expressed as

$$\int_{\partial\Sigma} \Big\{ \mathbf{T} - \mathbf{u} \otimes \mathbf{s} - (\nabla_s \mathbf{u})\mathbf{P}\mathbf{A} \Big\} \nu + \int_\Sigma \Big\{ \mathbf{t}_1 - \mathbf{d}_1 + \mathbf{t}_2 - \mathbf{d}_2 + \mathbf{b} + (s_1 + s_2)\mathbf{u} + (\nabla_s \mathbf{u})\mathbf{P}(\mathbf{a}_1 + \mathbf{a}_2) \Big\}$$

$$= d/dt \Big\{ \int_{\Sigma(t)} \sigma \mathbf{u} \Big\}, \tag{7.19}$$

where

$$\mathbf{T} := \mathbf{T}^- - \mathbf{D} \tag{7.20}$$

denotes the *interfacial stress tensor.*

Remarks

1. The decomposition (7.20) of interfacial stress into separate contributions from interactions and momentum transport is well-established (cf., e.g., equation (11.7) of Defay, Prigogine & Bellemans (1966) or equation (10.1) of March & Tosi (1976) and is the bidimensional analogue of a similar result for bulk stress (cf. $(3.10)_2$ in Murdoch (1985)). Given our observations that both s and $\mathbf{a}$ ($= \mathbf{A}\ \nu$) are expected to be very small (cf. Remark 2 of §5 and text following (6.12)) it follows from (4.11) that $\tilde{\mathbf{v}}_j$ in $(7.5)_2$ may be replaced, without significant change, by $\mathbf{v}_j$. Accordingly the contribution $-\mathbf{D}$ is *pressure-like* in that for all tangent vectors ν

$$-\mathbf{D}\nu.\nu = -\mathbf{d}.\nu < 0. \tag{7.21}$$

This follows from $(7.5)_2$ since $(\tilde{\mathbf{v}}_j.\nu)s_j^w > 0$ for each term in the sum: if P_j leaves the region, $\tilde{\mathbf{v}}_j.\nu > 0$ and $s_j^w = +1$ while if P_j enters, $\tilde{\mathbf{v}}_j.\nu < 0$ and $s_j^w = -1$.

If the interfacial stress is *tension-like*, that is

$$\mathbf{T}\nu.\nu > 0 \tag{7.22}$$

for all ν then (7.20,21) imply that

$$\mathbf{T}^{-}\nu.\nu \; > \; \mathbf{D}\nu.\nu > 0. \tag{7.23}$$

Suppose that $-\mathbf{D}$ is a surface pressure p and $\mathbf{T}$ a surface tension γ; that is,

$$-\mathbf{D} = -p\mathbf{1} \text{ and } \mathbf{T} = \gamma\mathbf{1},$$

where p, $\gamma > 0$ and $\mathbf{1}$ denotes the identity on the tangent space. Then

$$\mathbf{T}^{-} = \gamma'\mathbf{1}, \tag{7.24}$$

where

$$(p+\gamma) =: \gamma' > p > 0. \tag{7.25}$$

Inequalities (7.25) provide unequivocal evidence of the net cohesive effect of interfacial molecular interactions. Said differently, the existence of surface *tension* indicates that the resultant molecular interaction average represented by γ' is not only positive (evidence of net attraction) but exceeds the diffusive pressure p.

2. The mechanical effect of bulk phase "α" on the interface is represented by $\mathbf{t}_\alpha - \mathbf{d}_\alpha$ and is usually equated with $\mathbf{T}_\alpha\mathbf{n}_\alpha$ where $\mathbf{T}_\alpha$ denotes the limiting value of the bulk (Cauchy) stress tensor as S_α (with orientation $\mathbf{n}_\alpha$, directed into the bulk phase) is approached. Such identification is questionable, as seems first to have been noted by Slattery (1967). There are essentially three objections.

(i) Since $\mathbf{T}_\alpha$ is referred to S_α it should be the representation $\bar{\mathbf{t}}_\alpha - \bar{\mathbf{d}}_\alpha$, say, of $\mathbf{t}_\alpha - \mathbf{d}_\alpha$ on S_α which should be compared with $\mathbf{T}_\alpha\mathbf{n}_\alpha$. Specifically,

$$\mathbf{t}_\alpha - \mathbf{d}_\alpha = (\bar{\mathbf{t}}_\alpha \circ \mathbf{c}_\alpha - \bar{\mathbf{d}}_\alpha \circ \mathbf{c}_\alpha)J_\alpha,$$

where $\mathbf{c}_\alpha$ identifies with $\mathbf{x} \in S$ the point of intersection of S_α with the normal to S at $\mathbf{x}$ and has Jacobian (cf. Murdoch (1990a), A.16)

$$J_\alpha := (det\{(\mathbf{1} - (-1)^\alpha h_\alpha \mathbf{L}_n)^2 + \nabla_s h_\alpha \otimes \nabla_s h_\alpha\})^{\frac{1}{2}}.$$

Here h_α denotes the distance between corresponding points $\mathbf{x}$ and $\mathbf{c}_\alpha(\mathbf{x})$ of S and S_α, $\mathbf{1}$ the identity on tangent spaces of S, and $\mathbf{L}_n$ the curvature tensor of S with orientation

$\mathbf{n}$ directed from S to S_2. If S, S_α are plane parallel surfaces then $J_\alpha = 1$ and the above distinction is not necessary,

(ii) Quantities $\bar{\mathbf{t}}_\alpha - \bar{\mathbf{d}}_\alpha$ and $\mathbf{T}_\alpha \mathbf{n}_\alpha$ are computed as area-time averages in which different velocity fields are involved. The former are defined in terms of the motion of S prescribed by $\mathbf{u}_s$, while $\mathbf{T}_\alpha$ involves time averaging associated with the bulk motion prescribed by the bulk phase velocity field $\mathbf{v}_\alpha$. Further, the diffusive contributions $-\bar{\mathbf{d}}_\alpha$ and (for $\mathbf{T}_\alpha \mathbf{n}_\alpha$: cf. Murdoch (1985), $(3.10)_2$ and Remark 4 on p.310) $-\mathbf{D}_\alpha \mathbf{n}_\alpha$ involve different thermal velocities: the former are measured relative to $\mathbf{u}$ and the latter to $\mathbf{v}_\alpha$. Of course, all these differences disappear in the case of static equilibrium.

(iii) The interaction contribution to $\mathbf{T}_\alpha \mathbf{n}_\alpha$ represents the effect of "α" bulk phase molecules upon molecules on the side of S_α remote from this phase. Accordingly, unless these interactions are of range $< 10\mathring{A}$ they will involve interactions with molecules actually in the interface (rather than the *model* interfacial region which consists of the real interface sandwiched between two "layers" of thickness $\sim 10\mathring{A}$). Since the real interfacial region corresponds to a jump in two bulk densities whose ratio is of order 10^3:1, such interaction contributions will differ greatly from the situation within a bulk phase. Further, interactions of range $> 20\mathring{A}$ would imply that $\mathbf{T}_\alpha \mathbf{n}_\alpha$ should include the effect of "α" molecules on bulk phase molecules on the *opposite side of the interface.* Accordingly, unless the effect of interactions with range $> 10\mathring{A}$ is negligible, the identification of $\mathbf{T}_\alpha \mathbf{n}_\alpha$ with $\bar{\mathbf{t}}_\alpha - \bar{\mathbf{d}}_\alpha$ cannot be made.

3. The local form of (7.19) on using (7.18) and (7.10) is

$$div_s \mathbf{T} + (\mathbf{t}_1 - \mathbf{d}_1) + (\mathbf{t}_2 - \mathbf{d}_2) + \mathbf{b} = \sigma \dot{\mathbf{u}} + \mathbf{B} : \mathbf{A}. \quad (7.26)$$

8. Moment of Momentum Balance

A *generalised* moment of momentum balance is obtained by tensorial (pre-)multiplication of relation (4.2) by the displacement $(\mathbf{x}_i - \mathbf{x}_0)$ of the instantaneous location $\mathbf{x}_i$ of P_i from an (arbitrary) fixed point $\mathbf{x}_0$, summing over all particles in $R(\tau)$ at instant τ, and computing the Δ-time average of the resulting relation at instant t.

The continuum version of terms deriving from the left-hand side of (4.2) is

$$\int_{\Sigma} \{\mathcal{G} + \mathcal{C}_1 + \mathcal{C}_2 + \mathbf{G} + (\mathbf{x} - \mathbf{x}_0) \otimes (\mathbf{f} + \mathbf{t}_1 + \mathbf{t}_2 + \mathbf{b})\}, \tag{8.1}$$

where

$$\left.\begin{aligned}
\mathcal{G}(\mathbf{x}, t) &:= lim_\epsilon \Big\{ \overline{\sum_i{}' \sum_\ell (\mathbf{x}_i - \mathbf{x}) \otimes \mathbf{f}_{i\ell}}^{\Delta} / A_\epsilon \Big\}, \\
\mathcal{C}_1(\mathbf{x}, t) &:= lim_\epsilon \Big\{ \overline{\sum_i{}' \sum_r (\mathbf{x}_i - \mathbf{x}) \otimes \mathbf{f}_{ir}}^{\Delta} / A_\epsilon \Big\}, \\
\mathcal{C}_2(\mathbf{x}, t) &:= lim_\epsilon \Big\{ \overline{\sum_i{}' \sum_s (\mathbf{x}_i - \mathbf{x}) \otimes \mathbf{f}_{is}}^{\Delta} / A_\epsilon \Big\}, \\
\mathbf{G}(\mathbf{x}, t) &:= lim_\epsilon \Big\{ \overline{\sum_i{}' (\mathbf{x}_i - \mathbf{x}) \otimes \mathbf{b}_i}^{\Delta} / A_\epsilon \Big\}.
\end{aligned}\right\} \tag{8.2}$$

The sums in $(8.2)_{1,2,3,4}$, are taken over particles as delineated in $(7.2)_{1,2,3,4}$, respectively.

The right-hand sides of relations (4.2) yield the contribution

$$\frac{1}{\Delta} \int_{t-\Delta}^{t} \sum_i{}^{\tau} (\mathbf{x}_i - \mathbf{x}_0) \otimes \frac{d}{d\tau}(m_i \mathbf{v}_i) \; d\tau$$

which may be written as

$$\frac{1}{\Delta} \int_{t-\Delta}^{t} \sum_i{}^{\tau} \Big\{ d/d\tau((\mathbf{x}_i - \mathbf{x}_0) \otimes m_i \mathbf{v}_i) - \mathbf{v}_i \otimes m_i \mathbf{v}_i \Big\} d\tau. \tag{8.3}$$

At each instant τ in the time-averaging period the sum is over all P_i in $R(\tau)$.

The final term of (8.3) may be written (using A.1) as

$$-\frac{1}{\Delta} \int_{t-\Delta}^{t} \sum_{cells} \Big\{ \sum_i{}' (\tilde{\mathbf{v}}_i + \mathbf{u} + (\nabla_s \mathbf{u})\mathbf{P}(\mathbf{x}_i - \mathbf{x}) \otimes m_i(\tilde{\mathbf{v}}_i + \mathbf{u} + (\nabla_s \mathbf{u})\mathbf{P}(\mathbf{x}_i - \mathbf{x})) \Big\} d\tau. \tag{8.4}$$

Making the assumptions (cf. Remark 3 below)

$$A.3 \quad lim_\epsilon \Big\{ \overline{\sum_i{}' (\mathbf{x}_i - \mathbf{x}) \otimes m_i (\mathbf{x}_i - \mathbf{x})}^{\Delta} / A_\epsilon \Big\} \quad \text{is negligible} \tag{8.5}$$

and

$$T.M.2 \quad lim_\epsilon\left\{\overline{\sum_i{}'(\mathbf{x}_i - \mathbf{x}) \otimes m_i\tilde{\mathbf{v}}_i}^{\Delta} /A_\epsilon\right\} = \mathbf{0}, \tag{8.6}$$

(8.4) has the continuum representation (on invoking also A.2 and T.M.1)

$$-\int_\Sigma (\mathbf{K} + \mathbf{u} \otimes \sigma\mathbf{u}), \tag{8.7}$$

where the *interfacial thermal tensor*

$$\mathbf{K}(\mathbf{x}, t) := lim_\epsilon\left\{\overline{\sum_i{}'\tilde{\mathbf{v}}_i \otimes m_i\tilde{\mathbf{v}}_i}^{\Delta} /A_\epsilon\right\}. \tag{8.8}$$

The first term in (8.3) yields, in the manner of (5.2) and (6.1),

$$\frac{1}{\Delta}\left\{\sum_{P_i in R(t)}{}^t\phi_i - \sum_{P_i in R(t-\Delta)}{}^{(t-\Delta)}\phi_i\right\} + \frac{1}{\Delta}\left\{\sum_j{}^{(1)}\phi_j s_j^{(1)} + \sum_j{}^{(2)}\phi_j s_j{}^{(2)} + \sum_i{}^{w}\phi_j s_j^{w}\right\}, \tag{8.9}$$

where (k = i, j)

$$\phi_k := (\mathbf{x}_k - \mathbf{x}_0) \otimes m_k\mathbf{v}_k. \tag{8.10}$$

Appropriate cell partitions enable the second collection of sums in (8.9) to be expressed as cellular sums of quantities of the form (β = (1), (2), w)

$$\sum_j{}'^{\beta}(\mathbf{x}_j - \mathbf{x}_0) \otimes m_j\mathbf{v}_j s_j^\beta \tag{8.11}$$

taken over an individual cell. If this is centred at **x** then (8.11) may be written as (cf. (4.11))

$$\sum_k{}'^{\beta}((\mathbf{x}_k - \mathbf{x}) + (\mathbf{x} - \mathbf{x}_0)) \otimes m_k(\tilde{\mathbf{v}}_k + \mathbf{u} + (\nabla_s\mathbf{u})\mathbf{P}(\mathbf{x}_k - \mathbf{x}))s_k^\beta. \tag{8.12}$$

Accordingly the diffusive terms in (8.9) have the continuum counterpart

$$\int_{\Sigma}\left\{\mathbf{M}_1+\mathbf{M}_2-(\mathbf{a}_1+\mathbf{a}_2)\otimes\mathbf{u}+(\mathbf{x}-\mathbf{x}_0)\otimes(\mathbf{d}_1+\mathbf{d}_2-(s_1+s_2)\mathbf{u}-(\nabla_s\mathbf{u})\mathbf{P}(\mathbf{a}_1+\mathbf{a}_2))\right\}$$

$$+\int_{\partial\Sigma}\left\{\mathbf{M}+\mathbf{a}\otimes\mathbf{u}+(\mathbf{x}-\mathbf{x}_0)\otimes(\mathbf{d}+s\mathbf{u}+(\nabla_s\mathbf{u})\mathbf{Pa})\right\}, \tag{8.13}$$

where

$$\mathbf{M}_\alpha(\mathbf{x},t):=lim_{\epsilon,\Delta}\left\{\sum_j{}^{\prime(\alpha)}(\mathbf{x}_j-\mathbf{x})\otimes m_j\tilde{\mathbf{v}}_j s_j^{(\alpha)}/A_\epsilon\Delta\right\}$$

and (8.14)

$$\mathbf{M}(\mathbf{x},t):=lim_{\epsilon,\Delta}\left\{\sum_j{}^{\prime w}(\mathbf{x}_j-\mathbf{x})\otimes m_j\tilde{\mathbf{v}}_j s_j^{w}/\ell_\epsilon\Delta\right\}.$$

Here the term in each expression (8.12) involving products of ($\mathbf{x}_k-\mathbf{x}$) with itself has been neglected (cf. Remark 3 below).

The first term in (8.9), upon making a decomposition of form (8.12) (but without superscript (β) and diffusive transport factor $s_k^{(\beta)}$) is identifiable with

$$d/d\tau\left\{\int_{\Sigma(\tau)}(\mathbf{x}-\mathbf{x}_0)\otimes\sigma\mathbf{u}\right\}\Big|_{\tau=t} \tag{8.15}$$

on invoking T.M.1,2 and A.1,2,3.

From (8.1,7,13,15) we obtain the balance

$$-\int_{\partial\Sigma}\left\{\mathbf{M}+\mathbf{a}\otimes\mathbf{u}+(\mathbf{x}-\mathbf{x}_0)\otimes(\mathbf{d}+s\mathbf{u}+(\nabla_s\mathbf{u})\mathbf{Pa})\right\}$$

$$+\int_{\Sigma}\left\{\mathcal{G}+\mathcal{C}_1-\mathbf{M}_1+\mathcal{C}_2-\mathbf{M}_2+\mathbf{G}+\mathbf{K}+(\mathbf{a}_1+\mathbf{a}_2+\sigma\mathbf{u})\otimes\mathbf{u}\right.$$

$$\left.+(\mathbf{x}-\mathbf{x}_0)\otimes[\mathbf{f}+\mathbf{t}_1-\mathbf{d}_1+\mathbf{t}_2-\mathbf{d}_2+\mathbf{b}+(s_1+s_2)\mathbf{u}+(\nabla_s\mathbf{u})\mathbf{P}(\mathbf{a}_1+\mathbf{a}_2)]\right\}$$

$$=\frac{d}{d\tau}\left\{\int_{\Sigma(\tau)}(\mathbf{x}-\mathbf{x}_0)\otimes\sigma\mathbf{u}\right\}\Big|_{\tau=t} \tag{8.16}$$

$$= \int_{\Sigma} \Big\{ \mathbf{u}_s \otimes \sigma \mathbf{u} + (\mathbf{x} - \mathbf{x}_0) \otimes (\sigma \dot{\mathbf{u}} + [\dot{\sigma} + \sigma div_s \mathbf{u}_s]\mathbf{u}) \Big\}. \tag{8.17}$$

Relation (8.17), together with (5.13), (6.6) and (7.6), implies that if $\mathbf{M}$ is regular then there exists a rank-three tensor field $\mathcal{M}$ in terms of which

$$\mathbf{M} = \mathcal{M}\nu. \tag{8.18}$$

Substitution of (8.18) in (8.17) yields, after considerable manipulation and invocation of local balances (5.14), (6.8) and (7.10), the local form (cf. (A.2))

$$-div_s \mathcal{M} - \mathbf{A}(\nabla_s \mathbf{u})^T - \mathbf{I}\mathbf{A}^T\mathbf{I}(\nabla_s \mathbf{u})^T - \mathbf{I}\mathbf{D}^T + \mathcal{G} + \mathcal{C}_1 - \mathbf{M}_1 + \mathcal{C}_2 - \mathbf{M}_2 + \mathbf{G} + \mathbf{K} = \mathbf{0}. \tag{8.19}$$

Modulo the further interaction assumption

I.3 The net Δ-time averaged self-couple associated with particles in any microscopically-large interfacial region is negligible[24] (Δ macroscopically small)

we can deduce (in the manner of (7.14,15)) that

$$\int_{\Sigma} \Big\{ \mathcal{G} + (\mathbf{x} - \mathbf{x}_0) \otimes \mathbf{f} \Big\} = \int_{\partial\Sigma} \Big\{ \hat{\mathcal{C}} + (\mathbf{x} - \mathbf{x}_0) \wedge \mathbf{t} \Big\} + \int_{\Sigma} \Big\{ \mathbf{J} + (\mathbf{x} - \mathbf{x}_0) \sim \mathbf{f} \Big\}. \tag{8.20}$$

Here

$$\mathbf{J} := sym\ \mathcal{G}, \tag{8.21}$$

$$\hat{\mathcal{C}}(\mathbf{x}, t) := lim_\epsilon \Big\{ \overline{\sum_i{}' \sum_j (\mathbf{x}_i - \mathbf{x}) \wedge \mathbf{f}_{ij}}^{\Delta} / \ell_\epsilon \Big\} \tag{8.22}$$

and $\mathbf{a} \sim \mathbf{b}$ ($\mathbf{a} \wedge \mathbf{b}$) denotes the symmetric (skew) part of $\mathbf{a} \otimes \mathbf{b}$ for any pair of vectors $\mathbf{a}$, $\mathbf{b}$. The sum in (8.22) is taken over the same particles as in (7.16).

If $\hat{\mathcal{C}}$ is regular then (8.20) implies, on recalling (7.18), that there exists a rank-three tensor field $\hat{\mathbf{C}}^-$ such that

[24] Cf. Murdoch (1985),p.304 and Remark 1 below.

$$\hat{\mathcal{C}} = \hat{\mathbf{C}}^{-}\nu. \tag{8.23}$$

In such case (8.16) may be written, on using (8.18,20), (7.17,18), (5.13), (6.6) and (7.6), as

$$\int_{\partial\Sigma}\Big\{\bar{\mathbf{C}}\nu - \mathbf{A}\nu\otimes\mathbf{u} + (\mathbf{x}-\mathbf{x}_0)\wedge(\mathbf{T}-\mathbf{u}\otimes\mathbf{s}-(\nabla_s\mathbf{u})\mathbf{PA})\nu\Big\}$$

$$+\int_{\Sigma}\Big\{\mathcal{C}_1 - \mathbf{M}_1 + \mathcal{C}_2 - \mathbf{M}_2 + \mathbf{G} + \mathbf{J} + \mathbf{K} - sym(\mathbf{I}(\mathbf{T}^{-})^{\mathbf{T}}) + (\mathbf{a}_1 + \mathbf{a}_2 + \sigma\mathbf{u})\otimes\mathbf{u}$$

$$+(\mathbf{x}-\mathbf{x}_0)\otimes(\mathbf{t}_1 - \mathbf{d}_1 + \mathbf{t}_2 - \mathbf{d}_2 + \mathbf{b} + (s_1 + s_2)\mathbf{u} + (\nabla_s\mathbf{u})\mathbf{P}(\mathbf{a}_1 + \mathbf{a}_2))\Big\}$$

$$= d/dt\Big\{\int_{\Sigma(t)}(\mathbf{x}-\mathbf{x}_0)\otimes\sigma\mathbf{u}\Big\} \tag{8.24}$$

where

$$\bar{\mathbf{C}} := \hat{\mathbf{C}}^{-} - \mathcal{M}. \tag{8.25}$$

From (8.20,23) and (7.17,18)

$$\mathcal{G} = div_s\hat{\mathbf{C}}^{-} + \mathbf{J} + sk(\mathbf{I}(\mathbf{T}^{-})^{T}) \tag{8.26}$$

so by (8.19) the local form of (8.24) is

$$div_s\bar{\mathbf{C}} + sk(\mathbf{I}\mathbf{T}^{T}) - sym(\mathbf{I}\mathbf{D}^{T}) - (\mathbf{A} + \mathbf{I}\mathbf{A}^{T}\mathbf{I})(\nabla_s\mathbf{u})^{T} + \mathbf{J} + \mathbf{K} + \mathbf{G} + \mathcal{C}_1 - \mathbf{M}_1 + \mathcal{C}_2 - \mathbf{M}_2 = \mathbf{0}. \tag{8.27}$$

Remarks

1. The net self-couple associated with a set of particles whose interactions satisfy I.1 is

$$\sum_i\sum_k(\mathbf{x}_i - \mathbf{x}_0)\wedge\mathbf{f}_{ik}\ , \tag{8.28}$$

where $\mathbf{x}_0$ denotes any point: sum (8.28) is independent of $\mathbf{x}_0$ by I.1. Clearly

$$\textit{if } \mathbf{f}_{ik} + \mathbf{f}_{ki} = \mathbf{0} \textit{ and } \mathbf{f}_{ik} \textit{ is parallel to } (\mathbf{x}_i - \mathbf{x}_k) \tag{8.29}$$

then (8.28) simplifies to

$$\frac{1}{2}\sum_i \sum_k (\mathbf{x}_i - \mathbf{x}_k) \wedge \mathbf{f}_i k$$

and each term vanishes. Of course, I.3. is a much weaker restriction upon interactions than (8.29).

2. It is the skew part of (8.24) which corresponds to the usual skew tensor-valued (equivalently, axial vector-valued) balance of rotational momentum. Since **J** and **K** take symmetric values and div $\hat{\mathbf{C}}^-$ skew values, this balance has, from (8.27), the local form

$$div_s\hat{\mathbf{C}} + sk(\mathbf{IT}^T) + \hat{\mathbf{G}} + \hat{\mathcal{C}}_1 - \hat{\mathbf{M}}_1 + \hat{\mathcal{C}}_2 - \hat{\mathbf{M}}_2 - \mathbf{S} = \mathbf{0}, \tag{8.30}$$

where

$$\hat{\mathbf{C}}\nu := \hat{\mathbf{C}}^-\nu - sk(\mathcal{M}\nu),\ \hat{\mathcal{C}}_\alpha := sk\ \mathcal{C}_\alpha,\ \hat{\mathbf{M}}_\alpha := sk\ \mathbf{M}_\alpha,\ \hat{\mathbf{G}} := sk\,\mathbf{G}$$

and

$$\mathbf{S} := sk\Big\{(\mathbf{A} + \mathbf{IA}^T\mathbf{I})(\nabla_s\mathbf{u})^T\Big\}. \tag{8.31}$$

Terms $(\hat{\mathcal{C}}_\alpha - \hat{\mathbf{M}}_\alpha), (\alpha = 1, 2)$, represent couple-stresses exerted on the interface by bulk phases and $\hat{\mathbf{G}}$ denotes interfacial body couple stress; $\hat{\mathbf{C}}$ is the *interfacial couple-stress tensor*. While $\hat{\mathcal{C}}_\alpha - \hat{\mathbf{M}}_\alpha$ and $\hat{\mathbf{G}}$ might be negligible the great inhomogeneity associated with interfacial regions does not allow *a priori* neglect of $\hat{\mathbf{C}}$. Note further that even if *all* couple stresses are absent, in situations in which $\nabla_s\mathbf{u} \neq 0$ *and* **A** is non-negligible (so **S** is non-negligible) the surface stress tensor **T** will not map tangent vectors ν into tangent vectors in symmetric fashion in general (since $sk(\mathbf{IT}^T) \neq \mathbf{0}$ in general).

3. Thermal motion assumption T.M.2 expresses an aspect of the random nature of this quantity. Noting that $|\mathbf{x}_i - \mathbf{x}|$ is of order $O(\epsilon)$ for particles in an interfacial ϵ-cell, A.3 merely neglects terms of order $O(\epsilon^2)\sigma$. In manipulating expressions of form (8.12) terms of order $O(\epsilon^2)\mathbf{d}_\alpha$ and $O(\epsilon^2)\mathbf{d}_\alpha$ and $O(\epsilon^2)\mathbf{d}$ have similarly been neglected.

9. Energy Balance

On multiplying (4.2) scalarly by $\mathbf{v}_i$, summing each such equation for all particles in $R(\tau)$ at instant τ, and time averaging the result over the interval $t - \Delta \le \tau \le t$, the continuum form of energy balance is found to be

$$\int_{\Sigma} \left\{ Q + q_1 + q_2 + r + (\mathbf{f} + \mathbf{t}_1 + \mathbf{t}_2 + \mathbf{b}).\mathbf{u} + (\mathcal{G} + \mathcal{C}_1 + \mathcal{C}_2 + \mathbf{G}).\mathbf{I}(\nabla_s \mathbf{u})^T \right\}$$

$$= d/dt \left\{ \int_{\Sigma(t)} \sigma(h + \mathbf{u}^2/2) \right\} + \int_{\partial\Sigma} \left\{ k + \mathbf{d}.\mathbf{u} + s\mathbf{u}^2/2 + (\mathbf{a} \otimes \mathbf{u} + \mathbf{M}).\mathbf{I}(\nabla_s \mathbf{u})^T \right\}$$

$$+ \int_{\Sigma} \left\{ k_1 + k_2 - (s_1 + s_2)\mathbf{u}^2/2 + (\mathbf{d}_1 + \mathbf{d}_2).\mathbf{u} + (\mathbf{M}_1 + \mathbf{M}_2 - (\mathbf{a}_1 + \mathbf{a}_2) \otimes \mathbf{u}).\mathbf{I}(\nabla_s \mathbf{u})^T \right\}. \quad (9.1)$$

Here

$$\left.\begin{aligned}
Q(\mathbf{x}, t) &:= lim_\epsilon \left\{ \overline{\sum_i{}' \sum_\ell \mathbf{f}_{i\ell}.\bar{\mathbf{v}}_i}^{\Delta} / A_\epsilon \right\}, \\
q_1(\mathbf{x}, t) &:= lim_\epsilon \left\{ \overline{\sum_i{}' \sum_r \mathbf{f}_{ir}.\bar{\mathbf{v}}_i}^{\Delta} / A_\epsilon \right\}, \\
q_2(\mathbf{x}, t) &:= lim_\epsilon \left\{ \overline{\sum_i{}' \sum_s \mathbf{f}_{is}.\bar{\mathbf{v}}_i}^{\Delta} / A_\epsilon \right\}, \\
r(\mathbf{x}, t) &:= lim_\epsilon \left\{ \overline{\sum_i{}' \mathbf{b}_i.\bar{\mathbf{v}}_i}^{\Delta} / A_\epsilon \right\}, \\
(\sigma h)(\mathbf{x}, t) &:= lim_\epsilon \left\{ \overline{\sum_i{}' \tfrac{1}{2} m_i \bar{\mathbf{v}}_1^2}^{\Delta} / A_\epsilon \right\}, \left(= \tfrac{1}{2} tr\mathbf{K} \right) \\
k(\mathbf{x}, t) &:= lim_{\epsilon,\Delta} \left\{ \sum_j{}' \tfrac{1}{2} m_j \bar{\mathbf{v}}_j^2 s_j^w / \ell_\epsilon \Delta \right\}, \\
&\text{and} (\alpha = 1, 2) \\
k_\alpha(\mathbf{x}, t) &:= lim_{\epsilon,\Delta} \left\{ \sum_j{}'^{(\alpha)} \tfrac{1}{2} m_j \bar{\mathbf{v}}_j^2 s_j^{(\alpha)} / A_\epsilon \Delta \right\}.
\end{aligned}\right\} \quad (9.2)$$

In the derivation of (9.1) use is made of T.M.1,2 and A.1,2,3 and terms of order $|(\nabla_s \mathbf{u})\mathbf{P}|^2 s_\alpha \epsilon^2$ and $|(\nabla_s \mathbf{u})\mathbf{P}|^2 s \epsilon^2$ are neglected. The particles involved in definitions

$(9.2)_{1,2,3,4}$ are precisely those which appear in $(7.2)_{1,2,3,4}$, respectively and those in $(9.2)_{6,7}$ are the same as in (5.7,10), respectively. Provided k is regular, use of (5.13), (6.6), (7.6), (8.18) and relevant transport and surface divergence theorems imply there exists a tangential vector field **k** such that

$$k = \mathbf{k}.\nu. \tag{9.3}$$

Relation (9.1) can be made to resemble a more familiar form modulo the further interaction assumption

I.4 Time-averaged sums of the form [25]

$$\overline{\sum_i{}' \sum_p \mathbf{f}_{ip}.\tilde{\mathbf{v}}_i}^{\Delta}$$

are additive over microscopically-large [26] interfacial regions.

As a consequence of I.4 and I.2

$$\int_\Sigma Q = \int_\Sigma P + \int_{\partial\Sigma} q, \tag{9.4}$$

where

$$P(\mathbf{x},t) := lim_\epsilon \left\{ \overline{\sum_i{}' \sum_k{}' \mathbf{f}_{ik}.\tilde{\mathbf{v}}_i}^{\Delta} / A_\epsilon \right\}$$

and (9.5)

$$q(\mathbf{x},t) := lim_\epsilon \left\{ \overline{\sum_i{}' \sum_j \mathbf{f}_{ij}.\tilde{\mathbf{v}}_i}^{\Delta} / \ell_\epsilon \right\}.$$

In $(9.5)_1$ the sums are over pairs of particles both of which lie instantaneously within an interfacial ϵ-cell which is centred at **x** at instant t. Particle labels in $(9.5)_2$ are those employed in (7.16) and (8.22). If q is regular then from (9.4) there exists a tangential vector field $\mathbf{q}^-$ such that

---

[25] Here the sum, at any instant in the microscopically-long yet macroscopically-short time averaging period, is taken over all P_i within the (deforming) microscopically-large region and all P_p outside this region but within the interfacial zone. I.4 is the precise analogue of I.4 in Murdoch (1985) and essentially establishes the balance of heat conduction exchange rates between contiguous interfacial subregions.

[26] That is, an interfacial ϵ-cell($\epsilon \sim 10^{-5}m$) or union thereof.

$$q = -\mathbf{q}^{-}.\ \nu \tag{9.6}$$

and (9.4) yields

$$Q = P - div_s\mathbf{q}^{-}. \tag{9.7}$$

From (5.13), (6.6), (7.6,18), (8.18,26) and (9.3,7) balance (9.1) becomes

$$\int_{\partial\Sigma}\Big\{-\mathbf{q}.\nu - \mathbf{D}\nu.\mathbf{u} - (\mathbf{u}^2/2)\mathbf{s}.\nu - (\mathbf{A}\nu\otimes\mathbf{u} + \mathcal{M}\nu).\mathbf{I}(\nabla_s\mathbf{u})^T\Big\}$$

$$+\int_{\Sigma}\Big\{P + q_1 - k_1 + q_2 - k_2 + r(s_1 + s_2)\mathbf{u}^2/2$$

$$+(div_s\mathbf{T}^{-} + \mathbf{t}_1 - \mathbf{d}_1 + \mathbf{t}_2 - \mathbf{d}_2 + \mathbf{b}).\mathbf{u} + [\mathbf{J} + div_s\hat{\mathbf{C}}^{-} + sk(\mathbf{I}(\mathbf{T}^{-})^{\mathbf{T}})$$

$$+\mathcal{C}_1 - \mathbf{M}_1 + \mathcal{C}_2 - \mathbf{M}_2 + \mathbf{G} + (\mathbf{a}_1 + \mathbf{a}_2)\otimes\mathbf{u}].\mathbf{I}(\nabla_s\mathbf{u})^T\Big\}$$

$$= d/dt\Big\{\int_{\Sigma(t)}\sigma(h + \mathbf{u}^2/2)\Big\}. \tag{9.8}$$

Here the *interfacial heat flux vector*

$$\mathbf{q} := \mathbf{q}^{-} + \quad \mathbf{k}. \tag{9.9}$$

Using identities and surface divergence theorems (9.8) becomes

$$\int_{\partial\Sigma}\Big\{-\mathbf{q}.\nu + \mathbf{T}\nu.\mathbf{u} + (\mathbf{u}^2/2)\mathbf{s}.\nu + (\hat{\mathbf{C}}\nu - \mathbf{A}\nu\otimes\mathbf{u}).\mathbf{I}(\nabla_s\mathbf{u})^T\Big\}$$

$$+\int_{\Sigma}\Big\{-\sigma\hat{\alpha} + q_1 - k_1 + q_2 - k_2 + r + (s_1 + s_2)\mathbf{u}^2/2 + (\mathbf{t}_1 - \mathbf{d}_1 + \mathbf{t}_2 - \mathbf{d}_2 + \mathbf{b}).\mathbf{u}$$

$$+(\mathcal{C}_1 - \mathbf{M}_1 + \mathcal{C}_2 - \mathbf{M}_2 + \mathbf{G} + (\mathbf{a}_1 + \mathbf{a}_2)\otimes\mathbf{u}).\mathbf{I}(\nabla_s\mathbf{u})^T\Big\}$$

$$= d/dt\Big\{\int_{\Sigma(t)}\sigma(h + \mathbf{u}^2/2)\Big\}, \tag{9.10}$$

where

$$-\sigma\hat{\alpha} := P + (\mathbf{J} - sym(\mathbf{I}(\mathbf{T}^{-})^{T})).\mathbf{I}(\nabla_s\mathbf{u})^{\mathbf{T}} - \hat{\mathbf{C}}^{-}.\nabla_s(\mathbf{I}(\nabla_s\mathbf{u})^{T}). \tag{9.11}$$

If $\sigma\hat{\beta}$ is a solution (assumed unique up to an arbitrary constant) of

$$\dot{\widehat{\sigma\hat{\beta}}} + \sigma\hat{\beta} div_s \mathbf{u}_s = \sigma\hat{\alpha} \tag{9.12}$$

then, writing

$$\hat{e} := \hat{\beta} + h, \tag{9.13}$$

(9.10) assumes a more conventional appearance; namely,

$$\int_{\partial\Sigma}\Big\{-\mathbf{q}.\nu + (\mathbf{T} - \tfrac{1}{2}\mathbf{u}\otimes\mathbf{s} - \tfrac{1}{2}(\nabla_s\mathbf{u})\mathbf{PA})\nu.\mathbf{u} + (\hat{\mathbf{C}}\nu - \tfrac{1}{2}\mathbf{A}\nu\otimes\mathbf{u}).\mathbf{I}(\nabla_s\mathbf{u})^{T}\Big\}$$

$$+\int_{\Sigma}\Big\{q_1 - k_1 + q_2 - k_2 + r + (s_1 + s_2)\mathbf{u}^2/2$$

$$+(\mathbf{t}_1 - \mathbf{d}_1 + \mathbf{t}_2 - \mathbf{d}_2 + \mathbf{b} + \frac{1}{2}(\nabla_s\mathbf{u})\mathbf{P}(\mathbf{a}_1 + \mathbf{a}_2).\mathbf{u}$$

$$+(\mathcal{C}_1 - \mathbf{M}_1 + \mathcal{C}_2 - \mathbf{M}_2 + \mathbf{G} + \frac{1}{2}(\mathbf{a}_1 + \mathbf{a}_2)\otimes\mathbf{u}).\mathbf{I}(\nabla_s\mathbf{u})^{T}\Big\}$$

$$= d/dt\Big\{\int_{\Sigma(t)}\sigma(\hat{e} + \mathbf{u}^2/2)\Big\}. \tag{9.14}$$

Here we have also noted that

$$(\mathbf{A}\nu\otimes\mathbf{u}).\mathbf{I}(\nabla_s\mathbf{u})^{T} = (\nabla_s\mathbf{u})\mathbf{PA}\ \nu.\mathbf{u} \text{ and } (\mathbf{a}_\alpha\otimes\mathbf{u}).\mathbf{I}(\nabla_s\mathbf{u})^{T} = (\nabla_s\mathbf{u})\mathbf{P}\mathbf{a}_\alpha.\mathbf{u}$$

and written terms involving these quantities in a manner most simply to be compared with balances (7.19) and (8.24).

After considerable manipulation the local form of (9.14), on using balances (5.14), (6.8) and (7.26), is

$$r - div_s\mathbf{q} + q_1 - k_1 + q_2 - k_2 + \hat{\mathbf{C}}.\nabla_s(\mathbf{I}(\nabla_s\mathbf{u})^T)$$

$$+\Big\{\mathbf{I}\mathbf{T}^T + div_s\hat{\mathbf{C}} - \mathbf{A}(\nabla_s\mathbf{u})^T + \mathcal{C}_1 - \mathbf{M}_1 + \mathcal{C}_2 - \mathbf{M}_2 + \mathbf{G}\Big\}.\mathbf{I}(\nabla_s\mathbf{u})^T$$

$$= \dot{\widehat{\sigma\hat{e}}} + \sigma\,\hat{e}\,div_s\mathbf{u}_s.$$

$$(9.15)$$

$$= \sigma\;\dot{\hat{e}} + (s_1 + s_2 - div_s\mathbf{s})\hat{e}.$$

Equivalently, using (8.27) and (9.11,12)

$$r - div_s\mathbf{q} + q_1 - k_1 + q_2 - k_2 - \mathcal{M}.\nabla_s(\mathbf{I}(\nabla_s\mathbf{u})^T) + P$$

$$+\Big\{\mathbf{I}\mathbf{A}^T\mathbf{I}(\nabla_s\mathbf{u})^T - \mathbf{K}\Big\}.\mathbf{I}(\nabla_s\mathbf{u})^T = \sigma\dot{h} + (s_1 + s_2 - div_s\mathbf{s})h. \qquad (9.16)$$

This may be regarded as an evolution equation for the *(specific) heat content h.*

10. Alternative Energy and Moment of Momentum Balances

10.1 Preamble.

Following the methodology of Murdoch(1985) we now obtain alternative balances of energy and moment of momentum modulo a further interaction assumption, which is related to the existence of a binding energy density for the interface. Specifically we assume

I.5. Time-averaged [27] sums of the form [28]

$$\overline{\sum_i{}'\sum_k{}'\mathbf{f}_{ik}.\mathbf{v}_i}^{\Delta}$$

[27] Over intervals microscopically long.

[28] At any instant in the time-averaging period the sum is taken over all particles P_i, P_k within the region, whose walls are assumed to deform with the motion prescribed by $\mathbf{u}_s$.

are additive over microscopically-large (walled) interfacial regions.

10.2 Alternative Energy Balance

Consider the time average of

$$S(\tau) := \sum_i \sum_\ell{}^{\tau} \mathbf{f}_{i\ell}.\mathbf{v}_i \tag{10.1}$$

which[29] in §9 yielded the continuum representation

$$\int_\Sigma \left\{ Q + \mathbf{f}.\mathbf{u} + \mathcal{G}.\mathbf{I}(\nabla_s \mathbf{u})^T \right\}. \tag{10.2}$$

Equivalently, using (9.7), (7.18) and (8.26), this can be written as

$$\int_{\partial\Sigma} \left\{ -\mathbf{q}^-.\nu + \mathbf{T}^-\nu.\mathbf{u} + \hat{\mathbf{C}}^-\nu.\mathbf{I}(\nabla_s \mathbf{u})^T \right\}$$

$$+ \int_\Sigma \left\{ P + (\mathbf{J} - sym\ (\mathbf{I}(\mathbf{T}^-)^T)).\mathbf{I}(\nabla_s \mathbf{u})^T - \hat{\mathbf{C}}^-.\nabla_s(\mathbf{I}(\nabla_s \mathbf{u})^T) \right\}. \tag{10.3}$$

In view of I.5 the foregoing may be expressed differently. From (10.1)

$$S(\tau) = \sum_i \sum_k{}^{\tau} \mathbf{f}_{ik}.\mathbf{v}_i + \sum_i \sum_p{}^{\tau} \mathbf{f}_{ip}.\mathbf{v}_i, \tag{10.4}$$

where P_i, P_k denote any particles within $R(\tau)$ and P_p denotes any particle outside $R(\tau)$ but within the interfacial region. Upon breaking $R(t)$ into a system of disjoint walled interfacial ϵ-cells (so inducing a similar decomposition of $R(\tau)$ via the motion map) and taking the time average of (10.4), evaluated at t, the middle sum in (10.4) yields from I.5 the continuum representation

$$-\int_\Sigma \sigma\alpha,$$

where

[29] Sums are over all P_i in $R(\tau)$ and all particles P_ℓ in the interfacial zone, at instant τ.

$$(\sigma\alpha)(\mathbf{x},t) := -lim_\epsilon\left\{\overline{\sum_i{}' \sum_k{}' \mathbf{f}_{ik}.\mathbf{v}_i}^{\Delta} / A_\epsilon\right\}. \tag{10.5}$$

The last sum in (10.4) reduces, by I.2, to a sum over pairs of particles P_i, P_j, on opposite sides of $W(\tau)$ but very near thereto. Accordingly the "i" sum can be expressed as one over interfacial wall ϵ-cells (cf. §7), namely

$$\sum_{\substack{interfacial\\ wall\ cells}} \left\{\sum_i{}' \sum_j{}^{\tau} \mathbf{f}_{ij}.(\tilde{\mathbf{v}}_i + \mathbf{u} + (\nabla_s\mathbf{u})\mathbf{P}(\mathbf{x}_i - \mathbf{x}))\right\} \tag{10.6}$$

where $\mathbf{u}$, $\nabla_s\mathbf{u}$ and $\mathbf{P}$ are evaluated at the relevant cell centre $\mathbf{x}$. The continuum version of (10.6) is (cf. $(9.5_2, 6)$, (7.16,17))

$$\int_{\partial\Sigma}\left\{-\mathbf{q}^-.\nu + \mathbf{T}^-\nu.\mathbf{u} + \mathcal{C}.\mathbf{I}(\nabla_s\mathbf{u})^T\right\},$$

and (10.7)

$$\mathcal{C}(\mathbf{x},t) := lim_\epsilon\left\{\overline{\sum_i{}' \sum_j (\mathbf{x}_i - \mathbf{x}) \otimes \mathbf{f}_{ij}}^{\Delta} / \ell_\epsilon\right\}.$$

It follows from the equivalence of (10.3) with the sum of expressions $(10.5_1, 7_1)$ that if $\mathcal{C}$ is regular then there exists a rank three tensor field $\mathbf{C}^-$ such that

$$\mathcal{C} = \mathbf{C}^-\nu. \tag{10.8}$$

Of course, from $(4.27)_2$ and $(10.7)_2$

$$sk\ \mathcal{C} = \hat{\mathcal{C}} \tag{10.9}$$

so that for all tangent vectors ν (cf. (8.23), (10.8))

$$sk(\mathbf{C}^-\nu) = \hat{\mathbf{C}}^-\nu. \tag{10.10}$$

If $\sigma\beta$ is a solution of

$$\dot{\widehat{\sigma\beta}} + \sigma\beta\ div_s\mathbf{u}_s = \sigma\alpha \tag{10.11}$$

then the alternative form of energy balance, obtained by replacing (10.3) as it appears in (9.14) by the sum of $(10.5_1, 7_1)$ and using (10.8,11), is

$$\int_{\partial\Sigma}\Big\{-\mathbf{q}.\nu+(\mathbf{T}-\tfrac{1}{2}\mathbf{u}\otimes\mathbf{s}-\tfrac{1}{2}(\nabla_s\mathbf{u})\mathbf{PA})\nu.\mathbf{u}+(\hat{\mathbf{C}}\nu-\tfrac{1}{2}\mathbf{A}\nu\otimes\mathbf{u}).\mathbf{I}(\nabla_s\mathbf{u})^T\Big\}$$

$$+\int_{\Sigma}\Big\{q_1-k_1+q_2-k_2+r+(s_1+s_2)\mathbf{u}^2/2$$

$$+(\mathbf{t}_1-\mathbf{d}_1+\mathbf{t}_2-\mathbf{d}_2+\mathbf{b}+\tfrac{1}{2}(\nabla_s\mathbf{u})\mathbf{P}(\mathbf{a}_1+\mathbf{a}_2)).\mathbf{u}$$

$$+(\mathcal{C}_1-\mathbf{M}_1+\mathcal{C}_2-\mathbf{M}_2+\mathbf{G}+\tfrac{1}{2}(\mathbf{a}_1+\mathbf{a}_2)\otimes\mathbf{u}).\mathbf{I}(\nabla_s\mathbf{u})^T\Big\}$$

$$= d/dt\Big\{\int_{\Sigma(t)}\sigma(e+\mathbf{u}^2/2)\Big\}. \qquad (10.12)$$

Here the *generalised couple-stress tensor*

$$\mathbf{C} := \mathbf{C}^- - \mathbf{M} \qquad (10.13)$$

and the *(specific) internal energy*

$$e := \beta + h. \qquad (10.14)$$

Remarks

1. As discussed in Murdoch (1985) in the context of bulk continua, β has the interpretation of a binding energy[30] per unit mass when interactions are delivered by separation-dependent pair potentials. For this reason we prefer (10.12) to (9.14): e in (10.12) is simply the sum of the (specific) heat content h with the (specific) binding energy, when the latter is meaningful.[31]

2. The local form of (10.12) is merely (9.15) with $\hat{\mathbf{C}}$ and $\hat{e}$ replaced by $\mathbf{C}$ and e, respectively. (The equivalent relation (9.16) remains unmodified.)

3. The equivalence of (9.14) and (10.12) implies, using (8.25), (9.12), (10.11,13),

$$\int_{\partial\Sigma}(\mathbf{C}^- - \hat{\mathbf{C}}^-)\nu.\mathbf{I}(\nabla_s\mathbf{u})^T = \int_{\Sigma}\sigma(\alpha-\hat{\alpha}). \qquad (10.15)$$

---

[30] That is, the energy needed to assemble the interfacial particles from a state of infinite dispersion.

[31] The discussion involves a very general concept of interaction for which the notion of binding energy does not necessarily make sense.

Equivalently,[32]

$$\mathrm{div}_s\Big\{(\mathbf{C}^- - \hat{\mathbf{C}}^-)^\sim : \mathbf{I}(\nabla_s\mathbf{u})^T\Big\} = \sigma(\alpha - \hat{\alpha})$$

$$= \hat{\sigma}\overset{\cdot}{\overbrace{(\beta - \hat{\beta})}} + \sigma(\beta - \hat{\beta})div_s\mathbf{u}_s. \tag{10.16}$$

10.3 Alternative generalised moment of momentum balance

Balance (8.24) may formally be written as

$$\int_{\partial\Sigma}\Big\{\mathbf{C}\nu - \mathbf{A}\nu\otimes\mathbf{u} + (\mathbf{x}-\mathbf{x}_0)\otimes(\mathbf{T} - \mathbf{u}\otimes\mathbf{s} - (\nabla_s\mathbf{u})\mathbf{A})\nu\Big\}$$

$$+\int_{\Sigma}\Big\{\mathbf{Z} + \mathbf{G} + \mathcal{C}_1 - \mathbf{M}_1 + \mathcal{C}_2 - \mathbf{M}_2 + (\mathbf{a}_1 + \mathbf{a}_2 + \sigma\mathbf{u})\otimes\mathbf{u}$$

$$+(\mathbf{x}-\mathbf{x}_0)\otimes(\mathbf{t}_1 - \mathbf{d}_1 + \mathbf{t}_2 - \mathbf{d}_2 + (s_1 + s_2)\mathbf{u} + (\nabla_s\mathbf{u})\mathbf{P}(\mathbf{a}_1 + \mathbf{a}_2) + \mathbf{b}\Big\}$$

$$= d/dt\Big\{\int_{\Sigma(t)}(\mathbf{x}-\mathbf{x}_0)\otimes\sigma\mathbf{u}\Big\}, \tag{10.17}$$

where

$$\mathbf{Z} := \mathbf{J} + \mathbf{K} - sym(\mathbf{I}(\mathbf{T}^-)^T) - div_s(\mathbf{C}^- - \hat{\mathbf{C}}^-). \tag{10.18}$$

By virtue of (10.10), $div_s\hat{\mathbf{C}}^-$ is the skew part of $div_s\mathbf{C}^-$ so that $div_s(\mathbf{C}^- - \hat{\mathbf{C}}^-)$ takes symmetric values. Accordingly the symmetric natures of $\mathbf{J}$ and $\mathbf{K}$ ensure that, from (10.18), $\mathbf{Z}$ takes symmetric values. Since we have merely re-arranged terms the local form of (10.17) will be precisely (8.27).

---

[32] Here $(\mathbf{a}\otimes\mathbf{b}\otimes\mathbf{c})^\sim := \mathbf{c}\otimes\mathbf{b}\otimes\mathbf{a}$ and $(\mathbf{u}\otimes\mathbf{v}\otimes\mathbf{w}):(\mathbf{a}\otimes\mathbf{b}) := (\mathbf{w}.\mathbf{a})(\mathbf{v}.\mathbf{b})\,\mathbf{u}$ for simple tensors, with immediate analogues for general tensors of ranks three and two.

Acknowledgements

The author would like to thank H. Cohen, F. Dell'Isola, A. Romano, V. Veverka and G. Weiske for their encouragement, help and stimulation in his attempt to understand interfacial modelling, and to acknowledge support of this work by the British SERC, Canadian NSERC and Italian CNR in the past five years. The patience and care of A. Wells in word processing the original manuscript are gratefully acknowledged.

References

ADAM, N.K., *The Physics and Chemistry of Surfaces (third edition)*. Oxford University, London. (1941)

ADAMSON, A.W., *Physical Chemistry of Surfaces.* Interscience, New York - London. (1960)

BEDEAUX, D., Nonequilibrium thermodynamics and statistical physics of surfaces. In Advances in Chemical Physics (ed. I. Prigogine and Stuart A. Rice). Wiley, London - New York. (1986)

BIKERMAN, J.J., *Physical Surfaces.* Academic, New York - London. (1970)

BROWN, R.C., The surface tension of liquids. Contemp. Phys. **15**, 301-327. (1974)

BRUSH, S.G., A history of random processes I: Brownian movement from Brown to Perrin. Arch. Hist. Exact. Sc. **5**, 1-36. (1968)

BUFF, F.P., The Theory of Capillarity. In *Handbuch der Physik, Vol. X* (ed. S. Flugge). Springer, Berlin - Heidelberg - New York. (1960)

DAVIES, I.T. & RIDEAL, E.K., *Interfacial Phenomena.* Academic, New York - London. (1961)

DEFAY, R., PRIGOGINE, I. & BELLEMANS, A., *Surface Tension and Adsorption* (trans. D.H. EVERETT). Longmans, London. (1966)

DELL'ISOLA, F. & ROMANO, A. On the derivation of thermomechanical balance equations for continuous systems with a nonmaterial interface. Int. J. Eng. Sc. **25**, 1459-1468. (1987)

GIBBS, J.W., On the equilibrium of heterogeneous substances. Trans. Conn. Acad. **3**, 108-248 and 343-524. (1876) Reprinted in *The Scientific Papers of J. Willard Gibbs, Vol. I.* Dover, New York (1961).

GOPAL, E.S.R., *Statistical Mechanics and Properties of Matter.* Ellis Horwood, Chichester. (1974)

GURTIN, M.E., MURDOCH, A.I., A continuum theory of elastic material surfaces. Arch. Rational Mech. Anal **57**, 291-323 and **59**, 389-390. (1975)

HEER, C.V., *Statistical Mechanics, Kinetic Theory and Stochastic Processes.* Academic, New York - London. (1972)

HIRSCHFELDER, J.O., Study Week on Molecular Forces, Pontificiae Academiae Scientiarum Scripta Varia 31. North-Holland, Amsterdam; John Wiley, New York. (1967)

HUANG, K., *Statistical Mechanics.* Wiley, London - New York. (1963)

IRVING, J.H. & KIRKWOOD, J.G., The statistical mechanical theory of transport processes IV. The equations of hydrodynamics, J. Chem. Phys. **18**, 817-829.(1950)

ISHII, M. *Thermo-fluid Dynamic Theory of Two-phase Flow.* Collection de la Direction des Etudes et Recherches d'Electricite de France. Paris. (1975)

KOSIŃSKI, W., & ROMANO, A., Evolution balance laws in two-phase problems. Arch. Mech., to appear.

LANDAU, L.D. & LIFSCHITZ, E.M., *Fluid Mechanics.* Pergamon, London - Frankfurt. (1959)

LANDAU, L.D. & LIFSCHITZ, E.M., *Statistical Physics*, Part 1 (Third Edition). Pergamon, Oxford - New York-Frankfurt. (1980)

LEVICH, V.G. & KRYLOV, V.S., Surface tension-driven phenomena. In Annual Review in Fluid Mechanics **1**, Annual Review, New York. (1969)

MARCH, N.H. & TOSI, M.P., *Atomic Dynamics in Liquids.* Macmillan, London. (1976)

MOECKEL, G.P., Thermodynamics of an interface. Arch. Rational Mech. Anal. **57**, 255-280. (1975)

MURDOCH, A.I., A thermodynamical theory of elastic material interfaces. Q.J. Mech. Appl. Math, **29**, 245-275. (1976)

MURDOCH, A.I., The motivation of continuum concepts and relations from discrete considerations. Q.J. Mech. Appl. Math. **36**, 163-187. (1983)

MURDOCH, A.I., A corpuscular approach to continuum mechanics: basic considerations. Arch. Rational Mech. Anal. **88**, 291-321. (1985)

MURDOCH, A.I., A remark on the modelling of thin three-dimensional regions. In *Elasticity: Mathematical Methods and Applications.* The Ian N. Sneddon 70th Birthday Volume (ed. G. Eason & R.W. Ogden). Ellis Horwood, Chichester.(1990a)

MURDOCH, A.I., A coordinate-free approach to surface kinematics. Glasgow Math. J. **32**, 299-307. (1990b)

MURDOCH, A.I. & COHEN, H., On the continuum modelling of fluid interfacial regions I: kinematical considerations. Math. Departmental Report, Univ. Strathclyde. (1989)

NAGHDI, P.M., The Theory of Shells and Plates. In *Handbuch der Physik, Vol. VI a/2* (ed. C. TRUESDELL). Springer: Berlin - Heidelberg - New York. (1972)

NAPOLITANO, L.G., Thermodynamics and dynamics of surface phases. Acta Astronautica **6**, 1093-1112. (1979)

REIF, F., *Fundamentals of Statistical and Thermal Physics.* McGraw-Hill, New York. (1965)

ROWLINSON, J.S.& WIDOM, B., *Molecular Theory of Capillarity.* Oxford University, London. (1982)

RUSANOV, A.I., Recent investigations on the thickness of surface layers. In Progress in Surface and Membrane Science, Vol. 4 (ed. J.F. DANIELLI, M.D., ROSENBERG & D.A. CODENHEAD). Academic, New York - London. (1971)

SCRIVEN, L.E., Dynamics of a fluid interface. Chem. Eng., Sc. **12**, 98-108. (1960)

SLATTERY, J.C., Surfaces - I: momentum and moment of momentum balances for moving surfaces. Chem. Eng. Sc. **19**, 379-385. (1964)

SLATTERY, J.C., General balance equation for a phase interface. Ind. Eng. Chem. Fundamentals **6**, 105-115. (1967)

TRUESDELL, C. & NOLL, W., *The Non-Linear Field Theories of Mechanics.* Handbuch der Physik, *Vol. III/3* (ed. S. Flugge). Springer, Berlin - Heidelberg - New York. (1965)

YOUNG, T., An essay on the cohesion of fluids. Phil. Trans. Roy. Soc. 95, 65-87. (1805)

Appendix

Since the coordinate-free notation employed in these lectures will be unfamiliar, the geometrical interpretation of the associated terminology will here be outlined. (For a detailed development see Gurtin & Murdoch (1975), Murdoch (1976) or Murdoch (1990b)).

1. Surface gradients, curvature, and divergence theorems

Let S denote a smooth surface with unit normal field $\mathbf{n}$. The bidimensional space of vectors generated by displacements in the tangent plane at any point $\mathbf{x} \in S$ is termed the **tangent space** at $\mathbf{x}$ and will be denoted by $\mathcal{T}_{\mathbf{x}}$. The three-dimensional space of vectors generated by *all* spatial displacements will be labelled $\mathcal{V}$. The *perpendicular projection* of $\mathcal{V}$ upon $\mathcal{T}_{\mathbf{x}}$ will be written as $\mathbf{P}(\mathbf{x})$: that is, for any vector $\mathbf{u} \in \mathcal{V}$,

$$\mathbf{P}(\mathbf{x})\mathbf{u} = \mathbf{u} - (\mathbf{u}.\mathbf{n}(\mathbf{x}))\mathbf{n}(\mathbf{x}). \qquad (A.1)$$

The *inclusion* of $\mathcal{T}_{\mathbf{x}}$ in $\mathcal{V}$ is that linear transformation which identifies every element τ of $\mathcal{T}_{\mathbf{x}}$ as being also an element of $\mathcal{V}$:

$$\mathbf{I}(\mathbf{x})\tau = \tau. \qquad (A.2)$$

Notice that $\mathbf{I}(\mathbf{x})$, the identity $\mathbf{1}(\mathbf{x})$ on $\mathcal{T}_{\mathbf{x}}$, and the identity $\mathit{1}$ on $\mathcal{V}$ are all different. If an orthonormal basis for $\mathcal{V}$ is selected with two elements in $\mathcal{T}_{\mathbf{x}}$ then the matrices which represent $\mathbf{1}(\mathbf{x})$, $\mathbf{I}(\mathbf{x})$, $\mathbf{P}(\mathbf{x})$, and $\mathit{1}$ are, respectively, $(2 \times 2), (3 \times 2), (2 \times 3)$, and (3×3). Note also that

$$\mathbf{P}(\mathbf{x})\mathbf{I}(\mathbf{x}) = \mathbf{1}(\mathbf{x}), \;\; (\mathbf{P}(\mathbf{x}))^T = \mathbf{I}(\mathbf{x}) \text{ and } (\mathbf{I}(\mathbf{x}))^T = \mathbf{P}(\mathbf{x}). \qquad (A.3)$$

If ϕ denotes any scalar-valued field on S (for example, surface mass density) then its **surface gradient** at $\mathbf{x}$, written as $\nabla_s\phi(\mathbf{x})$, is a *tangent* vector (that is, $\nabla_s\phi(\mathbf{x}) \in \mathcal{T}_{\mathbf{x}}$) with the property that, for any *unit* vector $\tau \in \mathcal{T}_{\mathbf{x}}, \nabla_s\phi(\mathbf{x}).\tau$ is the rate of change of ϕ

(with respect to arc length s) along any oriented smooth curve Γ on S with unit tangent vector τ at $\mathbf{x}$. Symbolically,

$$\nabla_s\phi.\tau = d\phi/ds. \tag{A.4}$$

Similarly, if $\mathbf{v}$ represents a vector-valued field on S (for example, any surface velocity field) then its surface gradient at $\mathbf{x}$, $\nabla_s\mathbf{v}(\mathbf{x})$, is a linear transformation from $\mathcal{T}_\mathbf{x}$ into $\mathcal{V}$ such that

$$(\nabla_s\mathbf{v})\tau = d\mathbf{v}/ds. \tag{A.5}$$

In particular,

$$(\nabla_s\mathbf{n})\tau = d\mathbf{n}/ds. \tag{A.6}$$

If curve Γ above is chosen to be a plane section of S which contains the normal to S at $\mathbf{x}$ then its (signed)[1] curvature (at $\mathbf{x}$) is $\kappa := dt/ds.\mathbf{n}$ where $\mathbf{t}(= \tau$ at $\mathbf{x})$ denotes the unit tangent field on Γ. Consequently, since $\mathbf{t}.\mathbf{n} = 0$,

$$\kappa = -\mathbf{t}.d\mathbf{n}/ds = -\tau.\nabla_s\mathbf{n}(\mathbf{x})\tau = \mathbf{L}_\mathbf{n}(\mathbf{x})\tau.\tau \tag{A.7}$$

where[2]

$$\mathbf{L}_{\mathbf{n}(\mathbf{x})} := -\mathbf{P}(x)\nabla_s\mathbf{n}(\mathbf{x}). \tag{A.8}$$

The linear map $\mathbf{L}_\mathbf{n}(\mathbf{x})$, which takes vectors in $\mathcal{T}_\mathbf{x}$ into vectors in $\mathcal{T}_\mathbf{x}$, is termed the **curvature tensor** (corresponding to the choice $\mathbf{n}$ of unit normal: in the text suffix 'n' is omitted, for brevity) and has the property (A.7) of furnishing normal plane section curvatures at $\mathbf{x}$. The mean and Gaussian curvatures at $\mathbf{x}$ are, respectively,[3]

$$\kappa_\mathbf{n} := \frac{1}{2}\, tr\, \mathbf{L}_{\mathbf{n}(\mathbf{x})} \text{ and } G_\mathbf{n} := det\, \mathbf{L}_{\mathbf{n}(\mathbf{x})}. \tag{A.9}$$

---

[1] If Γ is a circle of radius R and $\mathbf{n}$ is directed radially *outward* then $\kappa = -1/R$.

[2] Since $\mathbf{n}.\mathbf{n} = 1, \mathbf{n}.d\mathbf{n}/ds = 0$ and so $\nabla_s\mathbf{n}(\mathbf{x})\tau.\mathbf{n}(\mathbf{x}) = 0$ ($cf.$(A.6)). Hence $\nabla_s\mathbf{n}(\mathbf{x})\tau \in \mathcal{T}_\mathbf{x}$. Accordingly, for all $\tau \in \mathcal{T}_\mathbf{x}$,

$$\mathbf{P}(\mathbf{x})\nabla_s\mathbf{n}(\mathbf{x})\tau = \nabla_s\mathbf{n}(\mathbf{x})\tau.$$

The difference between $\mathbf{P}\nabla_s\mathbf{n}$ and $\nabla_s\mathbf{n}$ (and hence between $\mathbf{L}_n$ and - $\nabla_s\mathbf{n}$) at $\mathbf{x}$ lies in their different co-domains. Said differently, their matrix representations are, respectively, (2×2) and (3×2).

[3] 'tr' and 'det' denote the *trace* and *determinant* of any linear transformation they prefix.

If *any* two mutually-orthogonal normal plane sections are taken at $\mathbf{x}$ and if κ_1, κ_2 denote the curvatures at $\mathbf{x}$ of the corresponding curves then

$$\kappa_{\mathrm{n}} := \frac{1}{2}(\kappa_1 + \kappa_2) \text{ and } G_{\mathrm{n}} := \kappa_1 \kappa_2. \tag{A.10}$$

In fact (cf.Gurtin & Murdoch (1975), Murdoch (1976) or Murdoch (1990b)) $\mathbf{L}_{\mathrm{n}}(\mathbf{x})$ is symmetric: its eigenvalues are termed *principal curvatures* and its eigenvectors define corresponding *principal directions* in the tangent plane.

If $\mathbf{u}$ denotes a vector field on S then its second surface gradient $\nabla_s^2 \mathbf{u}(\mathbf{x})$ at $\mathbf{x}$ is a symmetric bilinear map on $\mathcal{T}_{\mathbf{x}} \times \mathcal{T}_{\mathbf{x}}$ with values in $\mathcal{V}$. Indeed, with Γ and τ as above,

$$\Big((\nabla_s^2 \mathbf{u}(\mathbf{x}) \tau \Big) \tau = d^2 \mathbf{u}/ds^2, \tag{A.11}$$

the second derivative[4] (with respect to arc length) of $\mathbf{u}$ at $\mathbf{x}$ computed along Γ.

The **surface divergence** of any vector field $\mathbf{u}$ on S is defined as

$$div_s \mathbf{u} := tr(\mathbf{P} \nabla_s \mathbf{u}). \tag{A.12}$$

In particular, from $(A.8, 9_1)$,

$$div_s \mathbf{n} = -2\kappa_{\mathrm{n}}. \tag{A.13}$$

In general, writing

$$\mathbf{u} = \mathbf{u}_s + u_{\mathrm{n}} \mathbf{n}, \tag{A.14}$$

where

[4] The symmetric nature of $\nabla_s^2 \mathbf{u}\ (\mathbf{x}) =: \mathsf{S}$ implies that it is completely characterised by such second derivatives. Indeed, if two curves Γ_1, Γ_2 on S pass through $\mathbf{x}$ having unit tangents τ_1, τ_2 thereat then by the bilinearity and symmetry of S,

$$(\mathsf{S}\tau_1)\tau_2 = (\mathsf{S}\tau_2)\tau_1 = \frac{1}{2}\Big\{ (\mathsf{S}(\tau_1 + \tau_2))(\tau_1 + \tau_2) - (\mathsf{S}\tau_1)\tau_1 - (\mathsf{S}\tau_2)\tau_2 \Big\}.$$

This result implies that second derivatives $d^2\mathbf{u}/ds^2$ computed along three curves $\Gamma_1, \Gamma_2, \Gamma_3$ on S through $\mathbf{x}$ suffice to determine S, where two intersect orthogonally at $\mathbf{x}$ and the third "bisects" the angle between these at $\mathbf{x}$.

$$\mathbf{u}_s := \mathbf{P}\mathbf{u} \text{ and } u_{\mathbf{n}} := \mathbf{u}.\mathbf{n},$$

it turns out that if $\Sigma \subset S$ has piecewise-smooth boundary $\partial\Sigma$ with 'outward' unit normal [5] ν then

$$\int_{\partial\Sigma} \mathbf{u}.\nu \left(= \int_{\partial\Sigma} \mathbf{u}_s.\nu \right) = \int_{\Sigma} div_s \mathbf{u}_s. \qquad (A.15)$$

This is the **surface divergence theorem.**

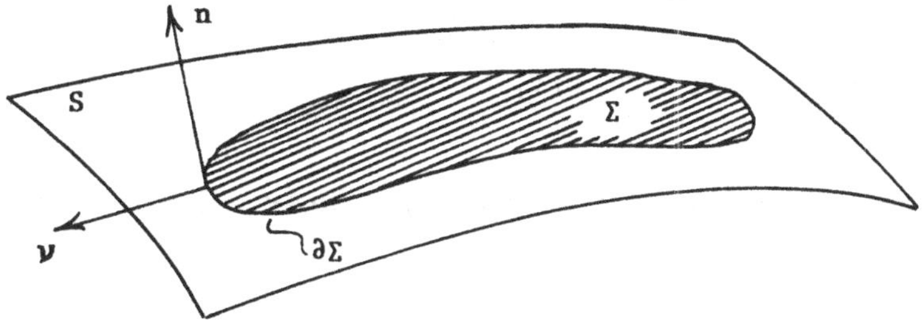

Figure A.1.

Further, since [6]

$$div_s \mathbf{u}_s = div_s \mathbf{u} - div_s(u_{\mathbf{n}}\mathbf{n}) = div_s \mathbf{u} + 2\kappa_{\mathbf{n}} u_{\mathbf{n}} \qquad (A.16)$$

we have

$$\int_{\partial\Sigma} \mathbf{u}.\nu = \int_{\Sigma} (div_s \mathbf{u} + 2\kappa_{\mathbf{n}} u_{\mathbf{n}}). \qquad (A.17)$$

---

[5] Notice that ν is tangential to S; that is, $\nu.\mathbf{n} = 0$.

[6] If ϕ, $\mathbf{v}$ denote scalar and vector fields on S then

$$\mathrm{div}_s(\phi\mathbf{v}) = \phi div_s \mathbf{v} + \nabla_s \phi.\mathbf{v}.$$

With $\phi = u_{\mathbf{n}}, \mathbf{v} = \mathbf{n}$ this yields

$$\mathrm{div}_s(u_{\mathbf{n}}\mathbf{n}) = u_{\mathbf{n}} div_s \mathbf{n} + \nabla_s u_{\mathbf{n}}.\mathbf{n} = -2\kappa_{\mathbf{n}} u_{\mathbf{n}}$$

using (A.13) and noting

$\nabla_s u_{\mathbf{n}}$ is a tangential vector field.

If $\mathbf{A}$ denotes a tensor field on S (rank ≥ 2) for which $\mathbf{A}(\mathbf{x})$ acts only on elements of $\mathcal{T}_{\mathbf{x}}$, then its surface divergence may be defined in such a way that

$$\int_{\partial\Sigma} \mathbf{A}\nu = \int_{\Sigma} div_s\mathbf{A}. \tag{A.18}$$

In particular, if $\mathbf{A}(\mathbf{x})$ maps $\mathcal{T}_{\mathbf{x}}$ linearly into $\mathcal{V}$ for all $\mathbf{x} \in S$ then $div_s\mathbf{A}$ is that vector field which, for any $\mathbf{k} \in \mathcal{V}$, satisfies

$$(div_s\mathbf{A}).\mathbf{k} := div_s(\mathbf{A}^T\mathbf{k}). \tag{A.19}$$

Using (A.19,3) and the result of footnote 6,

$$(div_s\,\mathbf{I}).\mathbf{k} = div_s(\mathbf{I}^T\mathbf{k}) = div_s(\mathbf{P}\mathbf{k}) = div_s(\mathbf{k} - (\mathbf{k}.\mathbf{n})\mathbf{n})$$

$$= -\mathbf{k}.\mathbf{n}\; div_s\mathbf{n} = 2\kappa_{\mathrm{n}}\mathbf{n}.\mathbf{k},$$

so that

$$div_s\mathbf{I} = 2\kappa_{\mathrm{n}}\mathbf{n}. \tag{A.20}$$

In similar fashion, if ϕ is a scalar field on S, we may show that

$$div_s(\phi\mathbf{I}) = 2\kappa_{\mathrm{n}}\phi\mathbf{n} + \nabla_s\phi. \tag{A.21}$$

2. Kinematics and the transport theorem

In Section 2.3 the concept of surface motion χ_0 prescribed by a velocity field $\mathbf{v}$ is introduced. If $\hat{\mathbf{x}} \in S_0$ then

$$\chi_0(.,t) : S_0 \to S(t) \subset \mathcal{E}$$

will map any oriented curve Γ_0 which lies on S_0, and passes through $\hat{\mathbf{x}}$, into an oriented curve $\Gamma_t \subset S(t)$ which passes through $\mathbf{x} := \chi_0(\hat{\mathbf{x}}, t)$.

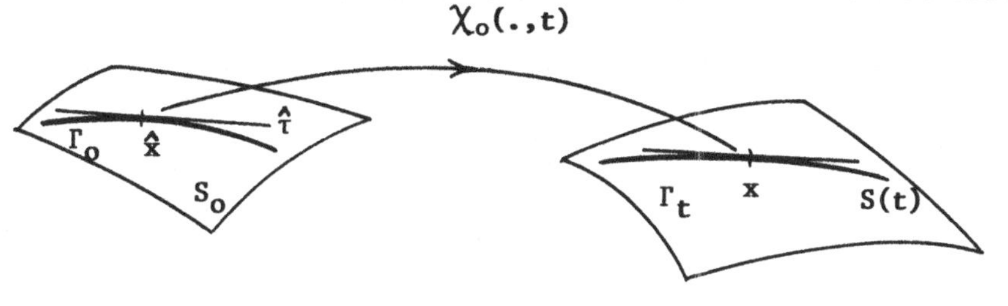

Figure A.2.

The surface gradient $\nabla_{S_0}\chi_0(\hat{\mathbf{x}},t)$ of $\chi_0(.,t)$ at $\hat{\mathbf{x}}$ delivers the rate of change of arc length along Γ_t with respect to arc length along Γ_0 (at $\hat{\mathbf{x}}$) as $\nabla_{S_0}\chi_0(\hat{\mathbf{x}},t)\hat{\tau}$ where $\hat{\tau}$ denotes the unit tangent to Γ_0 at $\hat{\mathbf{x}}$. Further, $\nabla_{S_0}\chi_0(\hat{\mathbf{x}},t)$ maps the tangent space to S_0 at $\hat{\mathbf{x}}$ linearly into the tangent space to $S(t)$ at $\mathbf{x}$.

Finally, we consider the appropriate transport theorem which, in conjunction with the relevant surface divergence theorem, allows local forms of balance relations to be formulated.

Let $\Sigma_t := \chi_0(\Sigma_0, t)$, where Σ_0 denotes a simply-connected subset of S_0 with piecewise smooth boundary $\partial\Sigma_0$. Then

$$\frac{d}{dt}\Big(\int_{\Sigma_t} f(.,t)\Big) := \frac{d}{d\tau}\Big(\int_{\Sigma_\tau} f(.,\tau)\Big)\Big|_{\tau=t} = \int_{\Sigma_t}\Big(\dot{f} + f\,div_s\mathbf{v}\Big). \qquad (A.22)$$

Here $\dot{f}$ denotes the time derivative of f following the motion prescribed by $\mathbf{v}$ ($cf.(2.11)_1$), and all fields in the last integrand (and $S(.)$) are evaluated at instant t.

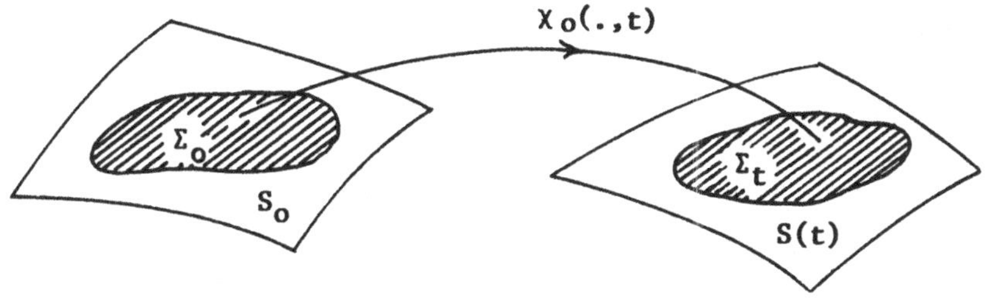

Figure A.3.

A PHENOMENOLOGICAL APPROACH TO FLUID-PHASE INTERFACES

W. Kosinski
Polish Academy of Sciences, Warsaw, Poland

Abstract. In these lectures an interface is modelled as a shell-like Eulerean region composed in principle of different material points at different instants. Consequent derivation of interfacial balance laws is performed in which true (not excess) quantities appear. Having (without any extrapolations) the exact integral relations for the true surface fields in terms of the bulk quantities from the layer, one can try to make constitutive assumption that take into account the interface and its interaction with the bulk phases as a whole, without retaining the macroscopically-irrelevant details of its structure. It is shown that one will in general have a nonsymmetric contribution to the interfacial stress tensor as well as nonvanishing normal component of it. This is true even for a non-polar fluid occupying the interface. Two cases of interfaces are considered, first with a constant thickness, and the other in which the thickness varies with time and space. A first correction to the Laplace relation is given. A more detailed description can be given in which higher order moments of the true fields appear. Relations between excess and true interfacial mass densities are given.

1. Introduction

Interfaces are well known to be regions of material bodies or systems formed of phase boundaries or phase transition layers where one observes - usually in a certain direction - abrupt variations of the properties of motion and physical quantities describing the system. These variations are not discontinuous, contrary to macroscopic appearance; they are rather the results of very high gradients of density, velocity, stress, *etc.*, which in a thin transition zone exceed essentially corresponding gradients outside the interface. Hence large gradients require that the analysis of motion of the system within the interface be based on equations more detailed than those sufficient in the surroundings, i.e. in the bulk. In particular these variations must be accounted for by appropriate constitutive relations capable of reflecting the characteristic features of the transition zone.

Defay et al.(1966) state that, due to experimental evidence, a satisfactory description of surface phenomena far from equilibrium is expected only if a structural model for the interface is developed, i.e. when the interface layer is modelled as a narrow 3D volume with special properties.

The use of the classical method of the dividing surface of Gibbs has been preferred by many authors (cf. Ono & Kondo (1960)) to avoid the necessity of assigning some thickness to the interfacial zone. In that method the surface tension is not the only quantity strictly defined with reference to the dividing surface since the excess contributions to the densities and fluxes (currents) appear there in the form of superficial quantities.

The above method for treating surface tension and description of the excess contributions to the densities and fluxes (currents) at the dividing surface in the form of superficial quantities, is simple, intuitive and useful but it is of an approximate nature from the molecular point of view (cf. Ono & Kondo (1960)).

Capillarity or surface phenomena, occurring when significant amounts of physical quantities are *localized* in a 3D region whose thickness is smaller than two others, have been described in the literature by means of three classes of models we shall describe hereafter. The region of localization is henceforth called a *layer*; its thickness varies from 10^{-9} m to 10^{-3} m, and even more, practically with no upper bound, depending on the interfacial phenomenon under considederation.

For example, a system of two immiscible fluid phases with their interphase layer (cf.Fig.1) can be modelled, at the macroscopic level, as two bulk phases separated by a surface Σ. If the interaction between

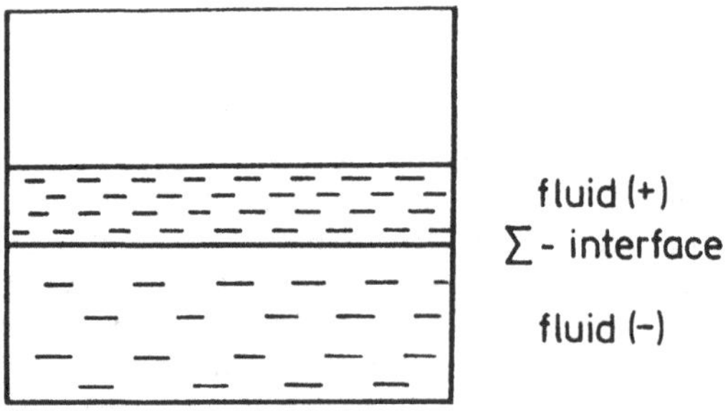

Fig.1, Singular surface as a fluid-fluid interface

the bulk phases and the layer is negligible, one can use a simple model in which the surface Σ is assumed to be a purely geometrical surface devoid of any thermodynamic property. Then the classical jump (so-called Rankine-Hugoniot) conditions of the bulk fields constitute the interface boundary conditions. When the interaction between the bulk phases and the layer cannot be neglected, more sophisticated models are needed. In the present lectures we confine ourselves to the phenomenological models leaving the corpuscular approach to Professor Murdoch's lectures.

The singular surface approach (cf. Scriven (1960), Barrère & Prud'homme (1973), Moeckel (1974), Lindsay & Straughan (1979), Napolitano (1978), Romano (1983), Iannece (1985), Kosiński (1986), Dell'Isola & Romano (1987)) to interfacial dynamics constitutes the first class of models. In this approach, which has limitations and some drawbacks, the interface is modelled as a '2D molecular world' (called the *singular surface* or the surface of singularities of physical fields), the dynamics of which is analogous to that of the ordinary 3D world of bulk phases; new physical fields appear on the surface and mechanical and thermodynamic balance laws with surface singularities are formulated in which these fields are related to discontinuities of the bulk fields at the surface.

In this framework one can observe the lack of the full interpretation of the surface fields in terms of 3D fields appearing in the description of the surrounding bulk phases. This drawback together with other inherent difficulties arising in formulating appropriate constitutive equations for these quantities can be overcome by constructing a structural model for the interface. This point of view is presented in Professor Hutter's lectures.

When the interface layer is modelled as a narrow 3D volume, two approaches are employed. In the first particular constitutive equations such as those of Korteweg (1901) or of Van der Waals fluids (cf. Slemrod (1983, 1989), Serrin (1983), Dunn & Serrin (1985)), are chosen for the whole system. They are chosen such that if both phases of the given material are present, the thermomechanical quantities show large gradients within the interfacial layer.

In the second approach the existence of a third narrow phase dividing two bulk phases is assumed *a priori*, while its properties are described by means of suitable constitutive equations. The resulting systems of universal balance laws of mechanics and thermodynamics have in principle similar forms. The interpretations of the terms appearing in the resulting balance equations differ from those appearing in the singular surface approach.

Formally, a 3D region already appeared in the approach of Gibbs, who employed the term "dividing surface" in his description of phase changes. In Gibbs' approach the layer is modeled as a surface on which an energy excess surface density and a uniform stress are distributed. It is a commonly accepted point of view that this model is suitable only for describing equilibrium phenomena and is not adequate for the description of irreversible and non-stationary phenomena (cf. Buff (1960), Defay et al.(1966)).

After the article of Scriven (1960) in which the motion of a deformable surface of a Newtonian fluid was first described and the book of Aris (1962) with its extended version, Slattery (1967) has described the motion of an interfacial surface together with that of the adjacent interfacial regions. Buff (1956), Slattery (1967), Ghez (1966), together with Ishii (1975), Bedeaux et al. (1976), Deemer & Slattery (1978), Albano et al.(1979), Napolitano (1978), Dumas (1980), Adamson (1982), as well as Alts (1986), Alts & Hutter (1989) and Professor Hutter's lectures belong to the group of authors, some of them from physical chemistry, who extend the work of Gibbs and correlate the fields on singular surfaces with mean values of *excess fields* defined over the 3D interface region of finite thickness. In this approach the interface region is a kind of boundary layer, in which the excess quantities represent the difference between the actual interfacial fields and averages of extrapolations of bulk fields into the layer.

In particular, in this approach the term surface concentration of substance *A*, say, is confusing since it seems to denote the total concentration of substance *A* in the interfacial layer, whereas in reality it represents the surface excess concentration (which can be positive or negative) of that substance with regard to its concentrations in the neighbouring bulk phases. A similar remark applies to the surface mass density defined in that model.

At this point one should underline the difference between two models based on the concept of the interface 3D layer: the first referring to the excess quantities and the second referring to the true ones. In the first model one introduces a dividing surface located somewhere in the transition (interface) zone and then the bulk quantities are extrapolated up to this surface by stipulating (cf.Dumais (1980), Alts & Hutter (1988) and Professor Hutter's lectures) that they must satisfy the typical 3D balance equations and the bulk constitutive relations (whatever these may be). The main problem of this model consists in introducing surface excess densities (quantities) to compensate the error introduced by replacing the exact (true) quantities by the extrapolated quantities in the transition zone.

In the second model no extrapolation is made (cf.Gogosov et al.(1983), Gatignol (1987), Gatignol & Seppecher (1986), Baillot et Prud'homme (1988), Kosiński & Romano (1989), Dell'Isola & Kosiński (1989, 1990)). Instead two dividing surfaces are introduced, which make the boundary between the single phase bulk media and the interface zone; in the latter multi-phase behaviour is observed, in which the confined matter possesses constitutive properties different from the surrounding bulk phases. Consequently, in dealing with this model the formulation of an initial-boundary value problem will be different from that in which excess quantities appear.

It is worthwhile to notice that in the first model the definition of excess quantities depends on the assumed form of the extensions of the true bulk fields into the multiphase region.

If a soap film stretched over a wire frame (cf.Fig. 2)

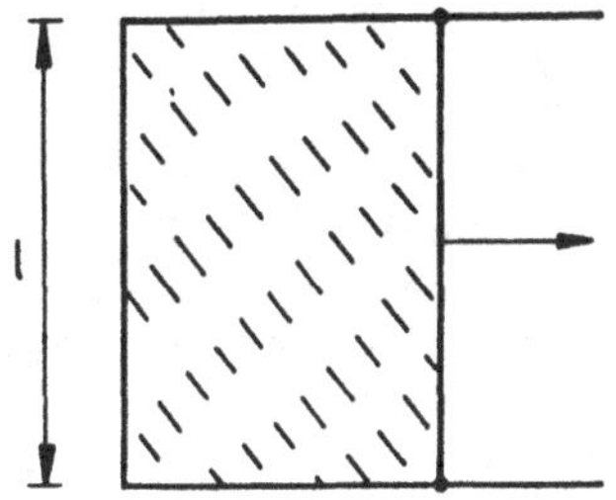

Fig. 2

is modelled then the film will appear between two interfacial regions. However, due to the small thickness of the film, an approximated picture can be obtained when the film is modelled by a layer of variable thickness, the equations of which can be supplied from those derived in Sec. 5.2.

In both models an averaging procedure is applied in which integration along the thickness is performed to get mean quantities defined as surface fields. In the first model one relates those quantities to the deviations between exact and extrapolated quantities in the layer, and in the second the mean quantities are defined as line integrals of the exact fields on some reference (e.g. mean) surface located in between the previous two. Here no physical meaning is ascribed to that surface: however, for convenience one can call it the dividing surface (as in the first model).

In the present lectures we follow the second model in which the interface is regarded as a shell-like Eulerean region composed in principle of different material points at different time instants[1].

Having, without any extrapolations, the exact integral relations for the true surface fields in terms of the bulk quantities from the layer, one tries to make constitutive assumptions that take into account the interface and its interaction with the bulk phases as a whole, without retaining the macroscopically - irrelevant details of its structure. The more detailed description[2] can be given by developing a theory in

---

1 This is evident in the the case of phase transition. However, adsorption may be also modelled by a material region with extra mass supply sources (cf. Ościk (1982)).

2 The other approach can be developed by introducing two different scales: micro- and macro- coordinates, the former responsible for the inner structure of the layer. The tool of nonstandard analysis could be helpful here.

which higher order moments of the true fields appear (cf. Dumais (1980) and Dell'Isola & Kosiński (1990)).

In the phenomenological approach we are presenting the interface is modelled as a finite slab and more detailed information about the structure of the dividing surface is introduced by relating the interfacial quantities to their 3D counterparts.

It is one of our aims to draw attention to the fact that in localizing surface phenomena to their carrier, namely to moving surfaces, we are losing some information necessary to constitutive modelling. To get this back we can explore the results of our exact derivation and the formulae in which the interfacial quantities appearing in the interfacial balance laws are defined in terms of the corresponding 3D quantities.

2. Moving surface in a continuum

Many physical and chemical processes occur at boundaries between two phases, while others are initiated at such interfaces. Adsorption is an example of processes of the first kind, while phase transition supplies an example for the second kind. Phenomenological modelling of both kinds of surface phenomena needs two tools: a mathematical structure composed of the universal balance laws and appropriate constitutive equations, the latter based on molecular physics, chemistry and mechanics. In the present lectures we shall be concerned with the preparation of the first tool, giving hints for constitutive modelling.

Surface effects in continuum physics are connected with singularities of volumetric energy densities. Although the notion of energy confined to a surface separating two phases is a theoretical concept, it has proved a convenient idealization in solving numerous problems of scientific value.

Surface phenomena play an essential role in the borderland between chemistry, physics and the mechanics of fluids and solids.

In the classical approach surface quantities were regarded as 2D analogues of those associated with a 3D material medium. It is one of our aims to draw attention to the fact that a moving surface carrying disturbances and physical properties different from those of the surrounding media can account for the structure of the interface region, i.e. its thickness, curvature and distributions of fields in it.

Dealing with the concept of dividing surface and excess quantities defined on it, one builds a model in which two well-behaved material media are separated by a singular surface which moves in the continuum and to which the physical properties of a phase change are attributed. This singular surface approach can be, of course, very helpful in modelling phenomena different from phase transition problems.

In the interfacial layer approach developed in Secs.3 and 4 a moving surface will be a carrier of a surface density function ψ^{s} expressed, however, in terms of mean values of 3D field ψ. Moreover the geometrical structure of the interfacial layer will be in some way preserved in dealing with true, not excess, quantities. In the model developed here, the interface is not treated as a singular surface and the bulk regions are not extended to the very edge of the moving surface (cf. Professor Hutter's lectures). Thus in formulating an initial boundary-value problem the geometry of the interface will influence the domains on which field equations are valid. This will be more evident in Sec. 3 and after the notation introduced there.

2.1 *Surface in a continuum*

A moving surface in a continuum is a mathematical model of a wave front or a thin interfacial layer which can be a carrier of discontinuities in 3D fields and/or in properties of the matter confined. No internal structure of the layer is introduced if one *a priori* assumes the existence of surface quantities together with discontinuities in the bulk fields on either side of the surface (cf. Kosiński (1986)). The model of an interface as a finite slab presented here uses the concept of a moving surface in a continuum, as a reference surface, similarly to that in shell theory. In what follows the basic geometrical concepts and the notation will first be settled.

The basic object is a hypersurface $\mathcal{S}$ in four dimensional space-time such that its spatial projection can be regarded as a one-parameter family of surfaces $\{\Sigma_t : t \in I\}$, where the time parameter t runs over $I = [t_0, t_f]$, a fixed interval.

We shall call $\mathcal{S}$ a *moving surface* in a continuum if the following condition are satisfied:

-there exists a nontrivial function g of C^2-class defined on $E^3 \times I$ such that the kernel of g contains the hypersurface $\mathcal{S}$, i.e.

$$\mathcal{S} \subseteq \left\{(x,t):\ g(x,t) = 0\right\},$$

-the spatial gradient of g exists at each point and is continuous and nonvanishing at points where g(x,t) vanishes, while the time derivative is continuous and not identically equal to zero.

One can assume weaker conditions than those above, if only an individual (material) surface Σ_t is considered (cf. Gurtin & Murdoch (1975) or Murdoch & Cohen (1989)).

If we denote the spatial gradient of g at a fixed t by grad g, then the *unit normal vector* to the surface Σ_t at x, denoted by **n**, will be

$$\mathbf{n} := \mathrm{grad}\, g(x,t) / \mathrm{grad}\, g \ , \ \text{when } g(x,t) = 0.$$

From the definition it follows that at any time t the surface Σ_t possesses a local parametrization called a *parametric representation*

$$x = \mathbf{r}(L^1, L^2, t) \tag{2.1}$$

which is of C^2-class, where the so-called *Gauss parameters* L^1, L^2 run over an open set in E^2, and has the property

$$g(\mathbf{r}(L^1, L^2, t), t) \equiv 0.$$

Both partial derivatives

$$\frac{\partial\, \mathbf{r}(L^\alpha, t)}{\partial\, L^1} =: \mathbf{a}_1\,, \qquad \frac{\partial\, \mathbf{r}(L^\alpha, t)}{\partial\, L^2} =: \mathbf{a}_2 \tag{2.2}$$

determine two linearly-independent vectors $\mathbf{a}_1$ and $\mathbf{a}_2$ *tangent* to Σ_t at $\mathbf{r}$, as well as the normal vector $\mathbf{n}$, by the relation

$$\mathbf{n} = \mathbf{a}_1 \times \mathbf{a}_2 \,/\, \|\mathbf{a}_1 \times \mathbf{a}_2\|. \tag{2.3}$$

From (2.3) it follows that $\mathbf{n} \cdot \mathbf{a}_\alpha = 0$, where the dot $\cdot$ denotes the inner product in E^3 and $\alpha = 1,2$.

The tangent vectors $\mathbf{a}_1$ and $\mathbf{a}_2$ together with the normal vector $\mathbf{n}$ form a basis in E^3, hence it is natural to ask about the reciprocal basis (or co-basis), i.e. the triple of vectors $\mathbf{a}^1$, $\mathbf{a}^2$ and $\mathbf{a}^3$ satisfying the relations

$$\mathbf{a}^1 \cdot \mathbf{a}_1 = \mathbf{a}^2 \cdot \mathbf{a}_2 = \mathbf{a}^3 \cdot \mathbf{n} = 1, \quad \mathbf{a}^1 \cdot \mathbf{a}_2 = \mathbf{a}^2 \cdot \mathbf{a}_1 = \mathbf{a}^3 \cdot \mathbf{a}_\alpha = 0. \tag{2.4}$$

From (2.4) it follows that $\mathbf{a}^3 = \mathbf{n}$, and $\mathbf{a}^1$, $\mathbf{a}^2$ are tangent to Σ_t. Hence the metric in E^3 can be split into a metric in the surface and a remaining one. If by $\mathbf{1}$ we denote the metric (unit) tensor in E^3, then the basis and its reciprocal $\mathbf{a}^1, \mathbf{a}^2$ and $\mathbf{n}$, give the identity

$$\mathbf{a}_1 \otimes \mathbf{a}^1 + \mathbf{a}_2 \otimes \mathbf{a}^2 + \mathbf{n} \otimes \mathbf{n} = \mathbf{1} = \mathbf{a}^1 \otimes \mathbf{a}_1 + \mathbf{a}^2 \otimes \mathbf{a}_2 + \mathbf{n} \otimes \mathbf{n}, \tag{2.5}$$

where the symbol $\otimes$ denotes the tensor product between vectors, and no distinction between the tangent and co-tangent spaces has been made. In the direct approach however, the distinction[3] between the tangent space $\mathcal{T}_{x,t}$ at x to Σ_t and the translation space $\mathcal{V}$ of E^3 is made. The difference between $\mathbf{1}$ and $\mathbf{n} \otimes \mathbf{n}$ is a tensor field, denoted here by $\mathbf{P}(x,t)$, which maps any spatial vector from E^3 into its component tangential to the surface: that is for any vector $\mathbf{u}$

$$\mathbf{P}(x,t)\mathbf{u} := (\mathbf{1} - \mathbf{n}(\mathbf{r}) \otimes \mathbf{n}(\mathbf{r}))\mathbf{u} = \mathbf{u} - (\mathbf{u} \cdot \mathbf{n}(\mathbf{r}))\mathbf{n}(\mathbf{r}). \tag{2.6}$$

It is evident that $\mathbf{P}(x,t)\mathbf{u}$ is 2D; hence the representation of $\mathbf{P}$ in the basis $\{\mathbf{a}_\alpha, \mathbf{n}\}$ (or in $\{\mathbf{a}^\alpha, \mathbf{n}\}$) is a (2×3) matrix.

---

[3] If the inclusion map $\mathbf{I}(x;t) \in \mathrm{Lin}(\mathcal{T}_{x,t}, \mathcal{V})$ is defined by

$$\mathbf{I}(x;t)\mathbf{t} = \mathbf{t}, \text{ for every } \mathbf{t} \in \mathcal{T}_{x,t},$$

then the perpendicular projection $\mathbf{P}(x;t) \in \mathrm{Lin}(\mathcal{V}, \mathcal{T}_{x,t})$ of $\mathcal{V}$ onto $\mathcal{T}_{x,t}$ the tangent space to Σ_t, is defined by $\mathbf{P}(x;t)\mathbf{I}(x,t) = \mathbf{1} - \mathbf{n} \otimes \mathbf{n}$, so that $\mathbf{I}(x;t)^T = \mathbf{P}(x;t)$ (cf. Professor Murdoch's lectures).

In the basis $\{\mathbf{a}_\alpha, \mathbf{n}\}$ the unit tensor $\mathbf{1}$ is represented as (3×3) matrix, while the *metric (unit) tensor* $\mathbf{1}_s$ of (the tangent space of) the surface Σ_t is represented as a 2×2 matrix. Hence in order to represent $\mathbf{1}_s$ as $\mathbf{a}_1 \otimes \mathbf{a}^1 + \mathbf{a}_2 \otimes \mathbf{a}^2$ or as $\mathbf{a}^1 \otimes \mathbf{a}_1 + \mathbf{a}^2 \otimes \mathbf{a}_2$, one needs to identify every tangent vector $\mathbf{d}$ (i.e. an element of the tangent space $\mathcal{T}_{x,t}$ of Σ_t at x) as being also a vector of E^3, i.e. en element of the translation space V of Σ_t. If we denote this identification (a linear transformation) by $I(x,t)$, and call it the *inclusion* of $\mathcal{T}_{x,t}$ in V, then for a tangent vector $\mathbf{t}$

$$I(x,t)\mathbf{t} = \mathbf{t},$$

and its representation in the basis $\{\mathbf{a}_\alpha, \mathbf{n}\}$ (or in $\{\mathbf{a}^\alpha, \mathbf{n}\}$) is a (3×2) matrix. Note also that

$$P(x,t)I(x,t) = \mathbf{1}(\mathbf{r}), \text{ and } (P(x,t))^T = I(x,t),$$

where the superscript ^{T} denotes transposition. At several places no visible distinction between $\mathbf{P}$ and $\mathbf{1}_s$ will be made; however, we can see that they are different objects

$$\mathbf{1}_s := (\mathbf{1} - \mathbf{n}\otimes\mathbf{n})I(x,t) = \mathbf{P}\, I(x,t) \tag{2.7}$$

The components of the metric tensor $\mathbf{1}_s$ are given by the relations

$$a_{\alpha\beta} := \mathbf{a}_\alpha \cdot \mathbf{a}_\beta, \quad a^{\alpha\beta} := \mathbf{a}^\alpha \cdot \mathbf{a}^\beta, \text{ with } a_{\alpha\beta}\, a^{\beta\vartheta} = \delta_\alpha^\vartheta, \tag{2.8}$$

where δ_α^ϑ is the Kronecker symbol and $\alpha, \beta, \vartheta = 1,2$. The four scalars[4] $a_{\alpha\beta}$ are also called the *components of the first fundamental form*, and moreover the determinant a of the matrix with its entries as $a_{\alpha\beta}$ gives the length squared of the cross (vector) product $\mathbf{a}_1 \times \mathbf{a}_2$, i.e.

$$a := \det[a_{\alpha\beta}] = \mathbf{n}\cdot(\mathbf{a}_1 \times \mathbf{a}_2) = \|\mathbf{a}_1 \times \mathbf{a}_2\|^2 . \tag{2.9}$$

If we introduce the permutation (fully antisymmetric) symbols

$$e^{11} = e^{22} = e_{11} = e_{22} = 0, \quad e_{12} = e^{12} = -e_{21} = -e^{21} = 1,$$

and define the *alternating tensor* ε by its components $\varepsilon_{\alpha\beta}$ in the basis $\{\mathbf{a}^\alpha \otimes \mathbf{a}^\beta\}$ as $\varepsilon_{\alpha\beta} := \sqrt{a}\; e_{\alpha\beta}$, then the vector products of basis vectors will be given by the very useful relations

---

[4] Note that $a_{\alpha\beta}$ are components of the metric tensor $\mathbf{1}_s$ in the tensor basis $\{\mathbf{a}^\alpha \otimes \mathbf{a}^\beta\}$, with $a^{\alpha\beta}$ the components in the basis $\{\mathbf{a}_a \otimes \mathbf{a}_\beta\}$ and δ_α^β the components in the 'mixed' basis $\{\mathbf{a}^\alpha \otimes \mathbf{a}_\beta\}$.

$$\mathbf{a}_\alpha \times \mathbf{a}_\beta = \varepsilon_{\alpha\beta}\mathbf{n}, \quad \mathbf{n} \times \mathbf{a}_\alpha = \varepsilon_{\alpha\beta}\mathbf{a}^\beta, \quad \mathbf{n} \times \mathbf{a}^\alpha = \varepsilon^{\alpha\beta}\mathbf{a}_\beta, \tag{2.10}$$

where $\varepsilon^{\alpha\beta} := (a)^{-0,5} e^{\alpha\beta}$ are components of ε in the basis $\{\mathbf{a}_\alpha \otimes \mathbf{a}_\beta\}$.

Let us differentiate the basis vector $\mathbf{a}_\beta$ with respect to the Gauss parameter L^α and express the result as a linear combination of basis vectors $\mathbf{a}_\vartheta$ and $\mathbf{n}$. We get

$$\frac{\partial\, \mathbf{a}_\beta(L^\alpha, t)}{\partial\, L^\alpha} =: \mathbf{a}_{\beta,\alpha} = \Gamma^\vartheta_{\beta\alpha}\mathbf{a}_\vartheta + b_{\beta\alpha}\mathbf{n}, \tag{2.11}$$

which immediately leads to the relation

$$b_{\alpha\beta} = \mathbf{a}_{\beta,\alpha} \cdot \mathbf{n}.$$

The coefficients $\Gamma^1_{\beta\alpha}$ and $\Gamma^2_{\beta\alpha}$ of the decomposition of $\mathbf{a}_{\beta,\alpha}$ onto tangent directions $\mathbf{a}_1$ and $\mathbf{a}_2$ are called *Christoffel symbols of the second kind.*

In what follows the comma followed by γ denotes partial differentiation with respect to the corresponding Gauss parameter L^γ.

From the orthogonality of $\mathbf{n}$ and $\mathbf{a}_\alpha$, and the unit length of $\mathbf{n}$, we obtain $\mathbf{n}\cdot\mathbf{n}_{,\alpha} = 0$ and

$$b_{\alpha\beta} = \mathbf{a}_{\beta,\alpha} \cdot \mathbf{n} = -\mathbf{a}_\beta \cdot \mathbf{n}_{,\alpha}, \quad \text{with } \mathbf{n}_{,\alpha} := \frac{\partial\, \mathbf{n}}{\partial\, L^\alpha}. \tag{2.12}$$

Hence, if we define the *surface gradient* grad_s by the formula

$$\mathrm{grad}_s \mathbf{u}(L^\beta, t) := \frac{\partial\, \mathbf{u}\ (L^\beta, t)}{\partial\, L^\alpha} \otimes \mathbf{a}^\alpha \tag{2.13}$$

for an arbitrary vector (or tensor)-valued smooth field $\mathbf{u}$, then (2.12) allows us to define a particular tensor field on (the tangent space of) the surface Σ_t (called *the curvature* or *second fundamental tensor*) by

$$\mathbf{b} := -\,\mathbf{P}(x,t)\ \mathrm{grad}_s \mathbf{n}\ , \tag{2.14}$$

for which $b_{\alpha\beta}$ will denote its components in the tensor basis $\{\mathbf{a}^\alpha \otimes \mathbf{a}^\beta\}$. The components of the curvature tensor in the basis $\{\mathbf{a}_a \otimes \mathbf{a}_\beta\}$ or in the mixed basis $\{\mathbf{a}_\alpha \otimes \mathbf{a}^\beta\}$ are given, respectively, by the relationships

$$b^{\alpha\beta} = b_{\gamma\vartheta}\, a^{\gamma\alpha}\, a^{\vartheta\beta}, \qquad b^\alpha_\beta = b_{\beta\vartheta}\, a^{\vartheta\alpha}.$$

At several places we do not distinguish between $\mathbf{P}\mathrm{grad}_s\mathbf{n}$ and $\mathrm{grad}_s\mathbf{n}$, since this difference lies in their matrix representations: the tensor $\mathbf{P}\mathrm{grad}_s\mathbf{n}$ has a (2×2) matrix, while $\mathrm{grad}_s\mathbf{n}$ has a (3×2) matrix.

From (2.2) and (2.11) it follows that the tensor $\mathbf{b}$ is symmetric; in the direct approach it is called the *Weingarten map* appropriate to $\mathbf{n}$, while (2.11) and (2.12) are called the Gauss and Weingarten formulae. Notice that $\mathbf{b}$ maps tangent vectors into tangent ones, its trace yields twice the *mean curvature* (at any point of Σ_t)

$$H := 0.5 \operatorname{tr} \mathbf{b} = 0.5\, b^{\alpha}_{\alpha} = 0.5\, b_{\alpha\beta} a^{\alpha\beta}, \tag{2.15}$$

and its determinant gives the so-called *Gauss curvature*

$$K := \det \mathbf{b} = \det [b^{\alpha}_{\beta}] = \det [b_{\alpha\beta}]/a.$$

Since the tensor **b** is symmetric its two eigenvalues, called *principal curvatures*, are real; if we denote them by k_1 and k_2, then $H = 0.5(k_1+k_2)$ and $K = k_1 k_2$; their reciprocals r_1 and r_2 are called the corresponding *radii of curvature*. For a cylindrical surface (and a cone) one of the curvatures vanishes together with the Gauss curvature. This is true for all developable surfaces, i.e. those which may be rolled out on a plane without stretching. Such surfaces can be obtained from a 3D (i.e. spatial) curve whose tangents will be generators of a surface. To finish the presentation of the second fundamental form of a surface, let us note that if the surface is given by

$$g(x,t) := x^3 - f(x^1, x^2, t) = 0$$

in Cartesian coordinates, then

$$b_{\beta\vartheta} = \frac{f_{,\beta\vartheta}}{(1 + f_{,1}^2 + f_{,2}^2)^{0.5}}, \quad \text{where} \quad \frac{\partial f}{\partial x^{\alpha}} =: f_{,\alpha} \quad \text{etc.}$$

This formula shows that the mean curvature of the surface is related to second derivatives of its representation. This is more evident from the formula

$$b_{\alpha\beta} a^{\alpha\beta} = 2H = \frac{f_{,11}(1 + f_{,2}^2) - 2f_{,1} f_{,2} f_{,12} + f_{,22}(1 + f_{,1}^2)}{(f_{,1}^2 + f_{,2}^2 + 1)^{1.5}}.$$

The tangential, or so-called covariant, derivative $D_s\mathbf{u}$ of a smooth field **u** can be defined by (cf. (2.13, 14))

$$D_s\mathbf{u}\, (L^{\beta}, t) := \mathbf{P}\operatorname{grad}_s \mathbf{u}(L^{\beta}, t)\cdot \tag{2.16}$$

It is time to recall the *Cayley-Hamilton theorem* for 2x2 matrices, from which follows the identity

$$\mathbf{d}^2 - \mathbf{d} \operatorname{tr}\mathbf{d} + \mathbf{1}_s \det\mathbf{d} = \mathbf{0} \tag{2.17}$$

for any surface tensor **d**; moreover, for any number λ (having the dimension of length if $\mathbf{d} = \mathbf{b}$) the following useful equation

$$\det(\mathbf{1}_s - \lambda\, \mathbf{d}) = 1 - \lambda \operatorname{tr}\mathbf{d} + \lambda^2 \det\mathbf{d} \tag{2.18}$$

holds. It is obvious that if $\mathbf{d} = \mathbf{b}$ and $\lambda = r_1$ or r_2, then the RHS of (2.18) vanishes. In (2.17) with **d** equal to **b** the first term $\mathbf{d}^2$ is called the *third fundamental tensor* (form).

Kinematics is described by time derivatives. The (nonvanishing) time derivatives of the function g or the function **r** in the parametric representation (2.1) characterize the mobility of the surface: at different instants of time different values of x (i.e. different points) satisfy the equation $g(x,t) = 0$. If we look at (2.1) as the function of motion of a (fictitious) point L^α = const., then the time derivative

$$\mathbf{c} := \partial\, \mathbf{r}(L^\alpha,\ t)/\ \partial\, t \tag{2.19}$$

will represent the *velocity of the displacement* of that point. It is obvious that in another parametric representation of the moving surface, $x = \mathbf{r}'(K^\alpha, t)$, say, the calculated time derivative will lead in general to another value

$$\mathbf{c}' := \partial\, \mathbf{r}'(K^\alpha,\ t)/\ \partial\, t.$$

There is, however, a common part of the both derivatives $\mathbf{c}$ and $\mathbf{c}'$, namely their projections on the normal direction

$$\mathbf{n}\cdot\mathbf{c} = \mathbf{n}\cdot\mathbf{c}' =: c_n \tag{2.20}$$

which are equal. This is easy to see, if one substitutes $\mathbf{r}$ or $\mathbf{r}'$ into the equation $g(x,t) = 0$ and performs the differentiation under identity sign. Hence also follows

$$\frac{\partial\, g(\mathbf{r}\ ,t)}{\partial\, t}\ /\ \|\mathrm{grad}\ g\| = -\ c_n, \tag{2.21}$$

due to the expression for the normal vector $\mathbf{n}$. The normal component c_n of the velocity of the displacement of the point L^α= const is called the *normal speed of displacement of the surface*, since it is independent of the parametrization. On the contrary the tangential component $c^\alpha := \mathbf{c}\cdot\mathbf{a}^\alpha$ is strictly related to the parametrization (L^α), for in the other parametrization $c'^\alpha = \mathbf{c}'\cdot\mathbf{a}^\alpha$. Let us notice that if both parametrizations are related by a time dependent transformation

$$K^\alpha = \hat{K}^\alpha(L^\beta, t) \qquad \text{then} \quad c'^\alpha = c^\alpha + \frac{\partial\, \hat{K}^\alpha(L^\beta,)}{\partial\, t}. \tag{2.22}$$

Now, if we choose the transformation $\hat{K}^\alpha(L^\beta, t)$ such that its time derivative is equal to $-c^\alpha$, then in the parametrization K^α the tangential velocity c'^α vanishes. That particular parametrization of the moving surface is known as the *convected parametrization* (cf. Bowen & Wang (1971), Kosiński (1986)). In such a parametrization[5] $\mathbf{c}'$ is equal to $c_n\mathbf{n}$. The integral curve of the field $\mathbf{u} = c_n\mathbf{n}$, i.e. the spatial projection of a solution of the vector differential equation

$$\frac{\mathrm{d}x}{\mathrm{d}s} = c\ \mathbf{n}(x), \quad \frac{\mathrm{d}t}{\mathrm{d}s} = 1, \tag{2.23}$$

with s as a curve parameter, is called the *normal trajectory* of the moving surface $\{\Sigma_t\}$ if at $t_0 = t(0)$ it begins at certain point of the initial surface Σ_{t_0}; each point of Σ_{t_0} is the starting point for a

---

[5] In his lectures Professor Hutter has assumed (ξ^β, t) be a convective parametrization.

certain normal trajectory, moreover through different points of Σ_t pass different normal trajectories. In the convected parametrization the normal trajectory is a geometrical locus of the surface point K^α =const. The choice of the convected parametrization in the description is of particular convenience in the derivation of any formula of general nature. Moreover no final formula should depend on the particular choice of the parametrization of the moving surface. Consequently it should be independent of the tangential component of the velocity field **c**: only its normal component c_n has geometrical meaning.

In the literature, however, one can find a discussion concerning the form of $c^\alpha \mathbf{a}_\alpha$. In our opinion this velocity has no physical meaning, unless a fictitious point L^α = const. is equipped with additional structure.

2.2 *Displacement time derivative*

In the derivation of the local balance laws in the singular surface approach as well as in the interfacial layer one, an invariant time derivative has to appear, which is independent of the chosen parametrization of the moving surface Σ_t. Introducing the velocity of displacement **c** of the moving surface (cf. (2.19) we have noticed that the invariant part of this velocity is the normal component $c_n\mathbf{n}$. Moreover, this component is uniquely determined by the time derivative of the representation of the surface in the particular parametrization, namely in the convected parametrization. Following this observation we introduce (after Thomas (1957, 1961) and Hayes (1957)) the so-called *displacement time derivative* $\frac{\delta}{\delta t}\mathbf{u}$ of a quantity (a C^1-smooth field) **u** defined on the hypersurface $\mathcal{S}$ as the time derivative of **u** at fixed convected parameters of the moving surface $\{\Sigma\}$, i.e. by the formula

$$\frac{\delta}{\delta t}\mathbf{u} := \frac{\partial \mathbf{u}(k^\alpha, t)}{\partial t}, \tag{2.24}$$

where (k^α, t) is the convected parametrization of $\{\Sigma_t\}$. We recall that in this parametrization the locus of the points $\{\mathbf{r}(k^\alpha, t),\ t \in I,\ k^\alpha = \text{const.}\}$ is a normal trajectory of Σ_t, given as integral curves of (2.23). It is a direct consequence of (2.24) and the comments following Eq.(2.22) that in any (not necessarily convected) parametrization (L^α, t) the (operational) formula for calculating this derivative is

$$\frac{\delta}{\delta t}\mathbf{u} = \frac{\partial \mathbf{u}(L^\alpha, t)}{\partial t} - \mathrm{grad}_s \mathbf{u}\, \mathbf{c}. \tag{2.25}$$

Additional to the displacement derivative another time derivative will be introduced in Sec.3.

We could close this section with the derivation of the local balance equations for a material system with surface singularities. It is already done in several places (e.g. Kosiński (1986)) and, moreover, the interfacial layer approach presented in the next two sections will contain as a particular case balance equations of the singular surface approach. We omit the repetition.

3. Moving shell-like Eulerean region in a continuum

Let us now assume that the effect of the interface in a continuous material system $\mathcal{B}$ occupying at time t in a motion χ a simply-connected region $\mathcal{B}_t$ in 3D space E^3 may be attributed to a three-dimensional *moving* region of finite thickness $\mathcal{Z}_t$. In Fig. 3 we see a typical a sub-region $\mathcal{P}_t$ which is occupied by a material subbody $\mathcal{P}$ of the body $\mathcal{B}$ and consists of two phases $\mathcal{P}_t \subseteq \mathcal{B}_t^+$ and $\mathcal{P}_t \subseteq \mathcal{B}_t^-$. In addition there exists a narrow layer $\mathcal{Z}_t$ which divides the volume phases $\mathcal{B}_t^{\pm}$ and at the same time the phases $\mathcal{P}_t^{\pm}$.

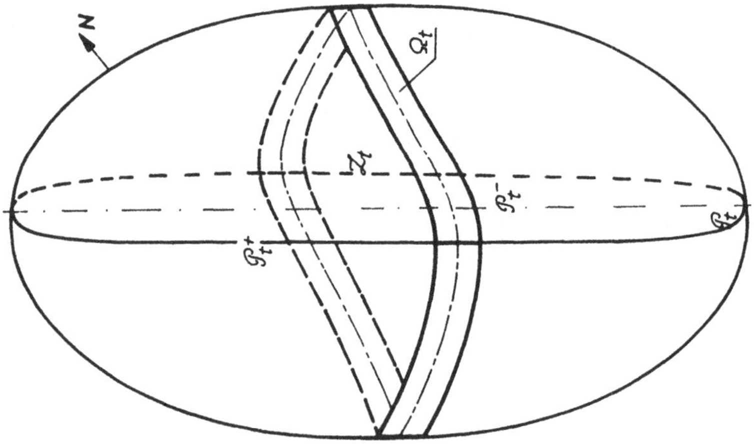

Fig. 3, Subbody with an interface zone

The boundaries between $\mathcal{Z}_t$ and both $\mathcal{B}_t^{\pm}$ are regular surfaces which will be denoted by Σ_t^+ and Σ_t^-; between them a reference surface Σ_t is located (cf. Fig. 4), to which the mean, interfacial, fields will be referred. (Note the subscript t appearing in the sets introduced, it reflects the fact that the sets are not fixed in the time, they change[6]

[6] Only Eulerean formalism is possible in modelling an interface.

with time.) In the case when excess quantities are used the reference surface Σ_t plays the role of the dividing surface (cf. Deemer and Slattery (1978) or Alts and Hutter (1989)).

In the case of a model with a shell-like interfacial region the geometry described above allows one to formulate an initial boundary-value problem in terms of bulk field equations valid in the regions $\mathcal{B}_t^{\pm}$ and the interfacial field equations valid on Σ_t. Then the region $\mathcal{Z}_t$ will be taken into account by the appearance, on the right hand sides (RHS) of the interfacial equations, one-sided limits of the true bulk fields on either $\Sigma_t^{\pm}$ and of those boundary conditions, which primitively (and physically) have been formulated on the lateral boundary of $\mathcal{Z}_t$, i.e on $\Omega_t := \partial\mathcal{Z}_t \setminus \Sigma_t^{\pm}$, and which now should be recalculated (read - integrated along the thickness) and put as boundary conditions for the interface equations.

3.1 *Constant thickness region*

First the case of the region of constant thickness $z = z^+ - z^-$, will be considered. Then the boundary surfaces $\Sigma_t^{\pm}$ will be equidistant (parallel) and the parallel surface coordinate system (Naghdi (1972), Napolitano (1982)) is most convenient for the description of an arbitrary point in the layer $\mathcal{Z}_t$. If the position of the reference surface is given by

$$y = \mathbf{r}(L^1, L^2, t), \tag{3.1}$$

then an arbitrary point x in $\mathcal{Z}_t$ can be represented as

$$x = \mathbf{r}(L^1, L^2, t) + L\,\mathbf{n}(\mathbf{r}(L^1, L^2, t)) \tag{3.2}$$

where $L \in [z^-, z^+]$ is the third coordinate, measuring the distance of the point x from Σ_t. It means that the zone $\mathcal{Z}_t$

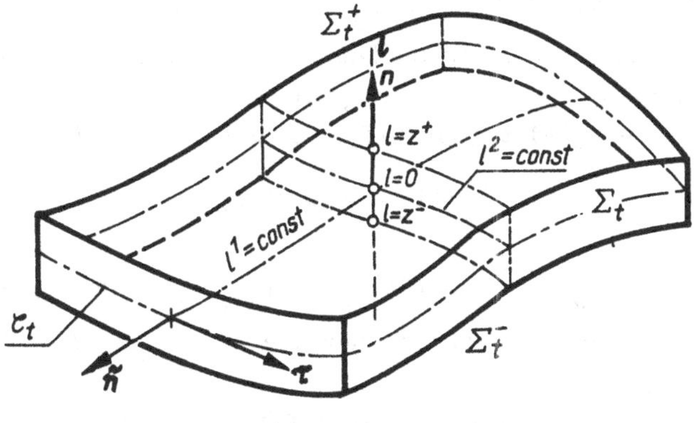

Fig. 4, A constant thickness region

is *delineated* by the surfaces Σ_t^L, distant L from Σ_t and represented by (3.2) with fixed L, and is delimited by the surfaces Σ_t^+ and Σ_t^-, distant $L = z^+$ and $L = z^-$ from Σ_t, respectively, where to the surfaces lying between Σ_t and $\mathcal{B}_t^-$ negative values of the coordinate L are attributed (cf. Fig. 4). Here points of the region $\mathcal{Z}_t$ are referred to a fixed rectangular Cartesian coordinate system.

3.2 *Geometry of the delineated layer*

To underline that except for the field of the normal vector $\mathbf{n}$, the inner as well as outer geometries, of individual Σ_t^L can be different, we shall explicitly write for the objects $\mathbf{a}^\alpha$, $\mathbf{a}_\beta$ and $\mathbf{b}$ their dependence on the third coordinate L, if L differs from zero. Using the representation (3.2), together with (2.2, 2.11) we have, after performing the necessary differentiation, the relation

$$\mathbf{a}_\alpha(L) = (\mathbf{1}_s - L\,\mathbf{b}\,)\mathbf{a}_\alpha\ . \tag{3.3}$$

In what follows we shall need the expansion formula for the second invariant of the surface tensor $(\mathbf{1}_s - L\,\mathbf{b}\,)$, denoted by $j(L)$, i.e. for the determinant of the matrix $[\,\delta_\alpha^\beta - L\,\mathrm{b}^\beta]$,

$$j(L):\ = \det\,[\delta_\alpha^\beta - L\,\mathrm{b}_\alpha^\beta] = 1 - 2HL + KL^{\,2}, \tag{3.4}$$

where we should point out to the reader that b_α^β are the mixed components of the curvature tensor $\mathbf{b}$ of the surface Σ_t, while H and K are its invariants, i.e. the mean and Gauss curvatures, respectively. If we denote by $k_1(\mathbf{r})$ and $k_2(\mathbf{r})$ the principal curvatures at $\mathbf{r} \in \Sigma_t$, then to avoid loss of regularity of the representation of the layer we have to restrict the thickness z of the layer to values satisfying

$$z < \sup\,\{\ \max(k_1(\mathbf{r})^{-1},\ k_2(\mathbf{r})^{-1})\ :\ \mathbf{r} \in \Sigma_t\}.$$

The relationships (3.3) between the basis of a typical surface Σ_t^L and Σ_t allows to calculate its surface metric tensor components $\mathrm{a}_{\alpha\beta}(L)$, and the area element $(a(L))^{1/2}$. We obtain

$$\mathrm{a}_{\alpha\beta}(L) := \mathbf{a}_\alpha(L)\cdot\mathbf{a}_\beta(L) = (\mathbf{1}_s - L\,\mathbf{b}\,)\mathbf{a}_\alpha\cdot(\mathbf{1}_s - L\,\mathbf{b}\,)\mathbf{a}_\beta$$
$$= \mathbf{a}_\alpha\cdot(\mathbf{1}_s - L\,\mathbf{b}\,)^2\,\mathbf{a}_\beta.$$

Hence

$$a(L) := \det\,[\mathrm{a}_{\alpha\beta}(L)] = \det\,[\mathrm{a}_{\alpha\beta}]\ \det(\mathbf{1}_s - L\,\mathbf{b}\,)^2 = a\ j(L)^2.$$

It means that the ratio of the surface area elements of Σ_t^L to those of Σ_t is given by $j(L)$

$$j(L) = (a(L)a^{-1})^{1/2}.$$

The last formula is particularly useful in splitting the volume measure $d\vartheta$ in the layer Z_t into the product of two measures: dL and da_L. Here dL represents the line measure (element) of a typical segment $\{\mathbf{r} + L\mathbf{n}(\mathbf{r}):\ L \in [z^-, z^+],\ \mathbf{r} \in \Sigma_t\}$ in the layer, orthogonal to each surface Σ_t^L, while da_L is the the surface measure (element) of Σ_t^L. The last measure can be written as $da_L = (a(L))^{1/2}\ dL^1 dL^2$. From the orthogonality of segments and the surfaces Σ_t^L, and the formula for $j(L)$, follows

$$d\vartheta = dL\ da_L = j(L)dL\ da\ , \tag{3.5}$$

where da is the the surface measure (element) of Σ_t. We proceed further to get relations for other geometrical objects in the layer. From (2.11, 13, 3.2) follow the useful relations

$$\mathbf{n},_\alpha = -\ \mathbf{b}\ \mathbf{a}_\alpha = -\mathbf{b}(L)\ \mathbf{a}_\alpha(L).$$

Hence, due to (3.3), we obtain

$$\mathbf{b}\ \mathbf{a}_\alpha = \mathbf{b}(L)\ (\mathbf{1}_s - L\ \mathbf{b}\ )\mathbf{a}_\alpha\ .$$

The last relation and the fact that both $\mathbf{b}$ and $\mathbf{b}(L)$ are surface tensors, lead to

$$\mathbf{b}(L) = \mathbf{b}\ (\mathbf{1}_s - L\ \mathbf{b}\ )^{-1}, \tag{3.6}$$

where $(\mathbf{1}\ \ - L\ \mathbf{b}\ )^{-1}$ is an inverse surface tensor so that

$$(\mathbf{1}_s - L\ \mathbf{b}\ )\ (\mathbf{1}_s - L\ \mathbf{b}\ )^{-1} = \mathbf{1}_s. \tag{3.7}$$

Repeated application of the Cayley-Hamilton theorem (2.14), first with $\mathbf{d}$ equal to $\mathbf{1}_s - L\ \mathbf{b}$ and then with $\mathbf{d}$ equal to $\mathbf{b}$, leads to

$$(\mathbf{1}_s - L\ \mathbf{b}\ )^{-1} = \frac{\mathbf{1}_s - L\ K\ \mathbf{b}^{-1}}{\det\ (\mathbf{1}_s - L\ \mathbf{b}\ )} = j(L)^{-1}(\mathbf{1}_s + L\ (\mathbf{b} - 2H\ \mathbf{1}_s)).$$

Hence (3.6) will be

$$\mathbf{b}(L) = j(L)^{-1}(\mathbf{b} - L\ K\ \mathbf{1}_s) = j(L)^{-1}\mathbf{b}\ (\mathbf{1}_s + L\ (\mathbf{b} - 2H\ \mathbf{1}_s)) \tag{3.6$_a$}$$

The surface tensors appearing under the sign of the last bracket, will play an important role in the further derivation. They, due to (3.6, 7,8), satisfy the relation

$$j(L)\mathbf{1}_s = (\mathbf{1}_s - L\ \mathbf{b}\ )\mathbf{A}_s(L) \tag{3.8}$$

where

$$\mathbf{A}_s(L) := \mathbf{1}_s + L\ \tilde{\mathbf{b}}, \quad \text{and} \quad \tilde{\mathbf{b}} = \mathbf{b} - 2H\mathbf{1}_s. \tag{3.9}$$

Note that at a spherical point of the surface Σ_t, where $\mathbf{b} = H_s\mathbf{1}_s$, the tensor $\tilde{\mathbf{b}}$ is opposite to $\mathbf{b}$, i.e. $\tilde{\mathbf{b}} = -\ \mathbf{b}$.

From (3.6, 6$_a$) the simple relationships between the mean and Gauss curvatures of Σ_t^L and Σ_t (cf. Thomas (1965), Naghdi (1972), Napolitano (1982), Kosiński (1986)) follow

$$\operatorname{tr} \mathbf{b}(L) = j(L)^{-1}(H - L\,K),\ \det\mathbf{b}(L) = j(L)^{-1}K.$$

To determine $\mathbf{a}^{\beta}(L)$ we use, together with (3.7), the relation $\mathbf{a}_{\alpha}(L)\cdot\mathbf{a}^{\beta}(L) = \delta^{\beta}_{\alpha}$, valid for any L. We can then write

$$\delta^{\beta}_{\alpha} = \mathbf{a}_{\alpha}\cdot\mathbf{1}_{s}\mathbf{a}^{\beta} = \mathbf{a}_{\alpha}\cdot(\mathbf{1}_{s} - L\,\mathbf{b}\,)\,(\mathbf{1}_{s} - L\,\mathbf{b}\,)^{-1}\mathbf{a}^{\beta}.$$

Due to the symmetry of $\mathbf{b}$, the RHS of the last relation can be rearranged, with the help of (3.3) and the expression using the inverse of the tensor $\mathbf{1}_{s} - L\,\mathbf{b}$, as

$$\begin{aligned}\delta^{\beta}_{\alpha} &= \mathbf{a}_{\alpha}\cdot(\mathbf{1}_{s} - L\,\mathbf{b}\,)\,(\mathbf{1}_{s} - L\,\mathbf{b}\,)^{-1}\mathbf{a}^{\beta}\\ &= (\mathbf{1}_{s} - L\,\mathbf{b}\,)^{-1}\mathbf{a}^{\beta}\cdot(\mathbf{1}_{s} - L\,\mathbf{b}\,)\mathbf{a}_{\alpha} = j(L)^{-1}(\mathbf{1}_{s} - LK\mathbf{b}^{-1})\mathbf{a}^{\beta}\cdot\mathbf{a}_{\alpha}(L).\end{aligned}$$

Hence we are getting the required representation (cf.(3.6)) for the reciprocal basis of the surface Σ^{L}_{t}

$$\begin{aligned}\mathbf{a}^{\beta}(L) &= j(L)^{-1}(\mathbf{1}_{s} - LK\mathbf{b}^{-1})\mathbf{a}^{\beta} = j(L)^{-1}(\mathbf{1}_{s} + L\,\tilde{\mathbf{b}}\,)\mathbf{a}^{\beta}\\ &= j(L)^{-1}\mathbf{A}_{s}(L)\mathbf{a}^{\beta}.\end{aligned} \tag{3.10}$$

The expressions (3.3, 5, 8, 10) relate the geometry of a typical surface Σ^{L}_{t} to the geometry of Σ_{t}, quite similarly to thin shell theory. This derivation differs from that in shell theory in that the region $\mathcal{Z}_{t}$ is not material, in general.

To finish the derivation of geometrical relationships in the layer let us transform the oriented surface element $\mathbf{N}(L)$da of the ruled surface Ω_{t} formed of segments $\{\mathbf{r} + L\mathbf{n}(\mathbf{r}) : L \in (z^{-}, z^{+}),\ \mathbf{r} \in \mathcal{C}_{t} \subset \Sigma_{t}\}$, where $\mathcal{C}_{t}$ is a curve[7] on Σ_{t}. Here $\mathbf{N}(L)$ is the outward unit normal to Ω_{t}, given by

$$\mathbf{N}(L) := \mathbf{t}_{L} \times \mathbf{n}\ \|\mathbf{t}_{L} \times \mathbf{n}\|^{-1} \tag{3.11}$$

where $\mathbf{t}_{L}$ is a tangent vector to the curve $\mathcal{C}^{L}_{t}$ after lifting of the curve $\mathcal{C}_{t}$ to Σ^{L}_{t}; the tangent vector to the latter we shall denote by $\mathbf{t}_{o}$ (cf.Fig.4). Due to (3.2) and the fact that each of $\mathbf{t}_{L}$ and $\mathbf{t}_{o}$ are orthogonal to $\mathbf{n}$, and $\mathbf{n}$ has unit length, we get

$$\mathbf{t}_{L} = (\mathbf{1}_{s} - L\,\mathbf{b}\,)\mathbf{t}_{o} \text{ and } \mathbf{t}_{L} \times \mathbf{n} = (\mathbf{1}_{s} - L\,\mathbf{b}\,)\mathbf{t}_{o} \times \mathbf{n}$$

and

$$\|\mathbf{t}_{L} \times \mathbf{n}\| = \|\mathbf{t}_{L}\| = j(L)\|\mathbf{t}_{o}\|. \tag{3.12}$$

If by $d\sigma_{L}$ and $d\sigma$ we denote the line elements of the curves $\mathcal{C}^{L}_{t}$ and $\mathcal{C}_{t}$, respectively, then from (3.12) it follows that the ratio $d\sigma_{L}/d\sigma :=$

[7] If $\mathcal{C}_{t} = \Sigma_{t} \cap \bar{\mathcal{Z}}_{t}$, then the ruled surface will be the lateral boundary of $\mathcal{Z}_{t}$, i.e. $\partial\mathcal{Z}_{t}\backslash(\Sigma^{-}_{t} \cup \Sigma^{+}_{t})$.

$= \|\mathbf{t}_L\|/\|\mathbf{t}_a\|$ is equal to $j(L)$. If the tangent vector $\mathbf{t}_a$ has[8] the splitting $d^{\alpha}\mathbf{a}_{\alpha}$, then its lifting $\mathbf{t}_L = d^{\alpha}\,\mathbf{a}_{\alpha}(L)$, and denotingcomponents of the alternation tensor $\varepsilon(L)$ of the surface Σ_t^L by $\varepsilon_{\alpha\beta}(L) = j(L)\,\varepsilon_{\gamma\alpha}$ (cf.(2.9, 3.5)), we obtain

$$\begin{aligned}\mathbf{t}_L \times \mathbf{n} &= d^{\alpha}\mathbf{a}_{\alpha}(L) \times \mathbf{n} = d^{\alpha}\,\varepsilon_{\gamma\alpha}(L)\,\mathbf{a}^{\gamma}(L) = d^{\alpha} j(L)\,\varepsilon_{\gamma\alpha}\,\mathbf{a}^{\gamma}(L)\\ &= d^{\alpha} j(L)\,\varepsilon_{\beta\alpha}\delta^{\beta}_{\gamma}\,\mathbf{a}^{\gamma}(L) = d^{\alpha} j(L)\,\varepsilon_{\beta\alpha}(\mathbf{a}^{\beta}\cdot\mathbf{a}_{\gamma})\,\mathbf{a}^{\gamma}(L)\\ &= d^{\alpha}\,j(L)\,(\mathbf{a}_{\alpha} \times \mathbf{n})\cdot\mathbf{a}_{\gamma}\,\mathbf{a}^{\gamma}(L) = j(L)(\mathbf{a}^{\gamma}(L) \otimes \mathbf{a}_{\gamma})\mathbf{t}_a \times \mathbf{n}.\end{aligned}$$

If we put $\tilde{\mathbf{n}} := \mathbf{t}_a \times \mathbf{n}/\|\mathbf{t}_a\|$ for the unit normal to the ve $\mathscr{C}_t$ that is both tangent and outwardly directed with respect to Σ_t, then the last expression, together with (3.11) and (3.12), will lead to relation

$$\mathbf{N}(L)\mathrm{da} = \mathbf{N}(L)dL d\mathsf{a}_L = j(L)(\mathbf{a}^{\gamma}(L) \otimes \mathbf{a}_{\gamma})\,\tilde{\mathbf{n}}\,dL d\mathsf{a}. \tag{3.13}$$

One can find the componentwise derivation of the last formula in Alts & Hutter (1986).

3.3 *Kinematics of the delineated layer*

To describe the kinematics of $\mathcal{Z}_t$ we employ the velocity of displacement $\mathbf{c}$ introduced in Sec.2 for Σ_t. If we use for the velocity of Σ_t the same denotation, then due to the representation (3.2) of the surface Σ_t^L its velocity of displacement will be

$$\mathbf{c}(L) = \mathbf{c} - L(\mathrm{grad}_s c_n + \mathbf{b}\,\mathbf{c}). \tag{3.14}$$

Here only the explicit dependence of the fields on the coordinate L is written: the dependence neither on L^{α} nor on t is represented. To get (3.14) we have to differentiate (3.2) with respect to time t keeping L^{α} and L constant. Hence we need the time derivative of the normal vector field $\mathbf{n}$ (note that here $\mathbf{n}$ is the same for each L in the zone $\mathcal{Z}_t$). Due to (2.2, 4, 5, 19), we get

$$\frac{\partial\,\mathbf{n}}{\partial\,t} = -\,\mathbf{n}\cdot\frac{\partial\,\mathbf{c}(L^{\alpha},t)}{\partial\,L^{\beta}}\otimes\mathbf{a}^{\beta} = -\mathbf{n}\,\mathrm{grad}_s\mathbf{c} = -(\mathrm{grad}_s c_n + \mathbf{b}\,\mathbf{c}) \tag{3.15}$$

Let us recall from Sec.2 that only the normal component c_n of the velocity of the displacement of the surface is independent of the parametrization. Consequently the component of the velocity field $\mathbf{c}(L,L^{\alpha},t)$, independent of parametrization, is

---

[8] If λ is a parameter of the curve $\mathscr{C}_t$ and each of the curves $\mathscr{C}_t^L$, then $d^{\alpha} := d\ell^{\alpha}/d\lambda$.

$$c(L) \doteq c_n \mathbf{n} - L\, \mathrm{grad}_s\, c_n, \qquad (3.16)$$

where the dot over the equality sign means that the formula is only valid in the convected parametrization. The derived formula (3.14) for the velocity is exact and can be compared with that proposed by Gatignol & Seppecher (1986) and others.

3.4 *Time derivatives in the interfacial motion*

The kinematics and geometry of the Eulerean region $\mathcal{Z}_t$ described above allow us to define the Jacobian of the transformation between two space-time reference systems given by (x, t) and (L, L^α, t). Here x is understood as a triple of spatial coordinates. To make the derivation more clear we shall put a star over t appearing in the first system.

The starting point is the transformation formula

$$\begin{aligned} x &= \mathbf{r}(L^\alpha, t) + L\, \mathbf{n}(\mathbf{r}(L^\alpha, t)) \\ t^* &= t. \end{aligned} \qquad (3.17)$$

Due to the above derivation ((3.2, 3, 10, 13) the *gradient of the transformation* (Jacobian) will be

$$\frac{\partial\,(x,\ t^*)}{\partial\,(L,\ L^\alpha,\ t)} = \begin{bmatrix} \mathbf{n} & \mathbf{a}_1(L) & \mathbf{a}_2(L) & \mathbf{c}(L) \\ 0 & 0 & 0 & 1 \end{bmatrix} \qquad (3.18)_1$$

and its inverse

$$\frac{\partial\,(L,\ L^\alpha,\ t)}{\partial\,(x,\ t^*)} = \begin{bmatrix} \mathbf{n} & -\mathbf{c}(L)\cdot\mathbf{n} \\ \mathbf{a}^1(L) & -\mathbf{c}(L)\cdot\mathbf{a}^1(L) \\ \mathbf{a}^2(L) & -\mathbf{c}(L)\cdot\mathbf{a}^2(L) \\ 0 & 1 \end{bmatrix} \qquad (3.18)_2$$

Note, that in the first matrix of (3.18) the normal vector **n** is a column one, while in the second - a row one.

The material body composed of material points in the bulk moves with the field **v**. To this field a *material time derivative*, denoted further by D/Dt, is related, which acting on an arbitrary field **u** (x,t) leads to the rate of change of **u** at fixed material point (particle). Its operating formula is

$$(D/Dt)\mathbf{u}(x,t) := (\partial/\partial t^*)\ \mathbf{u} + (\mathrm{grad}\ \mathbf{u})\ \mathbf{v}, \qquad (3.19)$$

where the simple contraction of the vector **v** with the gradient of **u**, in the case of a scalar-valued field **u**, is the inner product operation. With the help of (3.15) one can calculate the material time derivative of the normal vector and the coordinates L and L^α. First the derivative of the 'surface' parameters will be calculated. It leads to the formulae

$$DL/Dt = \frac{\partial L}{\partial t}_{*} + \text{grad } L\cdot\mathbf{v} = -\mathbf{c}(L)\cdot\mathbf{n} + \mathbf{n}\cdot\mathbf{v} = (\mathbf{v} - \mathbf{c}(L))\cdot\mathbf{n},$$

$$DL^{\alpha}/Dt = \frac{\partial L^{\alpha}}{\partial t}_{*} + \text{grad } L^{\alpha}\cdot\mathbf{v} = -\mathbf{c}(L)\cdot\mathbf{a}^{\alpha}(L) + \mathbf{a}^{\alpha}(L)\cdot\mathbf{v}, \tag{3.20}$$

$$= (\mathbf{v} - \mathbf{c}(L))\cdot\mathbf{a}^{\alpha}(L),$$

$$Dt/Dt = 1.$$

Replacing $\mathbf{u}$ by $\mathbf{n}$ in (3.16) and remembering that the independent variables[9] (L^{α}, t) appearing under the form of $\mathbf{n}$ are functions of x and t^{*}, we obtain

$$(D/Dt)\mathbf{n} = \frac{\partial\,\mathbf{n}}{\partial\,L^{\alpha}}\,DL^{\alpha}/Dt + \frac{\partial\,\mathbf{n}}{\partial\,t}\cdot 1 = -\,\mathbf{b}\,\mathbf{a}_{\alpha}\,(\mathbf{v} - \mathbf{c}(L))\cdot\mathbf{a}^{\alpha}(L)$$

$$- (\text{grad } c_{n} + \mathbf{b}\,\mathbf{c}) = -j(L)^{-1}\Big\{\mathbf{b}((\mathbf{1}_{s} + L\,\tilde{\mathbf{b}})(\mathbf{v} + L\,\text{grad}c_{n})$$

$$- \mathbf{b}\,(\mathbf{1}_{s} - L\mathbf{b})\,\mathbf{A}_{s}(L))\mathbf{c}\Big\} - \mathbf{bc} - \text{grad } c_{n}$$

$$= -\,j(L)^{-1}\Big\{\,\mathbf{b}\,\mathbf{A}_{s}(L)(\mathbf{v} + L\text{grad}c_{n}) - j(L)\mathbf{b}\,\mathbf{c}\Big\} - \mathbf{bc} - \text{grad}c_{n}$$

$$= -j(L)^{-1}\Big\{\mathbf{b}\,\mathbf{A}_{s}(L)(\mathbf{v} + \text{grad}c_{n})\,\Big\} - \text{grad}c_{n}, \tag{3.21}$$

due to (3.9, 11, 12). The last formula for the material time derivative of the normal vector is used in the the derivation of consequences of the angular momentum balance law in the model with interfacial true quantities (cf. Dell'Isola & Kosiński (1989)).

In the derivation of the local interfacial balance laws in either model an invariant time derivative has to appear, which is independent of the chosen parametrization of the moving reference surface Σ_t. It will be the displacement time derivative introduced in the previous section by formula (2.24) or (2.25).

Additionally, however, having the representation for $\mathbf{c}(L)$ derived in (3.14)), we may introduce the time derivative following the displacement of the whole region (layer) $\mathcal{Z}_t$. We denote this derivative by $\frac{d_c}{dt}$ and define it by

$$\frac{d_c}{dt}\psi := \frac{\partial\psi(x, t^{*})}{\partial\,t^{*}} + \text{grad}\psi(x, t^{*})\;\mathbf{c}(x) \tag{3.22}$$

for an arbitrary (C^1-smooth field) ψ defined (at least) on

$$\bigcup\Big\{\,\mathcal{Z}_t \times \{t\} \;:\; t \in I\,\Big\},$$

where we put $x = \mathbf{r}(L^1, L^2, t) + L\;\mathbf{n}(\mathbf{r}(L^1, L^2, t))$, according to (3.2), under the sign of $\mathbf{c}$.

In order to derive the local interfacial balance equations a time

---

9 Note that $\mathbf{n}$ is independent of L.

differentiation $\frac{d_c}{dt}$ of the surface integral defined on a moving reference surface Σ_t must be performed. Instead of introducing fictitious particles, as was done by Moeckel (1974), we shall use the representation of the motion of the surface Σ_t identical to that in (2.22) in a convected parametrization (k^α, t).

Since the field[10] $\mathbf{c} \doteq c_n \mathbf{n}$ as a solution of (2.22) determines the transformation (flow), which maps $\left\{\Sigma_{t_o} \times \{t_o\}\right\}$ onto $\left\{\Sigma_t \times \{t\}\right\}$, we can use the well-known formula for the (time) derivative of the Jacobian of flow (cf. Truesdell and Toupin (1960, p.342)) in order to calculate the derivative of the surface element da of Σ_t (cf. (2.8)). Using that formula, the transformations (3.18), and the definition (2.13), we get

$$d_c(da)/dt = \mathrm{div}_s \mathbf{c}\, da = (\mathrm{div}_s(\mathbf{Pc}) - 2Hc_n)da, \tag{3.23}$$

where the perpendicular projection $\mathbf{P}$ of V upon the tangent space of Σ_t has been used (cf. (2.7)). The above expression is used in the derivation of the local balance equations in either model (cf. Scriven (1960), Moeckel (1974), Deemer & Slattery (1978), Kosiński (1986), Alts & Hutter (1988), Kosiński & Romano (1989), Dell'Isola & Kosiński (1989)). One can find the most general expressions in Dumais (1980), Dell'Isola & Romano (1986) and Petryk & Mróz (1986).

If, however, in a parametrization (L^α, t) the displacement velocity of the surface Σ_t is $\mathbf{c}$, then (3.23) written for the element da regarded as a function of (L^α, t), will take the form

$$\frac{\partial da}{\partial t} = \mathrm{div}_s \mathbf{c}\, da.$$

In convected parametrization the RHS is equal to $2Hc_n da$.

3.5 *Variable thickness region*

The assumption about the constant thickness of the layer is rather reasonable when a very thin layer of the interfacial medium is modelled. However, the constant in time (and in surface coordinates) interfacial layer restricts the class of physical problems successfully treated by the model proposed. To overcome this drawback in this section we shall try to omit this assumption and to check its consequences on the previously derived formulae.

As before a narrow layer $\mathcal{Z}_t$ divides the volume phases $\mathcal{B}_t^{\pm}$, and the boundaries between $\mathcal{Z}_t$ and both $\mathcal{B}_t^{\pm}$ are regular surfaces Σ_t^+ and Σ_t^-. Now, however, they are not equidistant (parallel). As before, however, we can use the parallel surface coordinate system describing an arbitrary point in the layer $\mathcal{Z}_t$. If the position of the reference

[10] We recall that the dot over the equality sign means that it holds in the convected parametrization, only.

surface is given by (3.1), then an arbitrary point x in $\mathcal{Z}_t$ can be represented as in (3.2), where the third coordinate L measuring the distance of the point x from Σ_t does not run over the fixed interval $[z^-, z^+]$; Now two scalar fields $\mathfrak{z}^-$ and $\mathfrak{z}^+$ on the hypersurface $\mathcal{S}$ are defined: their values give the distance of the boundary surfaces $\Sigma_t^{\pm}$ from Σ_t. The thickness z of the layer can then change according to the difference of both functions, depending on position and time, i.e.

$$z(\mathbf{r}, t) := \mathfrak{z}^+(\mathbf{r}, t) - \mathfrak{z}^-(\mathbf{r}, t) \tag{3.24}$$

The interfacial zone $\mathcal{Z}_t$ is *delineated* by the surfaces Σ_t^L, distant L from Σ_t and represented by (3.2) with fixed L, and is *delimited* (Fig.5.) by the surfaces Σ_t^+ and Σ_t^- given by

$$\Sigma_t^{\pm} \equiv \left\{ \mathbf{y} \in \mathcal{Z}_t \ : \ \mathbf{y} = \mathbf{r} + \mathfrak{z}^{\pm}(\mathbf{r}, t)\mathbf{n}(\mathbf{r}), \ \mathbf{r} \in \Sigma_t \right\} \tag{3.25}$$

which are not parallel to Σ_t, unless $\mathfrak{z}^{\pm}(\mathbf{r}, t)$ is independent of position $\mathbf{r}$; (the single time dependence will describe a variable with time region bounded by equidistant surfaces). Let us notice that under the present weaker assumptions the layer can shrink locally to a surface if $\mathfrak{z}^{\pm}$ vanishes. Moreover, it is now possible to describe the situation when the lateral boundary of the whole layer is not a ruled surface.

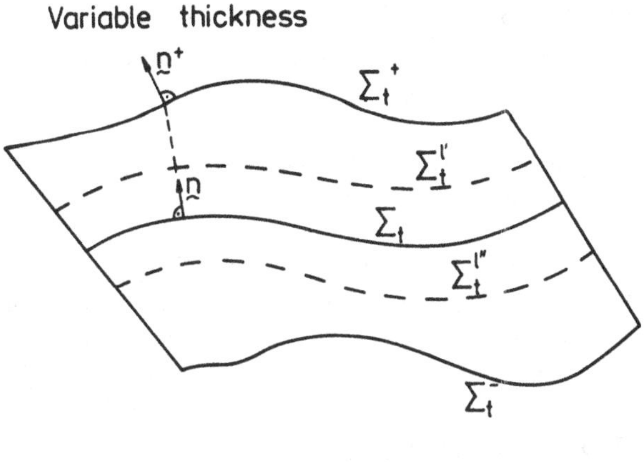

Fig.5 Cross-section of a variable thickness region

To avoid any singularities in the representation of the layer described in terms of the parallel surface coordinates (a similar assumption had to be made in Sec.3.2) and concerning the maximal thickness of the layer

$$\mathfrak{z} := \sup \{\max (\mathfrak{z}^-(\mathbf{r}, t), \mathfrak{z}^+(\mathbf{r}, t)) : (\mathbf{r}, t) \in \mathcal{S}\}$$

should be done.

The geometry of the boundary surfaces $\Sigma_t^{\pm}$ will be related to that of Σ_t as follows. If $\mathbf{a}_\alpha^{\pm}$, $\alpha = 1,2$, denote the natural basis vectors of either surface, then due to (3.25), we get

$$\mathbf{a}_\alpha^{\pm} = (\mathbf{1}_s - z^{\pm}\,\mathbf{b}\,)\mathbf{a}_\alpha + \frac{\partial z^{\pm}(\mathbf{r},t)}{\partial L^\alpha}. \tag{3.26}$$

For the components of the metric tensor we then obtain

$$a_{\alpha\beta}^{\pm} := \mathbf{a}_\alpha^{\pm}\cdot\mathbf{a}_\beta^{\pm} = \mathbf{a}_\alpha^{\pm}\cdot(\mathbf{1}_s - z^{\pm}\mathbf{b}\,)^2\,\mathbf{a}_\beta^{\pm} + \frac{\partial z^{\pm}}{\partial L^\alpha}\frac{\partial z^{\pm}}{\partial L^\beta}. \tag{3.27}$$

Normal (not unit) vectors to either surfaces will be given by

$$\begin{aligned}\mathbf{a}_1^{\pm}\times\mathbf{a}_2^{\pm} &= \mathbf{a}_1\times\mathbf{a}_2 - z^{\pm}(\mathbf{b}\mathbf{a}_1\times\mathbf{a}_2 + \mathbf{a}_1\times\mathbf{b}\mathbf{a}_2) + (z^{\pm})^2\mathbf{b}\mathbf{a}_1\times\mathbf{b}\mathbf{a}_2\\ &\quad + \mathbf{a}_1\times\mathbf{n}\frac{\partial z^{\pm}}{\partial L^2} + \mathbf{n}\times\mathbf{a}_2\frac{\partial z^{\pm}}{\partial L^1} - z^{\pm}(\mathbf{b}\mathbf{a}_1\times\mathbf{n}\frac{\partial z^{\pm}}{\partial L^2} + \mathbf{n}\times\mathbf{b}\mathbf{a}_2\frac{\partial z^{\pm}}{\partial L^1})\\ &= \varepsilon_{12}(1 - z^{\pm}\mathrm{tr}\mathbf{b} + (z^{\pm})^2\det\mathbf{b})\,\mathbf{n} + \varepsilon_{12}\,z^{\pm}(b_1^1\frac{\partial z^{\pm}}{\partial L^2} - b_2^1\frac{\partial z^{\pm}}{\partial L^1})\mathbf{a}^2\\ &\quad - \varepsilon_{12}\,z^{\pm}\,(b_1^2\frac{\partial z^{\pm}}{\partial L^2} - b_2^2\frac{\partial z^{\pm}}{\partial L^1})\,\mathbf{a}^1 - \varepsilon_{12}\mathrm{grad}_s z^{\pm}.\end{aligned} \tag{3.28}$$

Inspecting the coefficient of ε_{12} in the first bracket we can see the value of the Jacobian j at $L = z^{\pm}$, while the sum of the next two brackets give negative $z^{\pm}(\mathbf{b} - 2H\mathbf{1}_s)\mathrm{grad}_s z^{\pm}$ equal (cf. (3.8)) to $- z^{\pm}\tilde{\mathbf{b}}\,\mathrm{grad}_s z^{\pm}$. With this at hand and remembering the tensor $\mathbf{A}_s(L)$ from (3.9), we write the final relationship

$$\mathbf{n}^{\pm}(\mathbf{r})da^{\pm} = (j(z^{\pm},\mathbf{r})\mathbf{n}(\mathbf{r}) - \mathbf{A}_s(z^{\pm})\mathrm{grad}_s z^{\pm})da \tag{3.29}$$

where the surface element $da = a^{1/2}dL^1 dL^2$. In what follows we denote the ratio $da^{\pm}/da$ by $j^{\pm}$ (a function of $\mathbf{r}$ and t).

The boundary surfaces $\Sigma_t^{\pm}$ now move with different velocities to those prescribed by (3.14). In fact, we obtain the velocities $\mathbf{c}^{\pm}$ of displacement of $\Sigma_t^{\pm}$ from definitions (3.25) as

$$\mathbf{c}^{\pm} = \mathbf{c} - z^{\pm}(\mathrm{grad}_s c_n + \mathbf{b}\,\mathbf{c}) + \frac{\partial z^{\pm}}{\partial t}\mathbf{n}. \tag{3.30}$$

To finish the geometrical preparation to the next section concerning the general balance law for 3D fields and referred to the reference surface, we make the final calculation of the product of $j^{\pm}$ and the normal speed of displacement of either boundary surface $\Sigma_t^{\pm}$. It will proceed as follows $j^{\pm}\mathbf{c}^{\pm}\cdot\mathbf{n}^{\pm}(\mathbf{r}) =$

$$(\mathbf{c} - z^{\pm}(\mathrm{grad}_s c_n + \mathbf{b}\,\mathbf{c}) + \frac{\partial z^{\pm}}{\partial t}\mathbf{n})\cdot(j(z^{\pm},\mathbf{r})\mathbf{n}(\mathbf{r}) - \mathbf{A}(z^{\pm})\mathrm{grad}_s z^{\pm})$$

$$= j(z^{\pm},\mathbf{r})(c_n + \frac{\partial z^{\pm}}{\partial t} - \mathrm{grad}_s z^{\pm}\mathbf{c}) + z^{\pm}\mathrm{grad}_s z^{\pm}\mathbf{A}_s(z^{\pm})\cdot\mathrm{grad}_s c_n.$$

Using the formula (2.18) for the displacement derivative we can rewrite the last expression in the form

$$j^{\pm}\mathbf{c}^{\pm}\cdot\mathbf{n}^{\pm}(\mathbf{r}) = j(z^{\pm},\mathbf{r})(c_n + \frac{\delta z^{\pm}}{\delta t}) + z^{\pm}\mathrm{grad}_s z^{\pm}\mathbf{A}_s(z^{\pm})\cdot\mathrm{grad}_s c_n.$$

From (3.28, 29) follows the expression for $j^{\pm} := da^{\pm}/da$

$$j^{\pm} = \frac{\|a_1^{\pm} \times a_2^{\pm}\|}{a^{1/2}} = (j^2((\mathfrak{z}^{\pm}, \mathbf{r}) + \mathrm{grad}_s \mathfrak{z}^{\pm} \cdot \mathbf{A}_s^2(\mathfrak{z}^{\pm}) \mathrm{grad}_s \mathfrak{z}^{\pm})^{\frac{1}{2}} \tag{3.31}$$

which can be useful in some derivations.

4. Balance laws for a moving non-material shell-like region

The tools developed in the previous sections will be applied here in deriving the general and particular balance laws of thermomechanics. This will be done for the case of interfacial layers with nonvanishing thickness and for true, not excess, quantities.

4.1 *Surface mass density*

In a previous paper (cf. Kosiński & Romano (1989)) we have assumed that the lateral boundary of the whole interface (transition) region, i.e. $\partial \mathcal{Z}_t \setminus (\Sigma_t^+ \cup \Sigma_t^-)$, is a ruled surface. It turns out that this assumption can be dropped out on the global level (since liquid drops do not possess ruled lateral boundaries), keeping, however, this assumption, on the local level, i.e. during the passage from global to local form of balance laws. It can be done by assuming the integral form of the laws is valid for any sub-layer which is a proper subset of the whole layer bounded by subsurfaces of $\Sigma_t^{\pm}$ and a lateral boundary which is a ruled surface. In such case the natural boundary conditions given on the lateral boundary of $\mathcal{Z}_t$ need to be recalculated in an appropriate way.

According to our denotation the normal unit vector of $\Sigma_t^{\pm}$ will point from the region "-" to "+". The question of how to choose the surfaces $\Sigma_t^{\pm}$ and some reference surface Σ_t lying between $\Sigma_t^{\pm}$ will be discussed after the concept of a surface mass density is introduced.

The surface field ρ^s defined by

$$\rho^s(\mathbf{r}(L^1, L^2, t), t) = \int_{\mathfrak{z}^-}^{\mathfrak{z}^+} \rho(\mathbf{r}(L^\alpha, t) + L\mathbf{n}(\mathbf{r}), t) j(L, \mathbf{r}(L^\alpha, t)) \mathrm{d}L \tag{4.1}$$

with $j(L, \mathbf{r}(L^\alpha, t))$, $\alpha=1,2$, given by (3.4, 5), will be called the *surface mass density*.

Let us notice that the definition of the surface mass given by (4.1) does not take into account the type and the form of the 3D motion governed by the particle velocity field $\mathbf{v}$ in the layer (especially its tangential components). That motion is not known in advance, and consequently we should, in general, expect an extra term in the mass flux in the 2D continuity equation (cf. (4.21).

To make this statement clearer, let us assume that the motion in the interface region is such that any material particle entering that region moves along the normal to the reference surface. Then, in each period of time no segment of a line tangent to the normal is crossed by material particles: in other words, the line is a material particle path. If, additionally, the normal component of the material velocity is the same for all particles on one line, then the (normal) segment becomes a material segment, with a unique normal speed of propagation. Under the above circumstances the (mathematical) definition of the surface material particle (4.1) corresponds to the physically adequate one, and consequently no extra flux of the surface mass can be expected in the final form of the mass balance law (4.20). A nonconservative form of that law can be only due to influx or outflow of bulk mass and the difference between the normal speeds of the interface and the material segment.

The motion considered reflects (or better to say-corresponds to) the so-called Kirchhoff-Love hypothesis known in shell theory. Consequently, the interface with such a motion was called by dell'Isola and Kosiński (1991) a *Kirchhoff-Love type interface*. Particular examples of 3D motions in a material zone are discussed by Kosiński & Wąsowski (1991).

Let us notice that in working with excess quantities, as do Professors Hutter and Bedeaux, the situation is worst, since additional to the true mass density the extrapolated bulk densities from either regions appear under the integral sign, namely

$$\tilde{\rho}^s(L^1,L^2,t) := \int_{\mathfrak{z}^-}^{0} (\rho - \rho^-) j(L,L^\alpha,t)\,dL + \int_{0}^{\mathfrak{z}^+} (\rho - \rho^+) j(L,L^\alpha,t)\,dL$$

where $\tilde{\rho}^s$ denotes the excess mass density, while ρ^- and ρ^+ are the extrapolated bulk mass density limits from the regions "-" and "+" , respectively.

If one assume that the extrapolations are of particular forms, namely they are constant along the normal segments, then between the surface mass density ρ^s and the excess mass density $\tilde{\rho}^s$ the following relation

$$\tilde{\rho}^s = \rho^s - (\rho^- J^+ + \rho^+ J^+) \tag{4.2}$$

holds, where

$$J^- := \int_{\mathfrak{z}^-}^{0} j(L,L^\alpha,t)\,dL \quad \text{and} \quad J^+ := \int_{0}^{\mathfrak{z}^+} j(L,L^\alpha,t)\,dL.$$

Moreover, if the reference surface is a mean surface, i.e. $\mathfrak{z}^- = \mathfrak{z}^+$, then $J^- = -J^+$, and (4.2) simplifies to

$$\tilde{\rho}^s = \rho^s + (\rho^- - \rho^+) J^+.$$

Now we shall briefly list some possible methods of choosing the reference surface Σ_t known from the literature. First this might be done by the requirement that the volumetric mass density ρ attains prescribed values on them (cf. Seppecher (1987)). If ρ^+ is the value of the field ρ on the surface Σ_t^+, which may vary from point to point, then by extending the surface field ρ^+ into the zone $\mathcal{Z}_t$ in such a way that it is constant along lines orthogonal to Σ^+, the equation for Σ^+ could be written in the form:

$$\Sigma_t^+ = \{(x,t)\in \mathcal{B}_t \mid \rho(x,t) - \hat{}\rho^+(x,t) = 0\}.$$

Since the field[11] $\hat{}\rho^+$ does not increase in the direction normal to the surface, the gradient of the field ρ is collinear with $\mathbf{n}$. In the same way one could proceed with Σ_t^- and the surface field ρ^- on it to define Σ_t^-. The reference surface Σ could then be defined by the condition:

$$\Sigma_t = \{(x,t)\in E^3 \mid \rho(x,t) - \mathfrak{z}^{-1}\hat{}\rho^s(x,t) = 0\},$$

where $\hat{}\rho^s$ is the extension into $\mathcal{Z}_t$ of the surface field ρ^s.

The choice of reference surface Σ_t could be made by requiring that the center of mass of each segment

$$L_{\mathbf{r}} := \{\, \mathbf{r} + L\mathbf{n}(\mathbf{r}) : L \in [\mathfrak{z}^-, \mathfrak{z}^+] \,\} \tag{4.3}$$

be at $\mathbf{r}$. This will be satisfied if the integral equality

$$\int_{\mathfrak{z}^-}^{\mathfrak{z}^+} \rho(\mathbf{r}(L^\alpha,t)+L\mathbf{n}(\mathbf{r}),t)\, j(L,\mathbf{r}(L^\alpha,t))\, L\, dL = 0$$

holds at every $\mathbf{r}$ (cf. Alts & Hutter (1988), Murdoch & Cohen (1990)).

Notice that in terms of the surface mass density ρ^s the mass confined at instant t in the zone $\mathcal{Z}_t$ is

$$\mathcal{M}(\mathcal{Z}_t) := \int_{\mathcal{Z}_t} \rho(x,t)\, dv = \int_{\mathcal{Z}_t} \rho^s da .$$

Therefore our definition of 'surface' density is such that the introduction of the 2D continuum Σ_t as a model for the interface $\mathcal{Z}_t$ does not causes a *loss* of the quantity $\mathcal{M}$.

In the literature the reference surface Σ_t is mainly defined by means of suitable averages of excess quantities, all stemming from those used by Gibbs (1961). Moreover, each physical quantity leads to a particular mean (reference) surface.

In the definition of Levine (1978), when $\mathcal{Z}_t$ is the union of parallel planes, Σ_t corresponds to L_1, where the value L_1 satisfies the equation:

$$\int_{\mathfrak{z}^-}^{\mathfrak{z}^+} \phi(L)\, dL = \phi(\mathfrak{z}^+)\left[L_1 + \mathfrak{z}^-\right] + \phi(\mathfrak{z}^-)\left[\mathfrak{z}^+ - L_1\right],$$

---

11 Here $\hat{}\rho^\pm$ (and $\hat{}f$ - in the further part) denotes the extension of the field $\rho^\pm$ (of a field f) primitively defined on one of the parallel surfaces: Σ_t or $\Sigma_t^\pm$, to the field defined on the whole zone and constant along lines orthogonal to Σ_t.

where ϕ is the volume density of the physical quantity under consideration such as densityof volume mass or internal energy. In such case Σ_t is called the *Gibbs' dividing surface* for ϕ. We emphasize that L_1 does not belong to the interval $[\mathfrak{z}^-,\mathfrak{z}^+]$ when the integral of LHS is suitably large, as may occur in non-equilibrium phenomena.

Another choice could be made for the position of the reference surface Σ_t by assuming the condition:

$$\Sigma_t := \left\{ x \in \mathcal{Z}_t \,\middle|\, j(x)\rho - z^{-1}\rho^s(x) = 0 \right\} \tag{4.4}$$

where $z = \mathfrak{z}^+ - \mathfrak{z}^-$. If, along some segment $L_{\mathbf{r}}$, the density ρ is not a monotone[12] function of L, then definition (4.4) is not well-framed.

In the approach presented at the end of these lectures and dealing with the interface quantities defined as mean values of true 3D fields (and with nonvanishing thickness of the layer, when dealing with an initial boundary-value problem, cf. Introduction) no physical meaning is ascribed to Σ_t. The choice of its position should be based on mathematical convenience in the process of modelling and solving; this convenience is especially evident in the case of variable thickness of the layer (cf.(4.14, 15)).

After choosing the moving reference surface Σ_t as a geometrical object the remaining material structure of $\mathcal{Z}_t$ is preserved by equipping the surface Σ_t with the structure of a 2D continuum. This is done by defining surface densities and fluxes of physical quantities as suitable integrals of the corresponding volume ones along the thickness of $\mathcal{Z}_t$. Following the definition of the mass density, we can define for a 3D density field f - representing a density of a bulk quantity or a production term on the layer $\mathcal{Z}_t$ - the corresponding surface field f^s ($= \psi^s$ or p^s) by

$$\psi^s := \int_{\mathfrak{z}^-}^{\mathfrak{z}^+} j(L,\mathbf{r})\psi(\mathbf{r}+L\mathbf{n}(\mathbf{r}),t)\mathrm{d}L =: \langle j\ \psi\rangle . \tag{4.5}$$

The procedure makes possible the identification of surface quantities which appear in surface balance laws even when dealing with non-material continua. Moreover, this allows for a more careful discussion of the *Galilean invariance* of the balance laws.

To the *geometrical* surface speed field c_n, describing how $\mathcal{Z}_t$ moves in E^3, no physically meaningful tangential component can be added; in the literature, however, such a component is searched for, which is "reasonable" from a physical point of view, thus getting a "complete" velocity field to be used in the balance of linear momentum (cf. Ishii (1975), Gatignol & Seppecher (1986), Moeckel (1974)).

---

[12] This is likely for nonequilibrium phenomena (cf. Adamson (1982) and Ościk (1982) in equilibrium case).

In phase transition problems, for example, the material particles constituting the interfacial matter at instant t differ from those at another instant t'. Hence together with the field $c_n\mathbf{n}$, we introduce an average velocity $\mathbf{V}^s$ of particles belonging to the layer. In terms of the 3D material (particle) velocity field $\mathbf{v}$, the densities ρ and ρ^s, the *'surface' material point velocity* $\mathbf{V}^s$ is given by the relation for the *surface momentum density*

$$\rho^s\mathbf{V}^s = \int_{\mathfrak{z}^-}^{\mathfrak{z}^+} \rho\mathbf{v}j(L,\mathbf{r})dL, \tag{4.6}$$

together with (4.1). A continuous 2D system modelled by Σ_t will be called *non-material* if[13]

$$c_n :\equiv \mathbf{c}\cdot\mathbf{n} \neq \mathbf{V}^s\cdot\mathbf{n},$$

which means that in the mean the material points (particles) occupying the interface layer will not all the time stay in it. The difference

$$d_n := (\mathbf{c} - \mathbf{V}^s)\cdot\mathbf{n} \tag{4.7}$$

is Galilean-invariant and is relevant to phase transition and adsorption processes if it does not vanish. It can be regarded as a quantity which needs to be determined by a constitutive equation (cf. Kosiński (1989)).

4.2 *General balance laws*

The classical balance law for a quantity ψ with its Galilean invariant flux (current) $\mathbf{w}$ and the source (supply plus production) term p in the material volume $\mathcal{P}_t \subseteq \mathcal{B}_t$ is of the form

$$\frac{d}{dt}\int_{\mathcal{P}_t} \psi dv = -\int_{\mathcal{P}_t} \mathbf{w}\cdot\mathbf{N}da + \int_{\mathcal{P}_t} pdv, \tag{4.8}$$

where $\mathbf{N}$ is the outward unit normal to $\partial\mathcal{P}_t$. Using the derivative $\frac{d_c}{dt}$, we get the following integral balance law for ψ in the non-material, in general, region $\mathcal{Z}_t$ moving with the velocity $\mathbf{c}$

$$\frac{d_c}{dt}\int_{\mathcal{Z}_t} \psi\, dv = -\int_{\partial\mathcal{Z}_t}(\psi\,(\mathbf{v} - \mathbf{c}) + \mathbf{w})\cdot\mathbf{N}\, da + \int_{\mathcal{Z}_t} p\, dv. \tag{4.9}$$

After partition of both volumetric and surface measures into product measures (cf. (3.5, 15) we get

$$\frac{d_c}{dt}\int_{\Sigma_t}\psi^s da = -\int_{\mathcal{C}_t}(\mathbf{W}^{s'}\{\psi\} - \psi^s\mathbf{Pc})\tilde{\mathbf{n}}\, ds - \int_{\Sigma_t}((\tilde{\mathbf{w}}j\mathbf{n})^- + (\tilde{\mathbf{w}}j\mathbf{n})^+)da + \int_{\Sigma_t} p^s da.$$

Here the 'weighted' limiting bulk-field values $(\tilde{\mathbf{w}}j\mathbf{n})^{\pm}$ are

[13] It is p-material if equality holds (Kosiński (1986)).

$$(\tilde{w}jn)^{\pm} := \tilde{w}(r + \mathfrak{z}^{\pm}n(r), t)(jn)^{\pm}, \tag{4.10}$$

where $\tilde{w} := \psi\,(v_{+} - c) + w$ is the new flux appearing under the surface integral over $\Sigma_t^{\pm}$, and we have used the fact that $N da = (jn)^{\pm} da$ on $\Sigma_t^{\pm}$. Let us notice that in the case of constant thickness $\mathfrak{z}^{\pm} = z^{\pm}$ is independent of r and t, and

$$(jn)^{\pm} = \pm j(z^{\pm}, r)n, \tag{4.11}$$

while the case of variable thickness is governed by (3.29) with $\mathfrak{z}^{\pm}$ depending on r and t, in general. Since the definition of a general surface density has been given by (4.5) we give the definition of the surface flux in place of w. The surface flux corresponding to the flux $\tilde{w}$ in (4.9) is the sum of $-\psi^{s}Pc$ and $W^{s'}\{\psi\}$ (note the prime over s)

$$W^{s'}\{\psi\} :\equiv \langle \psi \otimes (v + L\, \mathrm{grad}_s c_n)\, A_s(L)\rangle + \langle w\, A_s(L)\rangle. \tag{4.12}$$

The above definition together with (4.5) give at the same time the only possible relationships between the surface quantities and their bulk counterparts (better to say - their primitives), in order for the interfacial balance law localized on the surface Σ_t to be compatible with and derivable from, the 3D law. The latter is postulated for ψ in the integral form (4.8). Let us notice that the surface flux $W^{s}\{\psi\}$ is Galilean invariant (due to the Cayley-Hamilton identity (2.17)). Now we can get the final form of the integral balance law for layer $\mathcal{Z}_t$:

$$\int_{\Sigma_t}\Big(\frac{d_c}{dt}\psi^{s} + \psi^{s}\mathrm{div}_s c\Big)da + \int_{\mathscr{C}_t}\big(W^{s'}\{\psi\} - \psi^{s}c_{\tau}\big)\tilde{n}\, ds =$$

$$= \quad - \int_{\Sigma_t}\big((\tilde{w}jn)^{-} + (\tilde{w}jn)^{+}\big)da + \int_{\Sigma_t} p^{s} da.$$

Here we have used c_{τ} to denote the tangential part of c. However, in order to obtain the local, differential form of the law we have to perform a localization procedure by applying the integral law to an arbitrary *subzone* $\mathcal{Z}'_t$ of $\mathcal{Z}_t$. Here by a *subzone* we mean an arbitrary (shell-like) subregion $\mathcal{Z}'_t$ of $\mathcal{Z}_t \subset \mathcal{P}_t$ bounded by subsurfaces $\Sigma'^{\pm}_t$ of $\Sigma^{\pm}_t$ with a subsurface Σ'_t of Σ_t and with a nonvanishing lateral boundary which is a ruled surface to which the Stokes and Green-Gauss theorems can be applied.

After calculating the time derivative of the first integral and applying such integral law obtained to an arbitrary Σ'_t which is a support of an arbitrary subzone $\mathcal{Z}'_t$ of the layer $\mathcal{Z}_t$, we get modulo continuity of the integrand[14],

$$\frac{d_c}{dt}\psi^s + \psi^s \mathrm{div}_s \mathbf{c} + \mathrm{div}_s(\mathbf{W}^{s'}\{\psi\} - \psi^s \mathbf{c}_\tau) =$$

$$= -\left\{(\{\psi(\mathbf{v}-\mathbf{c}) + \mathbf{w}\} j\mathbf{n})^- + (\{\psi(\mathbf{v}-\mathbf{c}) + \mathbf{w}\} j\mathbf{n})^+\right\} + p^s. \qquad (4.13)$$

Using the Thomas displacement derivative $\delta/\delta t$ (cf. (2.25)), together with the relationship

$$\frac{d_c}{dt}\psi^s + \psi^s \mathrm{div}_s \mathbf{c} - \mathrm{div}_s(\psi^s \mathbf{c}_\tau) = \frac{\delta}{\delta t}\psi^s - 2Hc_n\psi^s.$$

we arrive at

$$\frac{\delta}{\delta t}\psi^s - 2Hc_n\psi^s + \mathrm{div}_s\, \mathbf{W}^{s'}\{\psi\} =$$

$$\left\{(\{\psi(\mathbf{v}-\mathbf{c}) + \mathbf{w}\} j\mathbf{n})^- - (\{\psi(\mathbf{v}-\mathbf{c}) + \mathbf{w}\} j\mathbf{n})^+\right\} + p^s. \qquad (4.14)$$

The constant thickness case can be simplified in view of (4.11) and the form of the function $j(L,\mathbf{r})$, to

$$\frac{\delta}{\delta}\psi^s - 2Hc_n\psi^s + \mathrm{div}_s \mathbf{W}^{s'}\{\psi\} = [\![\psi(\mathbf{v}-\mathbf{c}) + \mathbf{w}]\!]\cdot\mathbf{n} + \tilde{p}^s \qquad (4.15)$$

where

$$[\![g]\!] := g(\mathbf{r} + \mathfrak{z}^-\mathbf{n}(\mathbf{r}), t) - g(\mathbf{r} + \mathfrak{z}^+\mathbf{n}(\mathbf{r}), t) \qquad (4.16)$$

for an arbitrary field g defined on $\mathcal{B}_t$.

The last equation is very well known in thermodynamics with singular surfaces, (cf. Moeckel (1974), Ishii (1975), Kosiński (1981, 85, 86), Romano (1983), Alts (1986), Dell' Isola & Romano (1987) and Professor's Hutter lectures). The term p^s, called there a *surface supply*, here is equal to

$$\tilde{p}^s = p^s + [\![h\{\psi(\mathbf{v}-\mathbf{c}) + \mathbf{w}\}]\!]\cdot\mathbf{n}, \qquad (4.17)$$

where $\qquad h(z^\pm, \mathbf{r}) := (K(\mathbf{r})z^\pm - 2H(\mathbf{r}))z^\pm,$

and as previously $\mathbf{r} \in \Sigma_t$.

Alts (1986), Alts & Hutter (1988), as well as Professor Hutter in his lectures, give a boundary layer model for curved phase boundaries and compare it with the model employing a singular surface. In their derivation, however, the surface quantities are identified with so-called *excess* interfacial quantities in contrast with the definition given in the present paper. Moreover, they made use of balance laws for fields which are 'extensions' of bulk-field values at the boundary layer edges, i.e. at $\Sigma_t^\pm$, in our notations. In this way their definition

[14] If weaker conditions are assumed, like measurability of the integrand, the derived equation (4.12) should hold except for a set which is at most a curve. This more general case leads to additional equations responsible for contact line effects.

of the surface quantities depends on the method of extrapolation of bulk fields into the layer. The same definition of surface quantities has been used by Dumais (1980), who performed the derivation for the case of a fixed material volume, not taking into account the diffusion terms. Partial results under similar definitions were obtained by Deemer and Slattery (1978) together with structural models for an interface employing local area averages dealing with excess quantities.

The present results can be compared with those of Gatignol & Seppecher (1986), where dimensional and quantitative analyses of an approximation were performed. The present derivations are rather close to the results of the 2D approximation theory of shells. A comparison will be made in another paper Kosinski & Wasowski (1991). In the next section we shall discuss some balance laws.

The variable thickness case ends with a relation similar to (4.14) in which, however, the surface supply term p^{s} is different and if we denote it by $p^{\mathrm{s}}_{\mathrm{v}}$, then due to (3.29) and (3.30) it is equal to

$$\tilde{p}^{\mathrm{s}}_{\mathrm{v}} = p^{\mathrm{s}} + [\![h(\psi(\mathbf{v} - \mathbf{c}) + \mathbf{w})]\!]\cdot\mathbf{n} - [\![\frac{\delta}{\delta t}\mathfrak{z}^{\pm} \psi j]\!] + \tag{4.18}$$
$$- [\![\psi\otimes(\mathbf{v} + \mathfrak{z}^{\pm}\mathrm{grad}_{\mathrm{s}}c_{\mathrm{n}})\mathbf{A}_{\mathrm{s}}(\mathfrak{z}^{\pm})\mathrm{grad}_{\mathrm{s}}\mathfrak{z}^{\pm}]\!] - [\![\mathbf{w}\mathbf{A}_{\mathrm{s}}(\mathfrak{z}^{\pm})\mathrm{grad}_{\mathrm{s}}\mathfrak{z}^{\pm}]\!].$$

A brief inspection of this term in comparison with the previous one shows the contribution of a new tangential part. This can be particularly important even in the case of the equilibrium equation for the interfacial stress tensor. The change of the thickness of a soap film on a bubble can be explained in those new terms.

4.3 *Particular balance laws*

Let us consider the particular quantities to be balanced by (4.14).

a) *Mass balance equation*: ψ is equal to ρ, and if mass is conserved in the bulk medium, the flux of mass $\mathbf{w}$ and supply p of ρ (compare denotation in (4.8)) are zero. The surface flux $\mathbf{W}^{\mathrm{s}}\{\rho\}$ is given by

$$\mathbf{W}^{\mathrm{s}'}\{\rho\} = \langle\rho\ (\mathbf{v} + L\ \mathrm{grad}_{\mathrm{s}}c_{\mathrm{n}})\ \mathbf{A}_{\mathrm{s}}(L)\rangle =: \langle\mathbf{m}(L)\rangle, \tag{4.19}$$

which can be split into two parts $\mathbf{W}^{\mathrm{s}'}\{\rho\} = \rho^{s}\mathbf{V}^{\mathrm{s}}_{\tau} + \mathbf{W}_{\rho}$, where

$$\mathbf{W}_{\rho} := \rho^{s}\mathbf{V}^{\mathrm{s}}_{\tau} + \langle\rho(\mathbf{A}_{\mathrm{s}}(L) - j(L)\mathbf{1}_{\mathrm{s}})\mathbf{v} + L\rho\ \mathrm{grad}_{\mathrm{s}}c_{\mathrm{n}}\mathbf{A}_{\mathrm{s}}(L)\rangle.$$

In the obvious way this equation leads to the definition of the extra mass flux $\mathbf{W}_{\rho}$. Hence the local balance equation for the mass is

$$\frac{\delta_{\mathrm{n}}}{\delta\ t}\rho^{\mathrm{s}} - 2Hc_{\mathrm{n}}\rho^{\mathrm{s}} + \mathrm{div}_{\mathrm{s}}(\rho^{\mathrm{s}}\mathbf{V}^{\mathrm{s}}_{\tau}) + \mathrm{div}_{s}\mathbf{W}_{\rho} = [\![j\ \rho(\mathbf{v} - \mathbf{c})]\!]\cdot\mathbf{n}$$
$$+ [\![\rho\otimes(\mathbf{v} + \mathfrak{z}^{\pm}\mathrm{grad}_{\mathrm{s}}c_{\mathrm{n}})\mathbf{A}_{\mathrm{s}}(\mathfrak{z}^{\pm})\mathrm{grad}_{\mathrm{s}}\mathfrak{z}^{\pm}]\!] - [\![\frac{\delta_{\mathrm{n}}}{\delta t}\mathfrak{z}^{\pm}\ \rho j]\!], \tag{4.20}$$

which from (4.19) yields the explicit form of the surface extra mass flux $\mathbf{W}_{\rho}$

$$\mathbf{W}_{\rho} := \langle L\rho\mathbf{v}\rangle\mathbf{b} - \mathrm{K}\langle L^{2}\rho\mathbf{P}\mathbf{v}\rangle + \{\langle L\rho\rangle\mathbf{1}_{s} + \langle L^{2}\rho\rangle\tilde{\mathbf{b}}\}\ \mathrm{grad}_{s}c_{\mathrm{n}}. \tag{4.21}$$

The last two terms on the RHS of (4.20) disappear in the *constant thickness case*. Simple inspection shows that the first two moments of mass (i.e. $\langle L\rho\rangle$ and $\langle L^2\rho\rangle$) and of momentum (i.e. $\langle L\rho\mathbf{v}\rangle$ and $\langle L^2\rho\mathbf{v}\rangle$) lead to nonvanishing, in general, extra flux of the mass.

Dealing with a p-material interface and the excess mass density in the constant thickness case, firstly Alts (1986) and then Alts & Hutter (1988) put the term corresponding to our $\mathbf{W}_\rho$ equal to zero. They chose, however, the surface coordinates as lines that are frozen to the motion of the surface 'particles'. Their particles, however, have been defined in terms of the excess mass density, which can be negative. It seems that their requirement means the motion, of the material points in the interface must not differ, in the mean sense, from the motion prescribed by the extension of the bulk fields into the interface. Moreover, their choice is local in L^β, and the disappearing of $\mathbf{W}_\rho$ cannot be interpreted as a constraint on the thickness of the layer which, in the constant thickness case, is a global quantity, independent of L^β.

Let us notice that the flux of the surface mass is different from the density of interfacial linear momentum. In his notes Professor Bedeaux has implicitly assumed that the mass current (read - flux) is the same as the interfacial linear momentum density, since he defines the surface velocity field $\vec{v}^s$ as the ratio of the mass current per unit surface area and the mass density per unit surface area.

b) *Linear momentum balance equation*: $\psi = \rho\mathbf{v}$, the Cauchy stress $\mathbf{T}$ with minus sign serves as the flux of linear momentum, and the body force $\rho\mathbf{b}$ is the supply term. For the surface flux $\mathbf{W}^s\{\rho\mathbf{v}\}$ we have

$$\mathbf{W}^{s'}\{\rho\mathbf{v}\} \equiv \langle\rho\mathbf{v}\otimes(\mathbf{v}+L\,\mathrm{grad}_s c_n)\,\mathbf{A}_s(L)\rangle - \langle\mathbf{T}\,\mathbf{A}_s(L)\rangle, \tag{4.22}$$

which can be split into two parts

$$\mathbf{W}^{s'}\{\rho\mathbf{v}\} = \mathbf{V}^s\otimes\mathbf{W}^{s'}\{\rho\} + \mathbf{T}_s\,.$$

The Galilean invariant *interfacial surface stress tensor* $\mathbf{T}_s$ can be written as the sum of two invariant parts

$$\mathbf{S}(\mathbf{r},t) := -\langle\mathbf{T}\mathbf{A}_s(L)\rangle + \langle\rho(\mathbf{v}-\mathbf{V}^s)\otimes(\mathbf{v}-\mathbf{V}^s)\mathbf{A}_s(L)\rangle$$

$$\mathbf{W}_{\rho\mathbf{v}}(\mathbf{r},t) := \langle\rho(\mathbf{v}-\mathbf{V}^s)\otimes((\mathbf{A}_s(L) - j(L)\mathbf{1}_s)\mathbf{V}^s + L\rho\,\mathrm{grad}_s c_n\mathbf{A}_s(L))\rangle$$

or as a sum of two other parts $\mathbf{S}^1$ and $\mathbf{S}^2$ contributing to $\mathbf{T}_s$, where we put

$$\mathbf{S}^1(\mathbf{r},t) := -\langle\mathbf{T}\mathbf{A}_s(L)\rangle, \quad \mathbf{S}^2(\mathbf{r},t) := \langle(\mathbf{v}-\mathbf{V}^s)\otimes\mathbf{m}(L)\rangle \tag{4.23}$$

$$\mathbf{T}_s = \mathbf{S}^1 + \mathbf{S}^2.$$

Using the last formula we can write the local balance equation for linear momentum as

$$\frac{\delta n}{\delta t}\left[\rho^s\mathbf{V}^s\right] - 2Hc_n\rho^s\mathbf{V}^s + \mathrm{div}_s(\mathbf{V}^s\otimes(\rho^s\mathbf{V}^s_\tau + \mathbf{W}_\rho) + \mathbf{T}_s) = \rho^s\mathbf{g}^s +$$

$$+ [\![(\rho\mathbf{v}\otimes(\mathbf{v}\cdot\mathbf{n} - c_n) - \mathbf{T}\mathbf{n})j]\!] - [\![\rho\mathbf{v}\otimes(\mathbf{v} + \mathfrak{z}\,\mathrm{grad}_s c_n)\mathbf{A}_s(\mathfrak{z})\,\mathrm{grad}_s\mathfrak{z}]\!] +$$

$$+ [\![\mathbf{TA}_s(\mathfrak{z})\mathrm{grad}_s\mathfrak{z}]\!] - [\![\frac{\delta n}{\delta t}\mathfrak{z}\, \rho \mathbf{v} j]\!]. \tag{4.24}$$

Let us consider a simplified version of the above equation under *quasi-static conditions*:

$$\mathrm{div}_s \mathbf{T}_s = [\![\mathbf{T}\ \mathbf{n} j]\!] + [\![\mathbf{TA}_s(\mathfrak{z})\ \mathrm{grad}_s \mathfrak{z}]\!] + \rho^s \mathbf{g}^s. \tag{4.25}$$

Then in the case of a fluid-fluid interface a first correction to the *Laplace equation* in the case of variable thickness can be given if p^+ and p^- denote the limit values of pressure of either bulk fluid phases, and $\gamma \mathbf{1}_s$ represents the isotropic surface stress tensor with γ as a surface tension coefficient. Let us notice in view of the representation (4.23), the surface stress $\mathbf{T}_s = - \langle \mathbf{TA}_s(L)\rangle$ will be isotropic if (cf. (3.8)) in the interface region the Cauchy stress $\mathbf{T}$ is proportional to $\mathbf{1}_s - L\mathbf{b}$. If we use ∇_s to denote grad_s, then from (4.25) we get

$$\nabla_s \gamma - 2H\gamma\mathbf{n} = (j(\mathfrak{z}^+)p^+ - j(\mathfrak{z}^-)p^-)\mathbf{n} +$$

$$+ \mathbf{A}_s(\mathfrak{z}^+)p^+\nabla\mathfrak{z}^+ - \mathbf{A}_s(\mathfrak{z}^-)p^-\nabla\mathfrak{z}^- + \rho^s\mathbf{g}^s. \tag{4.26}$$

In the non-equilibrium case, when quasi-static assumptions are not met, the above equation will contain the inertial term and, moreover, the representation formula (4.23) for the stress $\mathbf{T}_s$ will possess the extra part $\mathbf{S}^2$ due to diffusion forces.

Coming back to the general equation, the last three terms on the RHS of (4.24) disappear in the *constant thickness case*. The above expressions for the interfacial stress tensor $\mathbf{T}_s$ show that, when the diffusion terms are put equal to zero, the symmetry of the tangential components of $\mathbf{S}$ cannot hold automatically. An equilibrium case could allow us to put $\mathbf{S}^2$ and $\mathbf{W}_{\rho v}$ equal to zero. Moreover, the normal component $\mathbf{n}\mathbf{T}_s$ of the surface stress tensor $\mathbf{T}_s$ contains a contribution from the diffusion terms unless the tangential component $\mathbf{v}_\tau$ of the velocity field $\mathbf{v}$ is constant along each segment of the layer[15] i.e. is independent of L.

Note that in the case of a spherical interface with nonvanishing thickness under equilibrium conditions, when $\mathbf{v} - \mathbf{V}^s = \mathbf{0}$, and with the Cauchy stress written in terms of the negative hydrostatic pressure p, the surface stress will be

$$\mathbf{T}_s = \langle\, p(1 + L/r)\rangle \mathbf{1}_s,$$

---

[15] In that case $\langle f\mathbf{v}\rangle_\tau = \langle f\rangle\, \mathbf{v}_\tau$ for an arbitrary field f, and $\mathbf{v}_\tau = \mathbf{v}^{\pm}_\tau = \mathbf{V}^s_\tau$. Such condition has been admitted by Dell'Isola & Romano (1987) and interpreted as the perfect viscosity consequence of the viscosity of the 3D matter confined in the layer.

where r is the radius of curvature. We can interpret the term $\langle p \rangle$ as the classical surface tension. Here, however, the additional part $\langle pL/r \rangle$ appears, which is normally very small, unless the thickness of the interface is comparable with the radius r. This observation can be useful in modelling very fine soap bubbles.

Before the next balance law of angular momentum will be considered, we would like to point out to the reader that the 3D equilibrium equation

$$\operatorname{div} \mathbf{T} = \mathbf{0}$$

with the pure hydrostatic stress tensor $\mathbf{T} = -p\mathbf{1}$ cannot[16] be satisfied in the interface region with different limit values p^- and p^+ at the bottom and upper boundaries. To omit this contradiction one has either to take into account the nonvanishing body force $\rho\mathbf{g}$ on the RHS of the above equation or to admit a anisotropic stress in the interface layer.

c) *Angular momentum balance equation*: $\psi = \mathbf{x} \times \rho\mathbf{v}$. We restrict ourselves only to nonpolar continua. The master angular momentum balance law is well-known in the 3D theory; its interfacial counterpart requires, if the previous balance law is satisfied,

$$\langle\, j(\mathbf{1} \times \mathbf{T})\,\rangle = \mathbf{0},$$

which is automatically satisfied if $\mathbf{T}$ is symmetric. The derivation of this non-obvious result is given by Dell'Isola & Kosiński (1989).

Before closing discussion of the equations of motion we shall write explicit relations for the normal and antisymmetric parts of the surface stress tensor $\mathbf{T}_s$ in the natural basis $\{\, \mathbf{a}_\alpha, \mathbf{n}\,\}$. They are

$$\begin{aligned} T_s^{n\alpha} &= -\langle jT^{n\alpha}\rangle + \langle j\rho(\mathbf{v}\cdot\mathbf{n} - c_n)(v^\alpha - V^{s\alpha})\rangle \\ &\quad + \langle(\mathbf{v} - \mathbf{V}^s)\cdot\mathbf{n}\, w_\rho^\alpha\rangle - \langle jL\, b_\delta^\alpha\, T^{n\delta}\rangle, \end{aligned} \tag{4.27}$$

$$T_s^{12} - T_s^{21} = -\langle jL\,(b_2^1(T^{22} - T^{11}) + (b_1^1 - b_2^2)\, T^{12}\rangle,$$

where $T_s^{n\alpha} := \mathbf{n}\cdot\mathbf{T}_s\mathbf{a}^\alpha$, $T_s^{12} := \mathbf{a}^1\cdot\mathbf{T}\,\mathbf{a}^2$, etc.

d) *Energy balance equation*: $\psi = \rho(e + 0.5\mathbf{v}\cdot\mathbf{v}) =: \rho E$, where e represents the specific internal energy. The sum $\mathbf{q} - \mathbf{v}\mathbf{T}$ serves as the flux of the total energy, where $\mathbf{q}$ is the heat flux vector and the sum $\rho(\mathbf{g}\cdot\mathbf{v} + r)$ is the supply term, where r represents the body heat supply density. For the surface flux $\mathbf{W}^s\{\rho E\}$ we have

$$\mathbf{W}^{s\prime}\{\rho E\} \equiv \langle E\, \mathbf{m}(L)\rangle - \langle(\mathbf{v}\mathbf{T} - \mathbf{q})\mathbf{A}_s(L)\rangle.$$

If we define

$$\rho^s e^s := \langle j\rho\tilde{e}\rangle, \quad \rho^s\tilde{r}^s := \langle j\rho(r + \mathbf{g}\cdot(\mathbf{v} - \mathbf{V}^s))\rangle$$

$$\mathbf{q}^s := \langle(\rho\tilde{e}(\mathbf{v} - \mathbf{V}^s) + \mathbf{q} - (\mathbf{v} - \mathbf{V}^s)\mathbf{T})\mathbf{A}_s(L)\rangle,$$

$$\mathbf{W}_E := \langle\rho\tilde{e}(\mathbf{A}_s(L) - j(L)\mathbf{1}_s)\mathbf{V}^s + L\rho\tilde{e}\, \operatorname{grad}_s c_n\, \mathbf{A}_s(L)\rangle,$$

where $\tilde{e} := e + 0.5(\mathbf{v} - \mathbf{V}^s)^2$, then the local energy balance equation will

[16] The equilibrium equation requires the pressure p be constant along the thickness.

$$\frac{\delta n}{\delta t}\left[\rho^s(e^s + 0.5\,\mathbf{v}^s\cdot\mathbf{v}^s)\right] - 2Hc_n\rho^s\,(e^s + 0.5\,\mathbf{v}^s\cdot\mathbf{v}^s) +$$
$$+ \operatorname{div}_s(\rho^s(e^s + 0.5\mathbf{v}^s\cdot\mathbf{v}^s)\mathbf{v}^s_\tau + \mathbf{q}^s + \mathbf{W}_E + \mathbf{v}^s\mathbf{T}_s + 0.5\mathbf{v}^s\cdot\mathbf{v}^s\,\mathbf{W}_\rho)$$
$$= \rho^s\mathbf{g}^s\cdot\mathbf{v}^s + \rho^s\tilde{r}^s + [\![(\rho(e + 0.5\mathbf{v}\cdot\mathbf{v})(\mathbf{v} - \mathbf{c}) + \mathbf{q} - \mathbf{v}\mathbf{T})j]\!]\cdot\mathbf{n} +$$
$$- [\![\rho(e + 0.5\mathbf{v}\cdot\mathbf{v})\otimes(\mathbf{v} + \mathfrak{z}\,\mathrm{grad}_s c_n)\mathbf{A}_s(\mathfrak{z})\mathrm{grad}_s\mathfrak{z}]\!] +$$
$$+ [\![(\mathbf{v}\mathbf{T} - \mathbf{q})\mathbf{A}_s(\mathfrak{z})\mathrm{grad}_s\mathfrak{z}]\!] - [\![\frac{\delta n}{\delta t}\mathfrak{z}\,\rho(e + 0.5\mathbf{v}\cdot\mathbf{v})j\,]\!]. \tag{4.28}$$

The last three terms on the RHS of (4.28) disappear in the *constant thickness case*. The above expressions for the interfacial heat flux $\mathbf{Q}^s$ and the supply terms $\rho^s\tilde{r}^s$ lead to the following relations:

$$\mathbf{Q}^s = \mathbf{q}^s + \mathbf{W}_E, \qquad \rho^s r^s \neq \rho^s\tilde{r}^s,$$

which mean that even the case of a nonconductor of heat at the 3D level leads to nonvanishing interfacial heat flux, and vanishing heat supply term ρr at the 3D level leads to the interfacial heat supply $\rho^s r^s$ being equal to $\langle j\rho\, \mathbf{g}\cdot(\mathbf{v} - \mathbf{v}^s)\rangle$, which does not need to vanish if $\mathbf{g}$ is different from zero.

4.4 *Thermodynamic inequality*

The second law of thermodynamics for the 3D material continuum is assumed in the form of the entropy production inequality

$$\frac{d}{dt}\int_{\mathcal{P}_t}\rho\eta\, dv \geq -\int_{\partial\mathcal{P}_t}(\mathbf{q}/\vartheta + \mathbf{k})\cdot\mathbf{N}da + \int_{\mathcal{P}_t}\rho r/\vartheta dv, \tag{4.30}$$

where η and ϑ represent the specific entropy and the absolute temperature, respectively, while $\mathbf{k}$ is the so-called extra entropy flux. Performing the usual localization for the interfacial layer we get the inequality

$$\frac{\delta}{\delta t}\rho^s\eta^s - 2Hc_n\rho^s\eta^s + \operatorname{div}_s(\rho^s\eta^s\,\mathbf{v}^s_\tau + \tilde{\mathbf{k}}^s + \mathbf{W}_\eta) \geq$$
$$\geq \rho^s r^s_\eta + [\![(\rho\eta\,(\mathbf{v} - \mathbf{c}) + \mathbf{q}/\vartheta + \mathbf{k})j]\!]\cdot\mathbf{n} +$$
$$-[\![\,\rho\eta(\mathbf{v} + \mathfrak{z}\mathrm{grad}_s c_n)\mathbf{A}_s(\mathfrak{z})\mathrm{grad}_s\mathfrak{z}\,]\!] + [\![\,(\mathbf{q}/\vartheta + \mathbf{k})\mathbf{A}_s(\mathfrak{z})\mathrm{grad}_s\mathfrak{z}\,]\!] +$$
$$- [\![\,\frac{\delta}{\delta t}\mathfrak{z}\,\rho\eta j\,]\!], \tag{4.31}$$

where

$$\mathbf{W}^{s'}\{\rho\eta\} \equiv \langle\rho\eta\mathbf{m}(\ell)\rangle + \langle(\mathbf{q}/\vartheta + \mathbf{k})\mathbf{A}_s(\ell)\rangle,$$

and we have put

$$\rho^s\eta^s := \langle j\rho\eta\rangle,$$
$$\rho^s r^s_\eta := \langle j\rho(r/\vartheta)\rangle \tag{4.32$_1$}$$

$$\tilde{\mathbf{k}}^s := \langle(\rho(\mathbf{q}/\vartheta + \mathbf{k})\mathbf{A}_s(\ell) + \rho\eta(\mathbf{v} - \mathbf{V}^s))\mathbf{A}_s(\ell)\rangle, \qquad (4.32)_2$$
$$\mathbf{W}_\eta := \langle\rho\eta(\mathbf{A}_s(\ell) - j(\ell)\mathbf{1}_s)\mathbf{V}^s + \ell\rho\eta \operatorname{grad}_s c_n \mathbf{A}_s(\ell)\rangle.$$

The last three terms on the RHS of (4.31) disappear in the *constant thickness case*. As far as the expressions for the interfacial entropy flux $\mathbf{k}^s$ and $\mathbf{W}_\eta$ are concerned, no simple relation to the interfacial heat flux $\mathbf{Q}^s$ (even in the case of vanishing extra term $\mathbf{k}$) can be observed. However, under a particular set of assumptions concerning the kinematics and the constititive properties of the matter in the layer some simplification can be made in order to derive such a relation. This will be the subject of a further investigation.

5. Final remarks

Using the model developed in the last section one could not completely take into account the influence of the thickness of the layer on the thermomechanical behaviour of phase interfaces. This model we shall call a *0th order model*.

In order to resolve the aforementioned drawback an H-th order model was proposed for the case of constant thickness, by Dell'Isola & Kosiński (1990) the idea of which comes from Dumais (1980).

The main object in that model is the taking of higher order moments in the interface. One introduces the k-th moments ($k \le H$) of a typical quantity f by

$$^k f \equiv \ell^k(x,t^*)\, f, \quad \text{when } x \in \mathcal{Z}_t. \qquad (5.1)$$

Here ℓ^k means the k-th power of ℓ.

Then the following local balance equation for ψ can be easily derived by passing to the local form of (4.9), substituting fields by their k-th moments and by using the properties of the function $\ell(x,t^*)$ from (3.18):

$$\frac{\partial}{\partial t}{}^k\psi + \operatorname{div}({}^k\psi\otimes\mathbf{v} + {}^k\mathbf{w}) = {}^k p + k({}^{k-1}\psi\otimes(\mathbf{v}\cdot\mathbf{n} - c_n) + {}^{k-1}\mathbf{w}\cdot\mathbf{n}) \qquad (5.2)$$

Regarding (5.2) as the local form of a particular case of (4.13) and recalling the forms (4.12, 14), we get the following interfacial balance equation:

$$\frac{\delta}{\delta t}{}^k\psi^s - 2Hc_n\, {}^k\psi^s + \operatorname{div}_s(\langle{}^k\psi\otimes(\mathbf{v} + \ell \operatorname{grad}_s c_n)\mathbf{A}_s(\ell)\rangle + \langle{}^k\mathbf{w}\,\mathbf{A}_s(\ell)\rangle)$$
$$= [\![j({}^k\psi(\mathbf{v} - \mathbf{c}) - {}^k\mathbf{w}]\!]\cdot\mathbf{n} + {}^k p^s + \langle jk({}^{k-1}\psi\otimes(\mathbf{v}\cdot\mathbf{n} - c_n) + {}^{k-1}\mathbf{w}\cdot\mathbf{n})\rangle.$$

This equation is valid for any $k \ge 1$. If $k = 0$ the primitive interfacial balance law (4.15) is recovered.

A question can, however, arise concerning the mathematical

completeness of this kind of approach. To answer this question one should first notice that the k-th moment of a typical function f in (5.1) (regarded as a function of ℓ only) defines the projection of f on the polynomial ℓ^k belonging to the basis formed by all polynomials of the function space $L^2((z^-,z^+], d\mu = jd\ell)$. The measure μ is positive and absolutely continuous with respect to Lebesgue measure as long as j is positive and H and K are finite. This corresponds to the hypothesis on the thickness of the layer Z_t. Therefore the H-order theory deals with truncated expansions along the thickness of the layer of physical quantities to be balanced. A particular application of this model was presented in Dell'Isola & Kosiński (1991) for the case of a Kirchhoff-Love type interface.

In lecture notes no discussion of contact lines is given. In this respect we refer to Professor Slattery's notes as well as to his book Slattery (1990).

The phenomenological modelling of interfacial phenomena presented here is one of the approaches known in the literature. As a result of this we have obtained interfacial balance laws. However, interfacial constitutive relations also need to be discussed. This is one of subjects of Professor Hutter's lectures and Gogosov *et al.*(1983).

Finally, we would like to refer to some Polish contributions in this field, namely to Blinowski (1973), who discussed a particular higher (second) order model of liquid interface, and to Wilmański (1975), who discussed thermodynamics of interfaces, using singular surface approach. And last but not least let us recall the monograph of Ościk (1982) which was completely dedicated to adsorption problems.

Acknowledgements

This research was mainly sponsored by the Polish Academy of Sciences Research Project CPBP 02.01. A part of the work on these lecture notes was done while the author was a CNR-Visiting Professor at Dipartimento di Matematica e sue Applicazioni dell' Universita' di Napoli and Istituto di Matematica, Universita' di Potenza and Professeur Invité at Université Pierre et Marie Curie (Paris VI), Laboratoire de Modélisation en Mécanique, as well as during the author's tenure of an Alexander von Humboldt-Stiftung Fellowship at the Universities of Bonn, Heidelberg and TH Darmstadt. The final version of the notes was prepared at the Department of Mathematical Sciences, Loyola University, New Orleans. The possibility of collaboration and stimulating discussions in all these institutions are deeply appreciated.

6. References

A.W.Adamson (1982). Physical Chemistry of Surfaces, Interscience, New York-London.

A.M.Albano, D.Bedeaux and J.Vlieger (1979). On the description of interfacial properties using singular densities and currents at a dividing surface, Physica, **99 A**, 293-304.

T.Alts (1986). Some results of a boundary-layer theory for curved phase interfaces, in Lectures Notes in Physics, vol.249: Trends in

Applications of Pure Mathematics to Mechanics, Proc. of 6-th Symp. Bad Honnef, October 21-25, 1985, E.Kröner and K.Kirchgässner (eds.), Springer, Berlin-Heidelberg-New York- Tokyo, pp. 500-512.

T.Alts and K.Hutter (1986), Towards a theory of temperate glacier.Part I: Dynamics and thermodynamics of phase boundaries between ice and water, Mitteilungen No.82 der Versuchtsanstalt für Wasserbau, Hydrologie und Glaziologie, ETH, Zürich.

T.Alts and K.Hutter (1988). Continuum description of the dynamics and thermodynamics of phase boundaries between ice and water. Parts I and II, J.Non-Equilib. Thermodyn., **13**, 221-280.

T.Alts and K.Hutter (1989). Continuum description of the dynamics and thermodynamics of phase boundaries between ice and water. Parts III and IV, J.Non-Equilib. Thermodyn., **13**, 301-329, **14**, 1-22.

R.Aris (1962). Vectors, Tensors and the Basic Equations of Fluid Mechanics, Prentice-Hall, Englewood Cliffs, New York.

F.Baillot et R.Prud'homme (1988). Lois de comportement d'interface, Annales de Physique, Colloque # 2, suppl.au # 3,**13**, 23-34.

M.Barrère and R.Prud'homme (1973). Equations Fondamentales de l'Aérothermochimie, Masson et Cie, Paris.

D.Bedeaux (1991). Non-equilibrium thermodynamics and statistical physics of the liquid-vapour interface, CISM, Udine, Lecture Notes, Springer, in this volume.

A.Blinowski (1973). On the surface behaviour of gradient-sensitive liquids, Arch.Mechanics, **25**(2), 259-2268.

R.M.Bowen and C.-C.Wang (1971). On displacement derivatives, Quart.Appl. Math., **29**(1), 29-39.

F.P.Buff (1956). Curved fluid interfaces.I: Generalised Gibbs-Kelvin equation, J.Chem.Physics, **25**, 146-153.

F.P.Buff (1960). The theory of capillarity, *in* Handbuch der Physik, vol.X, Springer, Berlin-Heidelberg - New York, pp.281-304.

R.Defay, I.Prigogine and A.Bellemans (1966). Surface Tension and Adsorption (trans.D.H.Everett). Longmans, London.

F.Dell'Isola and W.Kosiński (1989). The interface between phases as a layer. Part I: Mechanical balance equations, Universita' di Napoli Reports, July, 1989, submitted for publication.

F.Dell'Isola and W.Kosiński (1991). The interface between phases as a layer. Part Ia, *in* Proc. Vth Meeting: Waves and Stability in Continuous Media, Sorrento, October 1989, in print.

F.Dell'Isola and A. Romano (1986). On a general balance law for continua with an interface, Ricerche di Matematica, **35**(2), 325-337.

F.Dell'Isola and A. Romano (1987). On the derivation of thermomechanical balance equations for continuous systems with a nonmaterial interface, Int.J. Engng Sci. 25(11/12), 1459-1468.

A.R.D.Deemer and J.C.Slattery (1978). Balance equations and structural models for phase interfaces, Int.J.Multiphase Flow, **4**, 171-192.

J.-F.Dumais (1980). Two and three-dimensional interface dynamics, Physica, **104 A**, 143-180.

J.E.Dunn and J.Serrin (1985). On the thermomechanics of interstitial working, Arch. Rational Mech. Analysis, **88**, 95,

R.Gatignol (1987). Liquid-vapour interfacial conditions, Rev.Romaine des Sci.Techn. Méc.Appl.,**3**.

R.Gatignol and P. Seppecher (1986). Modelization of fluid-fluid interfaces with material properties, J. Méc.Theor.et Appl., Numero special, 225-247.
R.Ghez (1966). A generalized Gibbsian surface, Surface Science, **4**, 125-140.
J.W.Gibbs (1961). The Scientific Papers of J.Willard Gibbs, vol.I. Dover Publications, New York.
V.V.Gogosov, V.A.Naletova, Chung Za Bing and G.A.Shaposhnikova (1983). Conservation laws for the mass, momentum, and energy on a phase interface for true and excess surface parameters, Izv.Akad.Nauk SSSR-Mekh. Zhidk. Gaz, **18**(6), 107, (Engl.transl. in Fluid Dynamics **6**, 923-930).
M.E.Gurtin and A.I.Murdoch (1975). A continuum theory of elastic material surfaces, Arch. Rational Mech. Analysis, **57**(4), 291-323.
D.Hayes (1957). The vorticity jump across a gasdynamic discontinuity, J.Fluid Mech., **2**(6),595-600.
K.Hutter, The physics of ice-water phase change surfaces, CISM, Udine, Lecture Notes, Springer, in this volume.
M.Ishii (1975). Thermo-fluid Dynamic Theory of Two Phase Flow, Eyroles, Paris.
W.Kosiński (1981). Wstęp do teorii osobliwości pola i analizy fal, PWN Warszawa-Poznań.
W.Kosiński (1983). Rownania ewolucji cial dyssypatywnych (in Polish), Habilitation Thesis, IFTR-Reports, 9/83.
W.Kosiński (1985). Thermodynamics of singular surfaces and phase transitions, *in* Free Moving Boundary Problems: Applications and Theory. vol.III, A. Bossavit, A. Damlamian and A.M. Fremond (eds.), Pitman, Boston-London-Melbourne, pp.140-151.
W.Kosiński (1986). Field Singularities and Wave Analysis in Continuum Mechanics, Ellis Horwood and PWN-Polish Scientific Publishers, Warsaw, Chichester.
W.Kosiński (1989). Characteristics in phase transition problems. Rend.Sem.Math.Univ.Pol.Torino, Fasciolo Speciale, Hyperbolic Problems, 137-144.
W.Kosiński and A.Romano (1989). Evolution balance laws in two-Phase problems, XVIIth IUTAM Congress in Grenoble, August 1988, also Arch. Mechanics, 41(2-3), 255-266.
W.Kosiński and A.O.Wąsowski (1991). On shell theory equations, in preparation.
I.N.Levine (1978). Physical Chemistry, McGraw-Hill, New York.
K.A.Lindsay and B.Straughan (1979). A thermodynamic viscous interface theory and associated stability problems, Arch.Rational Mech.Analysis, **71**(4), 301-326.
G.P.Moeckel (1974). Thermodynamics of an interface, Arch. Rational Mech. Analysis, **57**(3), 255-280.
A.I.Murdoch (1990). A remark on the modelling of thin three-dimensional regions. *in* Elasticity: Mathematical Methods and Applications. (The Ian Sneddon 70th Birthday Volume), G.E.Eason & R.W.Ogden (eds.), Chichester : Ellis Horwood, Chichester.
A.I.Murdoch (1991). On the physical interpretation of fluid interfacial concepts and quantities, CISM, Udine, Lecture Notes, Springer, in this volume.

A.I.Murdoch and H.Cohen (1989). On the continuum modelling of fluid interfacial regions I: Kinematical considerations. submitted for publication,
J.N.Murrel and E.A.Boucher (1982). Properties of Liquids and Solutions, John Wiley & Sons, Chichester.
P.M.Naghdi (1972). The theory of shells and plates, *in* Handbuch der Physik,vol.VIa/2, 425-640, Springer Verlag, Berlin-Heidelberg-New York.
L.G.Napolitano (1978). Thermodynamics and dynamics of pure interfaces, Acta Astronautica, **5**, 655-670.
L.G.Napolitano (1982). Properties of parallel-surface coordinate systems, Meccanica, **17**, 107-118.
S.Ono and S.Kondo (1960). Molecular theory of surface tension in liquids, in Handbuch der Physik, vol.X, 135-280, Springer Verlag, Berlin-Goettingen-Heidelberg.
J.Ościk (1982). Adsorption, PWN, Warszawa and Ellis Horwood, Chichester.
H.Petryk and Z. Mróz (1986). Time derivatives of integrals and functionals defined on varying volume and surface domains, Arch. Mech., **38** (5-6), 697-724.
A.Romano (1983). Thermodynamics of a continuum with an interface and Gibbs' rule, Ricerche di Matematica, **31**(2), 277-294.
L.E.Scriven (1960). Dynamics of a fluid interface, Chem.Engng.Sci.,**12**, 98-108.
P.Seppecher (1987). Étude d'une modélisation des zones capillares fluides interfaces et lignes de contact, Thèse, Univerisité Paris VI et E.N.S.T.A..
M.Slemrod (1983). Admissibility criteria for propagating phase boundaries in a Van der Waals fluid, Arch. Rational Mech. Analysis, **81**, 301.
M.Slemrod (1989). A limiting "viscosity" approach to the Riemann problem for materials exhibiting change of phase, Arch. Rational Mech. Analysis, **105** (4), 327-365.
J.Serrin (1983). The form of interfacial surfaces in Korteweg's theory of phase equilibria, Quart.Appl.Math., **41**,, 357-364.
J.C.Slattery (1967). General balance equation for a phase interface, Ind.Engng.Chem.Fundamentals, **6**(1), 108-115.
J.C.Slattery (1991). The prediction of static and dynamic contact angles, wetting and spreading, CISM, Udine, Lecture Notes, Springer, in this volume.
T.Y.Thomas (1957). Extended compatibility conditions for the study of surfaces of discontinuity in continuum mechanics, J.Maths.Mech.,**6**, 311-322, 907-908.
T.Y.Thomas (1961). Plastic Flow and Fructure in Solids, Academic Press, New York-London.
T.Y.Thomas (1965). Concepts from Tensor Analysis and Differential Geometry, 2nd Edition, Academic Press, New York-London.
C.A.Truesdell and R.A.Toupin (1960), The classical field theories, *in* Handbuch der Physik, vol.III/1, 226-793, Springer Verlag, Berlin-Göttingen-Heidelberg.
K.Wilmański (1975). Thermodynamic properties of singular surfaces in continuous media, Arch.Mechanics, **27**(3), 517-529.

NONEQUILIBRIUM THERMODYNAMICS AND STATISTICAL PHYSICS OF THE LIQUID-VAPOR INTERFACE

D. Bedeaux
University of Leiden, Leiden, The Netherlands

Abstract

In these lectures it is explained how the theory of nonequilibrium thermodynamics may be used to model dynamic phenomena at a fluid–fluid interface. Use is made of the concept of excess densities and currents introduced originally by Gibbs (1906) to describe the equilibrium properties of such a system. An explicit expression for the entropy production is given. Onsager relations are discussed. Furthermore it is shown how capillary waves give contributions to the equilibrium correlation functions of the system in the neighbourhood of the interface which have a new and unexpected behaviour. Finally fluctuating "forces" are introduced in the constitutive relations which make a calculation of the time dependence of the correlation functions possible. Fluctuation–dissipation theorems for these fluctuating forces are given.

1. INTRODUCTION

In these lectures I will first discuss the application of the theory of nonequilibrium thermodynamics, containing both the Onsager relations and an explicit expression for the entropy production, to the liquid–liquid (vapor) interface in a multi–component system. In the context of this description boundary conditions follow in a natural way from the interfacial entropy production. For a discussion of the use of such a description to model phenomena at liquid boundaries I refer to the lectures given by Golia (Napolitano). Subsequently I will discuss the influence of thermal fluctuations on the equilibrium correlation functions of the system. In particular I will discuss how capillary waves lead to contributions to the equilibrium auto–correlation functions which fundamentally differ from what one would expect on the basis of the behaviour of these correlation functions in the bulk regions away from the surface. Finally I will discuss the theory describing the time dependence of the correlation functions. In this description fluctuating "forces" are introduced in the constitutive relations. Such fluctuating forces appear not only in for instance the Navier–Stokes equation but also as fluctuating sources in the boundary conditions.

For a discussion of the general field of nonequilibrium thermodynamics and in particular its application to bulk phases we refer to the book on this subject by De Groot and Mazur (1962). The equilibrium thermodynamics of a socalled surface of discontinuity were already extensively discussed by Gibbs (1906). Following Gibbs the surface of discontinuity indicates a region of a small but finite thickness separating the two phases. A general method for the application of nonequilibrium thermodynamics to surfaces of discontinuity consistent with the equilibrium theory for surface thermodynamics formulated by Gibbs was given by Bedeaux, Albano and Mazur (1976). A major difficulty in such an analysis is that the surface of discontinuity may not only move through space but also has a time–dependent curvature. This makes it necessary to use time–dependent orthogonal curvilinear coordinates for the more difficult parts of the analysis. In section 2 we will discuss the mathematical description and the use of time–dependent curvilinear coordinates in this context in more detail.

The description in the context of nonequilibrium thermodynamics gives an explicit expression for the entropy production at the surface. One may identify the appropriate forces and fluxes from this expression. Onsager relations for the linear constitutive coefficients relating these forces and fluxes also follow. The expressions for the fluxes through the surface of discontinuity from one bulk phase to the other lead to what are usually called boundary conditions. Examples of the linear constitutive coefficients contained in these boundary conditions are for instance the slip coefficient and the temperature jump coefficient. Some fluxes give the flow from the bulk phases into the surface of discontinuity. An example is the flow of surface active material from the bulk to the interface. Finally some fluxes describe the flow along the interface. They are most similar to the usual fluxes in the bulk except for being two dimensional in the comoving reference frame. As we shall discuss, the entropy production at the interface contains the value of the fluxes into and through the interface in the comoving frame of reference. The fact that these fluxes should be calculated in a frame of reference in which the surface is at rest gave rise to considerable discussion among workers in the field of membrane transport, Mikulecky (1966).

An important property of the surface is that as a consequence of its existence the translational symmetry of the system is broken. As a consequence the tensorial nature of the fluxes and forces at the interface appearing in the interfacial entropy production differs from the corresponding behaviour in the bulk phases. At the interface there is still translational symmetry along the surface and rotational symmetry around the normal on the surface. As a consequence the force–flux pairs contributing to the interfacial entropy production are 2×2 tensors, 2–dimensional vectors and scalars. In the bulk regions one has 3×3 tensors, 3–dimensional vectors and scalars. At the interface the normal component of a 3–dimensional current, as for instance the asymptotic values of the bulk heat currents at the dividing surface, appear as scalar fluxes in the interfacial entropy production while the parallel parts appear as 2–dimensional vectors. Similarly a 3–dimensional tensor should be split up into a 2×2 tensor, two 2–dimensional vectors and a scalar. As a result the number of force–flux pairs in the interfacial entropy production is considerably larger than the corresponding number in the entropy production in the bulk. In view of Curie's symmetry principle, fluxes only depend on forces of the same tensorial nature. Even though this reduces the number of constitutive coefficients considerably the number of linear constitutive coefficients describing the behaviour of the system at the interface is still considerably larger than the number needed in the bulk regions. At the interface many processes may couple which do not couple with one another in the bulk.

The original paper by Bedeaux, Albano and Mazur (1976) considered the special case of an interface between two immiscible one–component fluids. A crucial element in their analysis was the description of the excess contributions to the densities and currents using δ–functions in the corresponding density and flow fields at the socalled dividing surface. Why the use of such δ–function singularities is mathematically sound and the proper definition of the excess densities and currents as a function of the position along the interface is not a trivial matter. On the most fundamental level one must start using the microscopic description and I refer to the lectures by Murdoch who addresses this problem. One may also use a continuous thermo–hydrodynamic description of the surface of discontinuity and then proceed to derive explicit expressions for the excess densities and currents as well as derive balance equations for the excess densities. Such an analysis is given by Albano, Bedeaux and Vlieger (1979). It should be pointed out that such a continuous thermo–hydrodynamic description is more detailed than the description in terms of the excess density and currents alone. The continuous description in particular contains more detailed information about the structure of the dividing surface. In the lectures by Hutter such details are discussed for the ice–water interface. Kosinski also addresses this problem in his lectures and in particular discusses the use of a finite slab model as an alternative for the use of excess quantities as suggested by Gibbs (1906). The finite slab model is in a way a model somewhere between a continuous description and the description using excess quantities. It my lectures I will show that the description using only the properly defined excess quantities gives a consistent and for many purposes adequate description of the nonequilibrium properties of the surface of discontinuity. While the continuous description and the description using the finite slab model may add some details they are completely compatible with the description using only excess quantities. In fact these compatibility requirements make it possible to analyse the behaviour of the continuous description in the surface of discontinuity; see Hutter's lectures for a discussion of this point.

A matter of considerable importance is the fact that there is no unique choice of the dividing surface. The problem which arises in this context is that the excess quantities depend on this choice. As the physics of the interface should be independent of the choice of a mathematical entity like the dividing surface it is clear that any prediction of a physical phenomenon must likewise be independent of this choice. Depending on the problem being considered it is usually convenient to wait with the choice until a point in the analysis where a specific choice simplifies the calculation.

Subsequent work extended the original analysis to multi-component systems with mass transport through the interface, Kovac (1977; 1981) and to systems in which electromagnetic effects play a role, Wolf (1979); Albano (1980; 1987). A review of much of this work and on some of the work on the description of the fluctuations of the interface has been given by Bedeaux (1986).

After the discussion of the mathematical description of the interface in section 2 we will proceed to discuss conservation laws in sections 3 – 5, entropy balance 6 and the resulting linear laws in section 7.

In section 8–11 we discuss the behaviour of the equilibrium fluctuations of the system. In particular we discuss the behaviour of the correlation functions close to the interface which is modified in a profound way from the known behaviour in the bulk , Bedeaux (1985; 1986). This is caused by the socalled capillary wave fluctuations of the position of the interface. In the last section we show how one may describe the time dependence of these correlation functions by the introduction of random fluxes in the constitutive relations, Zielinska (1982). Such random sources of fluctuations in fact also appear in the boundary conditions. Fluctuation–dissipation theorems for these random fluxes are given.

2. THE MATHEMATICAL DESCRIPTION OF INTERFACES

In order to give the time–dependent location of the dividing surface, it is convenient to introduce a set of time dependent orthogonal curvilinear coordinates: $\xi_i(\vec{r},t)$, $i = 1, 2, 3$, where $\vec{r} = (x, y, z)$ are the Cartesian coordinates and t the time, Morse (1953). These curvilinear coordinates are chosen in such a way that the time–dependent location of the dividing surface is given by

$$\xi_1(\vec{r},t) = 0 \tag{2.1}$$

The dynamical properties of the system are described using balance equations; for instance, balance equations for the mass densities of the various components. Consider as an example the balance equation for the density per unit of volume of an as yet unspecified variable $d(\vec{r},t)$

$$\frac{\partial}{\partial t} d(\vec{r},t) + \operatorname{div} \vec{J}_d(\vec{r},t) = \sigma_d(\vec{r},t) \tag{2.2}$$

Here $\vec{J}_d(\vec{r},t)$ is the current of the variable d and $\sigma_d(\vec{r},t)$ the production of d in the system, both at the position $\vec{r}$ and the time t. If $\sigma_d = 0$ the variable d, or more precisely the integral of the variable over the volume, is a conserved quantity. Examples of such variables are for instance the mass and the energy. In our description of interfaces d, $\vec{J}_d$ and σ_d vary continuously in the regions occupied by the bulk phases. At the dividing surface, however, these fields contain singular contributions equal to the properly defined excess of these fields near the surface of discontinuity, Albano (1979). Thus these fields have the following form

$$d(\vec{r},t) = d^-(\vec{r},t)\, \theta^-(\vec{r},t) + d^s(\vec{r},t)\, \delta^s(\vec{r},t) + d^+(\vec{r},t)\, \theta^+(\vec{r},t) \tag{2.3}$$

$$\vec{J}_d(\vec{r},t) = \vec{J}_d^-(\vec{r},t)\, \theta^-(\vec{r},t) + \vec{J}_d^s(\vec{r},t)\, \delta^s(\vec{r},t) + \vec{J}_d^+(\vec{r},t)\, \theta^+(\vec{r},t) \tag{2.4}$$

$$\sigma_d(\vec{r},t) = \sigma_d^-(\vec{r},t)\, \theta^-(\vec{r},t) + \sigma_d^s(\vec{r},t)\, \delta^s(\vec{r},t) + \sigma_d^+(\vec{r},t)\, \theta^+(\vec{r},t) \tag{2.5}$$

In these expressions θ^- and θ^+ are the time–dependent characteristic functions of the two bulk phases; 1 in one phase and 0 in the other. Using the time–dependent curvilinear coordinates these characteristic functions are defined by

$$\theta^\pm(\vec{r},t) \equiv \theta(\pm\xi_1(\vec{r},t)) \tag{2.6}$$

Here $\theta(s)$ is the Heaviside function; 1 if s is positive and 0 if s is negative. Furthermore, $\delta^s(\vec{r},t)$ is a δ–function on the dividing surface which is defined in terms of the curvilinear coordinates as

$$\delta^s(\vec{r},t) \equiv |\operatorname{grad} \xi_1(\vec{r},t)|\; \delta(\xi_1(\vec{r},t)) \tag{2.7}$$

The characteristic functions for the bulk phases restrict an integration over the volume to the regions occupied by the corresponding phase. Similarly $\delta^s(\vec{r},t)$ restricts an integration over the volume to a two-dimensional integral over the dividing surface.

It is clear from the above expressions for the fields that d^s, $\vec{J}^s_d$ and σ^s_d depend on the position along the dividing surface alone; thus one has for instance

$$d^s(\vec{r},t) = d^s(\xi_2(\vec{r},t),\xi_3(\vec{r},t),t) \tag{2.8}$$

and similarly for $\vec{J}^s_d$ and σ^s_d. An important consequence of this is that the spatial derivatives of d^s, $\vec{J}^s_d$ and σ^s_d normal to the dividing surface are zero.

When one substitutes the expressions (2.3) – (2.5) into the balance equation (2.1), one obtains spatial and temporal derivatives of the characteristic functions $\theta^\pm$ and of δ^s. For the spatial derivative of the characteristic functions one has

$$\text{grad}\ \theta^\pm(\vec{r},t) = \pm\, \vec{n}(\vec{r},t)\ \delta^s(\vec{r},t) \tag{2.9}$$

Here $\vec{n}$ is the normal on the dividing surface defined by

$$\vec{n}(\xi_2,\xi_3,t) \equiv \vec{a}_1(\xi_1 = 0,\xi_2,\xi_3,t) \tag{2.10}$$

where $\vec{a}_i$ is the unit vector in the direction of increasing ξ_i given by

$$\vec{a}_i(\vec{r},t) \equiv h_i(\vec{r},t)\ \text{grad}\ \xi_i(\vec{r},t) \quad \text{with} \quad h_i(\vec{r},t) \equiv |\text{grad}\ \xi_i(\vec{r},t)|^{-1} \tag{2.11}$$

These unit vectors which play an important role when one uses curvilinear coordinates are defined in each point in space and not only on the dividing surface like the normal $\vec{n}$. They form an orthonormal set for all $\vec{r}$ and t:

$$\vec{a}_i(\vec{r},t).\vec{a}_j(\vec{r},t) = \delta_{ij} \tag{2.12}$$

For the gradient of the surface δ-function one may show that, Bedeaux (1976)

$$\text{grad}\ \delta^s(\vec{r},t) = \vec{n}\,\vec{n}.\vec{\nabla}\ \delta^s(\vec{r},t) \tag{2.13}$$

where $\vec{\nabla} \equiv (\partial/\partial x,\ \partial/\partial y,\ \partial/\partial z)$ is the Cartesian gradient. An important aspect of both eq.(2.9) and eq.(2.13) is the fact that the gradient only has a component in the direction normal to the dividing surface. While this property is intuitively obvious, it nevertheless plays a crucial role. Even though this has not been explicitly indicated in the above equation, the direction of the normal depends on the position along the interface and the time. In the equations below we will similarly not always explicitly indicate the dependence on position and time as this dependence is usually clear from the context.

In order to give expressions for the time derivatives we introduce the velocity field of the curvilinear coordinate system relative to the fixed Cartesian coordinates

$$\vec{w}(\xi_1,\ \xi_2,\ \xi_3,\ t) \equiv \frac{\partial}{\partial t}\,\vec{r}(\xi_1,\ \xi_2,\ \xi_3,\ t) \tag{2.14}$$

The more physical velocity field of the dividing surface is given in terms of the velocity field of the curvilinear coordinate system by

$$\vec{w}^s(\xi_2,\xi_3,\ t) = \vec{w}(\xi_1 = 0,\ \xi_2,\ \xi_3,\ t) \tag{2.15}$$

Using the velocity of the dividing surface, one may show that the time derivative of the characteristic functions for the bulk phases is given by, Bedeaux (1976),

$$\frac{\partial}{\partial t}\,\theta^{\pm}(\vec{r},t) = \mp\, w_n^s(\vec{r},t)\ \delta^s(\vec{r},t) \tag{2.16}$$

Here the subscript n indicates the normal component. Similarly one may show that the time derivative of the surface δ-function is given by, Bedeaux (1976),

$$\frac{\partial}{\partial t}\,\delta^s(\vec{r},t) = -\,w_n^s(\vec{r},t)\ \vec{n}.\vec{\nabla}\ \delta^s(\vec{r},t) \tag{2.17}$$

It follows from the above equations that the time derivative of the characteristic functions and of the surface δ-function only contain the normal component of the surface velocity. The motion of the geometrical surface is in fact given in terms of the normal component of this velocity. The parallel component of the interfacial velocity should be related to material flow along the interface. This will become clear in its definition below.

The above formulae for the gradient and the time derivative of the characteristic functions and of the surface δ-function make it possible to substitute the explicit form of d, $\vec{J}_d$ and σ_d given in eqs.(2.3)–(2.4) into the balance equation (2.2) and one obtains, after some rearranging, the following more detailed formula for the balance equation

$$
\begin{aligned}
&[\frac{\partial}{\partial t} d^-(\vec{r},t) + \mathrm{div}\, \vec{J}_d^-(\vec{r},t) - \sigma_d^-(\vec{r},t)]\, \theta^-(\vec{r},t) \\
&+ [\frac{\partial}{\partial t} d^+(\vec{r},t) + \mathrm{div}\, \vec{J}_d^+(\vec{r},t) - \sigma_d^+(\vec{r},t)]\, \theta^+(\vec{r},t) \\
&+ [\frac{\partial}{\partial t} d^s(\vec{r},t) + \mathrm{div}\, \vec{J}_d^s(\vec{r},t) - \sigma_d^s(\vec{r},t) \\
&+ J_{d,n}^+(\vec{r},t) - J_{d,n}^-(\vec{r},t) - w_n^s(\vec{r},t)(d^+(\vec{r},t) - d^-(\vec{r},t))]\, \delta^s(\vec{r},t) \\
&+ [J_{d,n}^s(\vec{r},t) - w_n^s(\vec{r},t) d^s(\vec{r},t)]\, \vec{n}.\vec{\nabla}\, \delta^s(\vec{r},t) = 0 \qquad (2.18)
\end{aligned}
$$

This formula contains four different contributions, all of which are equal to zero. Thus, it follows from the first term that

$$
\frac{\partial}{\partial t} d^-(\vec{r},t) + \mathrm{div}\, \vec{J}_d^-(\vec{r},t) = \sigma_d^-(\vec{r},t) \qquad \text{for } \xi_1(\vec{r},t) < 0 \qquad (2.19)
$$

and it follows from the second term that

$$
\frac{\partial}{\partial t} d^+(\vec{r},t) + \mathrm{div}\, \vec{J}_d^+(\vec{r},t) = \sigma_d^+(\vec{r},t) \qquad \text{for } \xi_1(\vec{r},t) > 0 \qquad (2.20)
$$

The above two equations are the balance equations in the bulk regions and these equations have their usual form. As the above analysis shows these equations together with the restriction of the area where they are valid are contained in the general form of the balance equation (2.2) together with the explicit form of the fields given in eq.(2.3)–(2.5). The third term gives the balance equation for the interface

$$\frac{\partial}{\partial t} d^s(\vec{r},t) + \operatorname{div} \vec{J}_d^s(\vec{r},t) + J_{d,n}^+(\vec{r},t) - J_{d,n}^-(\vec{r},t) - w_n^s(\vec{r},t)(d^+(\vec{r},t) - d^-(\vec{r},t))$$

$$= \sigma_d^s(\vec{r},t) \qquad \text{for } \xi_1(\vec{r},t) = 0 \qquad (2.21)$$

In this equation the first two terms on the left hand side and the term on the right hand side are similar to the terms found in the bulk regions. The third term on the left hand side gives the flow of d leaving the surface of discontinuity into the + phase in the comoving frame. Similarly the fourth term gives the flow from the – phase into the surface of discontinuity in the comoving frame. Finally the fifth term in the above equation accounts for the rate of increase of d^s due to the motion of the dividing surface. The last term in eq.(2.18) gives

$$J_{d,n}^s(\vec{r},t) - w_n^s(\vec{r},t) d^s(\vec{r},t) = 0 \qquad \text{for } \xi_1(\vec{r},t) = 0 \qquad (2.22)$$

This condition expresses the important fact that in a reference frame moving along with the interface the excess current $\vec{J}_d^s$ is pointed along the interface. While of course it is difficult to see how it could be any other way, it is nevertheless important to realize that this property is contained in the general form of the balance equation (2.2) together with the explicit form of d, $\vec{J}_d$ and σ_d given in eqs.(2.3)–(2.5).

If one would contemplate describing a system in which the excess current is not along the interface in the comoving frame of reference one must necessarily modify eqs.(2.2)–(2.5). In this context one should then also add terms to d, $\vec{J}_d$ and σ_d which are proportional to normal derivatives of arbitrary order of the surface δ-function. Such terms would describe the time dependence of the detailed distribution of d inside the surface of discontinuity. In our description such details do not play a role. In Albano (1979) it is discussed why it is unnecessary to cary these details in our description along. We will restrict ourselves here to the observation that the proper choice of the variables is crucial in this context. This aspect is most apparent for a charge double layer. As the charge double layer has no excess charge the charge distribution has no contribution proportional to δ^s in this system. It does have a contribution proportional to $\vec{n}\vec{n}.\vec{\nabla}\,\delta^s$, however. As there is no excess charge one may, however, use the polarization density as a variable rather than the charge density. This polarization density only has a contribution proportional to δ^s which illustrates this point. If one would use the charge density anyway this is of course perfectly all right as long as one realises that the above equations should then be accordingly modified.

An important quantity of a surface of discontinuity is its curvature which is defined in the following way

$$C(\xi_2, \xi_3, t) \equiv -[\, h_1^{-1} \frac{\partial}{\partial \xi_1} \ln(h_2 h_3)]_s = \frac{1}{R_1} + \frac{1}{R_2} \tag{2.23}$$

The subindex s indicates that the value of the corresponding quantity is taken for $\xi_1 = 0$ [Morse 1953]. R_1 and R_2 are the so called radii of curvature. One may show that the curvature is minus the divergence of the normal on the surface, Bedeaux (1976),

$$C = -(\vec{\nabla}.\vec{n})_s \tag{2.24}$$

In order to make the dynamic description of the system complete one must also have an equation of motion for the surface. It may be shown that the normal on the surface satisfies the following equation, Albano (1980),

$$\frac{\partial}{\partial t}\vec{n}(\vec{r},t) = -(\vec{\vec{1}} - \vec{n}\vec{n})\,.\,(\vec{\nabla}\, w_n^s)_s \tag{2.25}$$

where $\overleftrightarrow{I}$ is a unit tensor. The time derivative of the normal is thus given in terms of the gradient of the normal component of the velocity of the surface.

As a matter of convenience we introduce the average of the extrapolated bulk values at the dividing surface:

$$d_+(r,t) \equiv \frac{1}{2}[d^+(\xi_1 = 0, \xi_2, \xi_3, t) + d^-(\xi_1 = 0, \xi_2, \xi_3, t)] \tag{2.26}$$

Another convenient combination is the jump of these extrapolated values across the surface

$$d_-(r,t) \equiv d^+(\xi_1 = 0, \xi_2, \xi_3, t) - d^-(\xi_1 = 0, \xi_2, \xi_3, t) \tag{2.27}$$

Similar definitions of the average and the jump of the extrapolated bulk values will be used for all bulk fields. Using the above definition of the jump one may write the balance equation for the interface (2.21) in the following form

$$\frac{\partial}{\partial t} d^s(\vec{r},t) + \operatorname{div} \vec{J}_d^s(\vec{r},t) + J^+_{d,n,-}(\vec{r},t) - w_n^s(\vec{r},t) d_-(\vec{r},t) = \sigma_d^s(\vec{r},t)$$

$$\text{for } \xi_1(\vec{r},t) = 0 \tag{2.28}$$

This is a little more compact then the original form.

3. CONSERVATION OF MASS

Consider a system consisting of n components. We will not consider chemical reactions so that the mass of each component is conserved. The balance equation for the mass density ρ_ℓ of component ℓ is given by

$$\frac{\partial}{\partial t} \rho_\ell + \operatorname{div} \rho_\ell \vec{v}_\ell = 0 \tag{3.1}$$

where $\vec{v}_\ell$ is the velocity of component ℓ. Both ρ_ℓ and $\rho_\ell \vec{v}_\ell$ have the form given in eqs.(2.3)-(2.5). In order to clarify the use of this notation we now give this form again explicitly

$$\rho_\ell(\vec{r},t) = \rho_\ell^-(\vec{r},t)\,\theta^-(\vec{r},t) + \rho_\ell^s(\vec{r},t)\,\delta^s(\vec{r},t) + \rho_\ell^+(\vec{r},t)\,\theta^+(\vec{r},t) \tag{3.2}$$

$$\rho_\ell(\vec{r},t)\vec{v}_\ell(\vec{r},t) = \rho_\ell^-(\vec{r},t)\,\vec{v}_\ell^-(\vec{r},t)\,\theta^-(\vec{r},t) + \rho_\ell^s(\vec{r},t)\,\vec{v}_\ell^s(\vec{r},t)\,\delta^s(\vec{r},t)$$

$$+\,\rho_\ell^+(\vec{r},t)\,\vec{v}_\ell^+(\vec{r},t)\,\theta^+(\vec{r},t) \tag{3.3}$$

The similarity of eq.(3.2) and eq.(2.3) is clear and needs no further discussion. We only note that while $\rho_\ell^\pm$ are mass densities per unit of volume that ρ_ℓ^s is a mass density per unit of surface area. For the mass current of component ℓ given in eq.(3.3) we have as usual written this current as the product of the mass density and a velocity. This procedure in fact defines the velocity field as the quotient of the mass current and the mass density. In the bulk regions this will be clear. The surface velocity field $\vec{v}^s$ is defined as the ratio of the mass current per unit of surface area and the mass density per unit of surface area both for component ℓ. Notice the fact that the surface velocity $\vec{v}_\ell^s$ and the bulk velocities $\vec{v}_\ell^\pm$ have the same dimensionality. This clearly is not the case for the surface and the bulk mass currents. The advantage of the use of specific quantities like the velocity is that one may consider for instance the difference of the velocity of a component in the bulk with the velocity of the same or in fact other components on the surface. This may not be done for the bulk and surface mass currents as there dimensionality is in fact different. It is important to realize, however, that an equation like eq.(2.4) may only be written down for the mass current and not for the velocity field.

In view of the fact that the mass of all components is conserved the total mass is also a conserved quantity. The total mass density is defined by

$$\rho \equiv \sum_{\ell=1}^{n} \rho_\ell \quad <=====> \quad \begin{cases} \rho^{\pm} \equiv \sum_{\ell=1}^{n} \rho_\ell^{\pm} & \text{for the bulk regions} \\ \rho^{s} \equiv \sum_{\ell=1}^{n} \rho_\ell^{s} & \text{for the interface} \end{cases} \tag{3.4}$$

The total mass density satisfies the following balance equation

$$\frac{\partial}{\partial t}\rho + \operatorname{div} \rho\vec{v} = 0 \tag{3.5}$$

where the barycentric velocity field is defined by

$$\rho\vec{v} \equiv \sum_{\ell=1}^{n} \rho_\ell \vec{v}_\ell \quad <=====> \quad \begin{cases} \rho^{\pm}\vec{v}^{\pm} \equiv \sum_{\ell=1}^{n} \rho_\ell^{\pm}\vec{v}_\ell^{\pm} & \text{for the bulk regions} \\ \rho^{s}\vec{v}^{s} \equiv \sum_{\ell=1}^{n} \rho_\ell^{s}\vec{v}_\ell^{s} & \text{for the interface} \end{cases} \tag{3.6}$$

The condition for the normal component of the mass current, cf. eq.(2.22), gives for the normal components of the various interfacial velocity fields

$$v_{\ell,n}^{s} = v_n^{s} = w_n^{s} \tag{3.7}$$

Thus the excess mass of the different components and as a consequence also the barycentric motion of the interface have the same normal velocity as the dividing surface. It is clear that a proper choice of the time–dependent position of the dividing surface is in fact only acceptable if $v_n^s = w_n^s$. The fact that the excesses of all components have the same velocity in the normal direction is necessary to assure that the interface does not split up in two or more different interfaces. This is not a phenomenon we want to describe. We now use the freedom in the choice of the curvilinear coordinate system to choose the velocity of the interface equal to the material flow velocity

$$\vec{v}^s = \vec{w}^s \tag{3.8}$$

The advantage of this is that also the velocity of the curvilinear coordinate system at the dividing surface in the direction parallel to the dividing surface has a direct physical meaning.

Using the general balance equation (2.21) for the excess mass density of component ℓ one obtains using also eq.(3.7)

$$\frac{\partial}{\partial t}\rho_\ell^s + \operatorname{div}\rho_\ell^s\vec{v}_\ell^s + [\rho_\ell(v_{\ell,n} - v_n^s)]_- = 0 \tag{3.9}$$

As is clear from this equation the rate of change of the excess mass of component ℓ is given in terms of the divergence of the interfacial mass current of component ℓ and the flow of mass of component ℓ from the bulk regions into the surface of discontinuity.

Similarly, one finds for the total excess mass

$$\frac{\partial}{\partial t}\rho^s + \operatorname{div}\rho^s\vec{v}^s - [\rho(v_n - v_n^s)]_- = 0 \tag{3.10}$$

It is convenient to define a barycentric time derivative for the dividing surface

$$\frac{d^s}{dt} \equiv \frac{\partial}{\partial t} + \vec{v}^s.\operatorname{grad} = \frac{\partial}{\partial t} + v_2^s\frac{1}{h_{2,s}}\frac{\partial}{\partial\xi_2} + v_3^s\frac{1}{h_{3,s}}\frac{\partial}{\partial\xi_3} \tag{3.11}$$

This barycentric time derivative gives the time rate of change of a variable in a reference frame which moves along with the excess surface mass. Using this barycentric time derivative in eq.(3.10) one finds

$$\frac{d^s}{dt}\rho^s + \rho^s\operatorname{div}\vec{v}^s - [\rho(v_n - v_n^s)]_- = 0 \tag{3.12}$$

for the mass balance at the interface. In the description of interfaces one often uses the freedom in the choice of the dividing surface and chooses the socalled Gibbs dividing surface for which $\rho^s = 0$. We will not do this at this point in the analysis as it is convenient to introduce specific densities which would then be impossible. It should be emphasised once again that any resulting prediction of a physical phenomenon must be independent of the choice of the dividing surface.

The interfacial and the bulk diffusion fluxes of component ℓ are defined by

$$\vec{J}^s_\ell \equiv \rho^s_\ell(\vec{v}^s_\ell - \vec{v}^s) \quad \text{and} \quad \vec{J}^\pm_\ell \equiv \rho^\pm_\ell(\vec{v}^\pm_\ell - \vec{v}^\pm) \tag{3.13}$$

It follows from the definition of the barycentric velocities that

$$\sum_{\ell=1}^{n} \vec{J}^s_\ell = 0 \quad \text{and} \quad \sum_{\ell=1}^{n} \vec{J}^\pm_\ell = 0 \tag{3.14}$$

which implies that as usual only n − 1 diffusion currents are independent. In view of the fact that the velocities of all the components in the direction normal to the interface are identical to v^s_n it follows that

$$J^s_{\ell,n} = 0 \tag{3.15}$$

The interfacial diffusion currents are therefore along the dividing surface. If we define the interfacial and bulk mass fractions by

$$c^s_\ell \equiv \rho^s_\ell / \rho^s \quad \text{and} \quad c^\pm_\ell \equiv \rho^\pm_\ell / \rho^\pm \tag{3.16}$$

one finds upon substitution in the interfacial balance equation (3.9) for component ℓ

$$\rho^s \frac{d^s}{dt} c^s_\ell + \text{div}\, \vec{J}^s_\ell + [(v_n - v^s_n)\rho(c_\ell - c^s_\ell) + J_{\ell,n}]_- = 0 \tag{3.17}$$

Of course eq.(3.9) or eq. (3.17) are completely equivalent and one should simply use the one that is most convenient for any particular application.

As an intermezzo we now return to the general balance equation discussed in section

2. At the interface this balance equation is given by

$$\frac{\partial}{\partial t} d^s + \operatorname{div} \vec{J}_d^s + [J_{d,n} - v_n^s d]_- = \sigma_d^s \tag{3.18}$$

Furthermore the normal component of the current satisfies

$$J_{d,n}^s - v_n^s d^s = 0 \tag{3.19}$$

The purpose of the intermezzo is to see how the introduction of specific densities

$$d \equiv \rho a \quad <=====> \begin{cases} a^{\pm} \equiv d^{\pm} / \rho^{\pm} & \text{for the bulk regions} \\ a^s \equiv d^s / \rho^s & \text{for the interface} \end{cases} \tag{3.20}$$

gives rise to alternative equations for the balance at the interface. For this purpose it is furthermore convenient to define currents in the comoving reference frame

$$\vec{J}_d \equiv \rho a \vec{v} + \vec{J}_a \quad <=====> \begin{cases} \vec{J}_d^{\pm} \equiv \rho^{\pm} a^{\pm} + \vec{J}_a^{\pm} & \text{for the bulk regions} \\ \vec{J}_d^s \equiv d^s \rho^s + \vec{J}_a^s & \text{for the interface} \end{cases} \tag{3.21}$$

Here $\rho a \vec{v} = d \vec{v}$ is the so-called convective contribution to the current. Upon substitution of these definitions into eq.(3.18) one obtains, using also the balance equation for the excess mass,

$$\rho^s \frac{d^s}{dt} a^s + \operatorname{div} \vec{J}_a^s + [(v_n - v_n^s)\rho(a - a^s) + J_{a,n}]_- = \sigma_d^s \tag{3.22}$$

Furthermore eq.(3.19) gives

$$J_{a,n}^s = 0 \tag{3.23}$$

These alternative equations, which are of course equivalent to the original ones, will be often used in the analysis below. It must be emphasised that this is entirely a matter of personal choice. Using the original form of these equations yields exactly the same results. This in particular also when the dividing surface can be chosen such that $\rho^s = 0$. We shall come back to this point below.

4. CONSERVATION OF MOMENTUM

The equation of motion of the system is given by

$$\frac{\partial}{\partial t}\rho\vec{v} + \vec{\nabla}.[\rho\vec{v}\vec{v} + \overleftrightarrow{P}] = \sum_{\ell=1}^{n} \rho_\ell \vec{F}_\ell \tag{4.1}$$

where the convective contribution to the momentum flow is equal to

$$\rho\vec{v}\vec{v} = \rho^-\vec{v}^-\vec{v}^-\theta^- + \rho^s\vec{v}^s\vec{v}^s\delta^s + \rho^+\vec{v}^+\vec{v}^+\theta^+ \tag{4.2}$$

Furthermore $\overleftrightarrow{P}$ is the pressure tensor that gives the rest of the momentum flow and $\vec{F}_\ell$ is an external force acting on component ℓ. The pressure tensor can be written in terms of the hydrostatic pressures p^- and p^+ in the bulk regions, the surface tension γ and the viscous pressure $\overleftrightarrow{\Pi}$ in the following way

$$\overleftrightarrow{P} = p^-\overleftrightarrow{I}\theta^- - \gamma(\overleftrightarrow{I} - \vec{n}\vec{n})\delta^s + p^+\overleftrightarrow{I}\theta^+ + \overleftrightarrow{\Pi} \tag{4.3}$$

The viscous pressure tensor may also have an excess in the dividing surface. It will be assumed that the surface tension is independent of the curvature. For a fluid–fluid interface such a dependence is unimportant, cf Gibbs (1906), as long as the surface tension is not too small. See in this context also section 6. As is usual in the discussion of the bulk regions we furthermore assume that the system possesses no intrinsic internal angular momentum, this neither in the bulk nor at the interface. Using invariance of the system under rotation the viscous contributions are then found to be symmetric, cf De Groot (1962) for the bulk phases and Waldmann (1967) for the interface. In fact Waldmann's argument assumes invariance of the surface under rotation around the normal on the dividing surface. While this is clearly correct for a fluid–fluid interface if the curvature radii are equal to each other it is not correct if these radii are not equal to each other. Thus one will in general have an asymmetric contribution to the excess viscous pressure tensor proportional to $(R_1 - R_2)$. I refer to the lectures by Kosinski for more details. As one may expect that this asymmetric contribution will usually be small for the fluid–fluid interface we will assume it to be zero. Eq.(3.23) then gives for this case

$$\vec{n}.\vec{\vec{\Pi}} = \vec{\vec{\Pi}}.\vec{n} = 0 \qquad \text{and} \qquad \vec{n}.\vec{\vec{P}} = \vec{\vec{P}}.\vec{n} = 0 \tag{4.4}$$

Consequently the excess of the viscous pressure tensor and of the total pressure tensor are symmetric 2×2 tensors. Notice the fact that the symmetric choice of the pressure tensor has limited the discussion to fluid–fluid interfaces. It is not difficult to give the alternative expressions for the case that one or both phases are solids.

We restrict the discussion to conservative forces which can therefore be written in terms of time–independent potentials:

$$\vec{F}_\ell = -\text{grad}\,\Psi_\ell \tag{4.5}$$

Because of these external forces the momentum is not conserved. The balance equation for the excess interfacial momentum (4.1) can now be written in a form analoguous to eq.(3.23) which gives

$$\rho^s \frac{d^s}{dt}\vec{v}^s + \vec{\nabla}.[-\gamma(\vec{\vec{1}} - \vec{n}\vec{n}) + \vec{\vec{\Pi}}^s] + \vec{n}.[(\vec{v} - \vec{v}^s)\rho(\vec{v} - \vec{v}^s) + \vec{\vec{P}}]_- = \sum_{\ell=1}^{n} \rho_\ell^s \vec{F}_{\ell,s} \tag{4.6}$$

Here the subscript s indicates as usual the value of the corresponding quantity at the dividing surface. Notice in this context the fact that the external force is continuous at the dividing surface. Notice in this context the fact that the external force is continuous at the dividing surface. Another point which should be realized is that after taking the gradient or the divergence of a quantity which is only defined on the surface, which when one differentiates merely implies that this quantity does not depend on ξ_1, one should take the result for $\xi_1 = 0$. The gradient operator may otherwise introduce spurious ξ_1 dependence. We will not indicate this fact explicitly in the formulae, by setting for instance $(\text{grad } \gamma)_s$ instead of grad γ, for ease of notation.

In order to study the force resulting from the surface tension we write

$$\vec{\nabla}.[\gamma(\vec{\vec{1}} - \vec{n}\vec{n})] = \vec{\nabla}\gamma - \gamma\vec{n}\,\text{div}\,\vec{n} = \vec{\nabla}\gamma - \gamma\vec{n}\,C \tag{4.7}$$

where we made use of the fact that the gradient of a scalar quantity like the surface tension which is only defined on the surface is always directed along the dividing surface. As is clear from this equation the gradient of the surface tension gives rise to a force along the interface while in the normal direction one obtaines a force equal to the curvature times the surface tension. For an interface in an equilibrium situation, ie when $\vec{v} = 0$ and $\vec{\vec{\Pi}} = 0$, eq.(4.6) reduces to

$$\vec{\nabla}\gamma - \gamma\vec{n}\,C + \vec{n}\,p_- = \sum_{\ell=1}^{n} \rho_\ell^s \vec{F}_{\ell,s} \tag{4.8}$$

The balance of forces for the equilibrium situation normal to the surface is thus given by

$$-\gamma C + p_- = -\gamma C + p_s^+ - p_s^- = \sum_{\ell=1}^{n} \rho_\ell^s F_{\ell,s,n} \tag{4.9}$$

This is a direct generalization of the Laplace equation for the hydrostatic pressure difference in terms of the surface tension and the curvature. The balance of forces for the equilibrium situation parallel to the surface is given by

$$-\vec{\nabla}\,\gamma = \sum_{\ell=1}^{n} \rho_\ell^s \vec{F}_{\ell,s}\cdot(\vec{\vec{1}} - \vec{n}\vec{n}) \qquad (4.10)$$

In the non–equilibrium case both the Laplace equation and the above equation for the forces parallel to the surface contain additional terms due to the inertial term, the viscous pressure and the convective contribution to the momentum current. The contributions may easily be found using eq.(4.6). As the surface tension will in general depend on the concentration of the various components and also on the temperature, gradients of these quantities along the interface will lead to a force due to the gradient of the surface tension along the interface. It is this force along the interface which leads to the socalled Marangoni effect where convection cells develop if there are such gradients. See the lectures by Golia for a more extensive discussion of the Marangoni effect.

In the analysis of the conservation of energy in the next section we need an equation for the time rate of change of the kinetic energy density. As we are in particular considering the behaviour of the interface we will restrict ourselves to the excess kinetic energy. In general we will refer to De Groot (1962) for a discussion of bulk contributions and of the relevant equations for the bulk regions. For the excess kinetic energy one finds using eq.(4.6)

$$\rho^s \frac{d^s}{dt}\frac{1}{2}|\vec{v}^s|^2 = \rho^s \vec{v}^s\cdot\frac{d^s}{dt}\vec{v}^s$$

$$= -\vec{\nabla}\cdot(\vec{\vec{P}}^s\cdot\vec{v}^s) + \vec{\vec{P}}^s : \vec{\nabla}\,\vec{v}^s - \vec{n}\cdot[(\vec{v}-\vec{v}^s)\rho(\vec{v}-\vec{v}^s) + \vec{\vec{P}}\,]_-\cdot\vec{v}^s + \sum_{\ell=1}^{n} \rho_\ell^s \vec{F}_{\ell,s}\cdot\vec{v}^s \qquad (4.11)$$

Similarly we need an expression for the time rate of change of the potential energy of the excess mass:

$$\rho^s \Psi^s \equiv \sum_{\ell=1}^{n} \rho_\ell^s \Psi_{\ell,s} \qquad (4.12)$$

Using eq.(3.17) one finds

$$\rho^s \frac{d^s}{dt}\Psi^s = \rho^s \frac{d^s}{dt}\sum_{\ell=1}^{n} \Psi_{\ell,s} c_\ell^s = \sum_{\ell=1}^{n} \Psi_{\ell,s}\rho^s \frac{d^s}{dt} c_\ell^s + \rho^s \sum_{\ell=1}^{n} c_\ell^s \vec{v}^s . \nabla \Psi_{\ell,s}$$

$$= -\sum_{\ell=1}^{n} \Psi_{\ell,s}\vec{\nabla}. \vec{J}_\ell^s - [(v_n - v_n^s)\rho(\Psi - \Psi^s) + \sum_{\ell=1}^{n} \Psi_\ell J_{\ell,n}]_- - \sum_{\ell=1}^{n} \rho_\ell^s \vec{F}_{\ell,s}.\vec{v}^s \tag{4.13}$$

It follows from the above equations that neither the kinetic nor the potential energy are conserved quantities. As we will discuss below only the total energy is conserved.

5. CONSERVATION OF ENERGY

The total energy of the system is conserved. As a consequence one has for the specific energy density e

$$\frac{\partial}{\partial t}\rho e + \mathrm{div}(\rho e\vec{v} + \vec{J}_e) = 0 \tag{5.1}$$

The total energy density can be written as the sum of the kinetic energy, the potential energy and the internal energy u:

$$e^\pm = \frac{1}{2}|\vec{v}^\pm|^2 + \Psi^\pm + u^\pm \qquad \text{for the bulk regions}$$
$$e^s = \frac{1}{2}|\vec{v}^s|^2 + \Psi^s + u^s \qquad \text{for the interface} \tag{5.2}$$

The energy current is similarly the sum of a mechanical work term, a potential–energy flux due to diffusion and a heat flow $\vec{J}_q$

$$\vec{J}_e = \vec{\vec{P}}.\vec{v} + \sum_{\ell=1}^{n} \Psi_\ell \vec{J}_\ell + \vec{J}_q \tag{5.3}$$

In fact one may use the above equations as a definition of internal energy and of the heat current. Notice the fact that we have not included the convective contribution $\rho e\vec{v}$ in the energy current $\vec{J}_e$ as is done in De Groot (1962).

Using energy conservation (5.1) one immediately finds a balance equation for the interfacial energy density e^s from the general form (3.22) of such a balance equation

$$\rho^s \frac{d^s}{dt} e^s + \mathrm{div}\, \vec{J}_e^s + [(v_n - v_n^s)\rho(e - e^s) + J_{e,n}]_- = 0 \tag{5.4}$$

The excess energy defined above does not contain a convective contribution and is as a consequence directed along the interface

$$J_{e,n}^s = 0 \tag{5.5}$$

The definition of the internal energy given in eq.(5.2) may then be used to find a balance equation for the internal energy. Together with the balance equations for the interfacial kinetic and potential energy given in the previous section and the definition of the heat current (5.3) one then obtains

$$\rho^s \frac{d^s}{dt} u^s = -\,\mathrm{div}\, \vec{J}_q^s - \overleftrightarrow{P}^s : \vec{\nabla}\vec{v}^s + \sum_{\ell=1}^{n} \vec{F}_{\ell,s} \cdot \vec{J}_\ell^s - \{(v_n - v_n^s)\rho[u - u^s - \tfrac{1}{2}|\vec{v} - \vec{v}^s|^2]$$

$$+ J_{q,n} + [(v_n - v_n^s)\rho(\vec{v} - \vec{v}^s) + \vec{n}.\overleftrightarrow{P}].(\vec{v} - \vec{v}^s)\}_- \tag{5.6}$$

Furthermore one finds that also the excess heat flow is directed along the dividing surface

$$J_{q,n} = 0 \tag{5.7}$$

The internal energy is not a conserved quantity, because of conversion of kinetic and potential energy into internal energy and vice versa. The above balance equation for the excess internal energy gives in fact the first law of thermodynamics for the interface.

6. ENTROPY BALANCE

In order to also formulate the second law of thermodynamics we now consider the balance equation for the entropy

$$\frac{\partial}{\partial t}\rho s + \mathrm{div}(\rho s\vec{v} + \vec{J}) = \sigma \tag{6.1}$$

Here s is the entropy density per unit of mass, $\vec{J}$ is the entropy current and σ the entropy source. Notice the fact that we do not use subscripts s to indicate that the current or the source are of the entropy. The reason for this is that this would lead to confusion with the subindex s which indicates that the value of a bulk variable should be taken at the dividing surface. For the interface the equation for the entropy balance becomes, cf again eq.(3.22),

$$\rho^s \frac{d^s}{dt} s^s = -\mathrm{div}\,\vec{J}^s - [(v_n - v_n^s)\rho(s - s^s) + J_n]_- + \sigma^s \tag{6.2}$$

If we compare this with the expression for the balance of entropy in the bulk regions, De Groot (1962),

$$\rho^\pm \frac{d^\pm}{dt} s^s = \rho^\pm \left(\frac{\partial}{\partial t} + \vec{v}^\pm.\vec{\nabla}\right) s^s = -\mathrm{div}\,\vec{J}^\pm + \sigma^\pm \tag{6.3}$$

one sees that the balance equation for the interfacial entropy density contains, in addition to the divergence of a current and a source term, a flow of entropy from the bulk phases into the dividing surface. It is this last term and similar terms in the balance of excess energy and mass which will be found to also lead to an excess of the entropy production. The origin of the usual boundary conditions, which will be discussed below, is related to the appearence of these terms.

The use of nonequilibrium thermodynamics to describe systems with two phases separated by an interface implicitly assumes that in volume elements small compared to the overall size of the system but large compared to the molecular size one may use the usual thermodynamic relations. Thus one assumes that in each volume element the second law of thermodynamics applies and concludes that the entropy production is non–negative

$$\sigma^- \geq 0, \qquad \sigma^s \geq 0 \qquad \text{and} \qquad \sigma^+ \geq 0 \tag{6.4}$$

While this property is well known in the bulk phases it is less obviously true for the interface. In Gibbs' discussion of the equilibrium thermodynamics of interfaces the interface is treated as a separate thermodynamic system. In the present context one may clearly argue that in volume elements located in the surface of discontinuity one may also assume the usual thermodynamic relations and it then follows that the excess entropy production is non–negative. Notice in this context that the surface of discontinuity as defined by Gibbs (1906) and as used here has a finite thickness, this contrary to the dividing surface also introduced by Gibbs which is a mathematically sharp two dimensional surface.

From thermodynamics we know that the entropy and any other thermodynamic function, such as the hydrostatic pressure, are well–defined functions of the thermodynamic parameters necessary to specify the macroscopic state of the system. For the system under consideration we use as parameters the internal energy, the specific volume ($1/\rho^\pm$) or the specific surface area ($1/\rho^s$), and the mass fractions. For the entropy we may thus write

$$s^- = s^-(u^-, 1/\rho^-, c_\ell^-), \quad s^s = s^s(u^s, 1/\rho^s, c_\ell^s) \quad \text{and} \quad s^- = s^-(u^-, 1/\rho^-, c_\ell^-) \tag{6.5}$$

We stress the important fact that the excess entropy only depends on u^s, $1/\rho^s$ and c_ℓ^s and not on the extrapolated values $u_s^\pm$, $1/\rho_s^\pm$ and $c_{\ell,s}^\pm$ of these variabes in the bulk as is sometimes done [Defay 1966]. Nowhere do we find that such a dependence on the extrapolated bulk values is needed. In the context of the above argument that volume elements in the surface of discontinuity can be treated as separate thermodynamic systems, this would imply a dependence on the variables in neighbouring volume elements; such a nonlocal dependence is clearly contrary to the usual assumptions made in the context of nonequilbrium thermodynamics. In Gibbs discussion of interfaces he also discusses a possible dependence of the excess thermodynamic variables on the curvature of the surface. As he argues, such a dependence is small for most systems and may therefore usually be neglected. Notable exceptions are systems in which the surface tension is very small due to the presence of sufficient surfactants like for instance microemulsions or in systems close to the critical point. We will in these lectures always neglect the dependence on the curvature.

At equilibrium the total differential of the entropy is given by the Gibbs relation. In the bulk regions this relation is given by

$$T^{\pm}ds^{\pm} = du^{\pm} + p^{\pm}d(1/\rho^{\pm}) - \sum_{\ell=1}^{n} \mu_{\ell}^{\pm} dc_{\ell}^{\pm} \tag{6.6}$$

where $T^{\pm}$ is the temperature and $\mu_{\ell}^{\pm}$ is the chemical potential of component ℓ in the bulk regions. For the interface one similarly has, Gibbs (1906),

$$T^{s}ds^{s} = du^{s} - \gamma\, d(1/\rho^{s}) - \sum_{\ell=1}^{n} \mu_{\ell}^{s} dc_{\ell}^{s} \tag{6.7}$$

where T^{s} is the interfacial temperature and μ_{ℓ}^{s} is the interfacial chemical potential of component ℓ. If one writes the surface tension as minus the hydrostatic surface pressure p^{s} the Gibbs relation for the interface is term by term analogous to the Gibbs relation in the bulk. If the surface tension depends on the curvature, a dependence which we neglected, one has additional contributions to eq.(6.7) proportional to the change of curvature.

In nonequilibrium thermodynamics the whole system is not in equilibrium. One assumes, however, that each volume element which is large compared to molecular sizes but small compared to the overall size of the system is in equilibrium. This is the socalled assumption of local equilibrium. In each volume element one may therefore write the entropy and other thermodynamic quantities as a function of the energy, the specific volume (surface area) and the mass fractions in the same volume element. Thus one has

$$s^{-}(\vec{r},t) = s^{-}(u^{-}(\vec{r},t),\, 1/\rho^{-}(\vec{r},t),\, c_{\ell}^{-}(\vec{r},t))$$

$$s^s(\xi_2,\xi_3,t) = s^s(u^s(\xi_2,\xi_3,t),\ 1/\rho^s(\xi_2,\xi_3,t),\ c_\ell^s(\xi_2,\xi_3,t))$$

$$s^-(\vec{r},t) = s^-(u^-(\vec{r},t),\ 1/\rho^-(\vec{r},t),\ c_\ell^-(\vec{r},t)) \tag{6.8}$$

The local equilibrium assumption implies that the Gibbs relation remains valid in the frame moving with the center of mass. We therefore have

$$T^\pm \frac{d^\pm}{dt} s^\pm = \frac{d^\pm}{dt} u^\pm + p^\pm \frac{d^\pm}{dt}(1/\rho^\pm) - \sum_{\ell=1}^{n} \mu_\ell^\pm \frac{d^\pm}{dt} c_\ell^\pm \tag{6.9}$$

in the bulk phases and

$$T^s \frac{d^s}{dt} s^s = \frac{d^s}{dt} u^s - \gamma \frac{d^s}{dt}(1/\rho^s) - \sum_{\ell=1}^{n} \mu_\ell^s \frac{d^s}{dt} c_\ell^s \tag{6.10}$$

for the interface.

The next step in the analysis is to substitute the expressions for the barycentric derivatives of the internal energy, the specific volume (surface area) and the mass fractions in the above Gibbs relations. This leads in particular for the interface to rather complicated expressions for the barycentric derivative of the entropy density. Comparing this expression for the barycentric derivative of the entropy density with eq.(6.3) one may then identify the entropy current and the entropy production. In this way one obtains as entropy current in the bulk phases, De Groot (1962]),

$$\vec{J}^\pm = (\vec{J}_q^\pm - \sum_{\ell=1}^{n} \mu_\ell^\pm \vec{J}_\ell^\pm)/T^\pm \tag{6.11}$$

For the interfacial entropy current one finds in a similar way

$$\vec{J}^s = (\vec{J}_q^s - \sum_{\ell=1}^{n} \mu_\ell^s \vec{J}_\ell^s)/T^s \tag{6.12}$$

This expression is in essence identical to the one in the bulk phases.

For the entropy production one finds in the bulk phases, De Groot (1962),

$$\sigma^\pm = -(T^\pm)^{-2} \vec{J}_q^\pm \cdot \text{grad}\, T^\pm + \vec{J}_\ell^\pm \cdot [(\vec{F}_{\ell,s}/T^\pm) - \text{grad}(\mu_\ell^\pm/T^\pm)] - (T^\pm)^{-1} \vec{\vec{\Pi}}^\pm : \text{grad}\, \vec{v}^\pm \tag{6.13}$$

For the excess entropy production one finds in this way after some tedious algebra

$$\sigma^s = -(T^s)^{-2} \vec{J}_q^s . \text{grad}\, T^s + \sum_{\ell=1}^{n} \vec{J}_\ell^s . [(\vec{F}_{\ell,s}/T^s) - \text{grad}(\mu_\ell^s/T^s)] - (T^s)^{-1} \vec{\vec{\Pi}}^s : \text{grad}\, \vec{v}^s$$

$$+ \{T[J_n + (v_n - v_n^s)\rho s][(1/T) - (1/T^s)]\}_- - (1/T^s)\{[\vec{n}.\vec{\vec{\Pi}} + (v_n - v_n^s)\rho\vec{v}] . (\vec{v} - \vec{v}^s)\}_-$$

$$-(1/T^s) \sum_{\ell=1}^{n} \{[J_{\ell,n} + (v_n - v_n^s)\rho_\ell][\mu_\ell - \mu_\ell^s - \tfrac{1}{2}|\vec{v}|^2 + \tfrac{1}{2}|\vec{v}^s|^2]\}_- \tag{6.14}$$

The above expression was first given by Bedeaux (1986). The expression in that paper contained $J_{q,n}/T$ rather then J_n which is incorrect, Albano (1987). The first four terms in the excess entropy production are similar to the contributions found in the bulk regions and are a consequence of transport processes along the interface. The other terms are new and are the result of transport processes from the bulk regions into and through the interface. One may rewrite the above form of σ^s in a somewhat different but equivalent forms. Our reason to choose the above form is the fact that, in the terms corresponding to the flow into and through the interface, one systematically finds as currents the sum of the diffusive and the convective part of the heat, momentum and mass flow in the comoving reference frame. This seems to be a reasonable choice and gives a relatively elegant form. In the treatment of membrane transport the question how the choice of the reference frame affected the entropy production gave rise to considerable discussion, Mikulecki (1966). Of course one may recombine the various terms in different combinations. In practice it is usually good to think about this choice carefully for every particular problem under consideration. Some contributions are usually more important than others and this must motivate the choice. In the analysis below we will in fact rewrite the above expression and combine all the contributions due to the convective part of the currents. This combined term will in fact turn out to couple the mass current through and into the interface to pressure differences.

In the derivation of the above expressions for the entropy production and current we have used the following thermodynamic identities:

$$T^{\pm}s^{\pm} = u^{\pm} + p^{\pm}/\rho^{\pm} - \sum_{\ell=1}^{n} \mu_{\ell}^{\pm} c_{\ell}^{\pm} \tag{6.15}$$

for the bulk phases and

$$T^{s}s^{s} = u^{s} - \gamma/\rho^{s} - \sum_{\ell=1}^{n} \mu_{\ell}^{s} c_{\ell}^{s} \tag{6.16}$$

for the interface. In the expression for the interface we have not used the freedom to make one of the excess densities equal to zero by the appropriate choice of the location of the dividing surface. If one would use this freedom to take ρ^{s} equal to zero it is better to write the last equation in the alternative form

$$T^{s}(\rho s)^{s} = (\rho u)^{s} - \gamma - \sum_{\ell=1}^{n} \mu_{\ell}^{s} (\rho c_{\ell})^{s} \tag{6.17}$$

It is important to realize that $\rho^{s} = 0$ does not imply that for instance $(\rho s)^{s}$ is also zero. The definition $s^{s}\rho^{s} \equiv (\rho s)^{s}$ is somewhat confusing in this case. If one prefers to avoid this problem one should rewrite the above analysis using densities per unit of surface area rather than using densities per unit of surface mass. The resulting expression for the entropy current and the entropy production are of course excacly the same. In a given problem certain excess densities or currents may be zero to a good approximation. In that case it is most convenient to eliminate these terms from the relevant balance equations and from the above expressions for the excess entropy production. In view of the large number of force–flux pairs in the excess entropy production, such a simplification is most conveniently made at that stage of the analysis.

The diffusion currents in the bulk regions as well as along the interface are not independent. In fact their sum is zero, cf eq.(3.16). Using this property one may eliminate one of these diffusion currents from the expression for the entropy production. In solutions it is most convenient to eliminate the diffusion current of the solvent. If one eliminates $\vec{J}_n^s$ from the excess entropy production one obtains

$$\sigma^s = -(T^s)^{-2}\,\vec{J}_q^s\cdot \text{grad}\,T^s + \sum_{\ell=1}^{n-1} \vec{J}_\ell^s\cdot[((\vec{F}_{\ell,s}-\vec{F}_{n,s})/T^s) - \text{grad}((\mu_\ell^s-\mu_n^s)/T^s)]$$

$$-(T^s)^{-1}\,\overleftrightarrow{\Pi}^s : \text{grad}\,\vec{v}^s + \{T[J_n + (v_n - v_n^s)\rho s][(1/T) - (1/T^s)]\}_-$$

$$-(1/T^s)\{[\vec{n}.\overleftrightarrow{\Pi} + (v_n - v_n^s)\rho\vec{v}]\,.(\vec{v}-\vec{v}^s)\}_-$$

$$-(1/T^s)\sum_{\ell=1}^{n-1}\{[J_{\ell,n} + (v_n - v_n^s)\rho_\ell][\mu_\ell - \mu_\ell^s - \mu_n + \mu_n^s]\}_-$$

$$-(1/T^s)\{(v_n - v_n^s)\rho(\mu_n - \mu_n^s)]\}_- \tag{6.18}$$

It is now convenient to combine the various contributions proportional to $\rho(v_n - v_n^s)$ and to write TJ_n in terms of the heat current and the diffusion currents, cf eq.(6.11). This yields the following expression

$$\sigma^{s} = -(T^{s})^{-2}\,\vec{J}^{s}_{q}\cdot \mathrm{grad}\,T^{s} + \sum_{\ell=1}^{n-1} \vec{J}^{s}_{\ell}\cdot [((\vec{F}_{\ell,s}-\vec{F}_{n,s})/T^{s}) - \mathrm{grad}((\mu^{s}_{\ell}-\mu^{s}_{n})/T^{s})]$$

$$-(1/T^{s})\,\vec{\vec{\Pi}}^{s} : \mathrm{grad}\,\vec{v}^{s} + \{J_{q,n}[(1/T)-(1/T^{s})]\}_{-} - (1/T^{s})\{\vec{\Pi}_{n,\parallel}\cdot(\vec{v}_{\parallel}-\vec{v}^{s}_{\parallel})\}_{-}$$

$$-\sum_{\ell=1}^{n-1}\{J_{\ell,n}[(\mu_{\ell}/T)-(\mu^{s}_{\ell}/T^{s})-(\mu_{n}/T)+(\mu^{s}_{n}/T^{s})]\}_{-}$$

$$-(1/T^{s})\{[\rho(v_{n}-v^{s}_{n})][s(T-T^{s}) + \sum_{\ell=1}^{n} c_{\ell}(\mu_{\ell}-\mu^{s}_{\ell}) + (\Pi_{n,n}/\rho) + \tfrac{1}{2}|\vec{v}-\vec{v}^{s}|^{2}]\}_{-} \tag{6.19}$$

The subindex $\parallel$ indicates the projection parallel to the dividing surface. Notice furthermore that the subindex n is usually used to indicate the normal component of a vector. If it appears as a subindex of the chemical potential, however, it indicates the n–th component.

In order to understand the driving force for the mass flow through and into the interface a little better we use the following thermodynamic relation for the bulk phases

$$(1/\rho^{\pm})dp^{\pm} = s^{\pm}dT^{\pm} + \sum_{\ell=1}^{n} c^{\pm}_{\ell}\,d\mu^{\pm}_{\ell} \tag{6.20}$$

A similar relation which we will not need, however, is valid for the interface

$$(1/\rho^{s})d\gamma = -s^{s}dT^{s} - \sum_{\ell=1}^{n} c^{s}_{\ell}\,d\mu^{s}_{\ell} \tag{6.21}$$

Now we use the fact that the system, in the context of nonequilibrium thermodynamics, is never very far from equilibrium. This implies in the bulk regions that the gradient of the temperature and the gradient of the chemical potentials are small. At the interface this similarly implies on the one hand that these gradients are also small along the interface and on the other hand that the temperature and chemical potential jumps are small. We may therefore write

$$\{[\rho(v_n - v_n^s)][s(T - T^s) + \sum_{\ell=1}^{n} c_\ell(\mu_\ell - \mu_\ell^s) + \Pi_{n,n} + \tfrac{1}{2}|\vec{v} - \vec{v}^s|^2]\}_-$$

$$= \{[\rho(v_n - v_n^s)][(p - p(T^s, \mu_\ell^s) + \Pi_{n,n})/\rho + \tfrac{1}{2}|\vec{v} - \vec{v}^s|^2]\}_- \qquad (6.22)$$

This expression implies that the mass flow through and into the interface couples directly to pressure differences. $p^{\pm}(T^s, \mu_\ell^s)$ is the pressure of the bulk phases at the temperature and chemical potentials of the interface.

7. THE PHENOMENOLOGICAL EQUATIONS

In equilibrium the entropy production is zero. This implies that all the thermodynamic forces and fluxes are zero. Close to equilibrium the fluxes are linear functions of the thermodynamic forces. It should be realized that the linear nature of this relation between the fluxes and forces does not imply that the resulting differential equations describing the time dependent state of the system are linear. These equations are non linear for a number of reasons. The first is the presence of convective contributions to all the currents. The second is the nonlinear nature of the equations of state relating the various thermodynamic quantities to each other. The third is the dependence of the linear constitutive coefficients on the variables. The only crucial limitation is the fact that these constitutive coefficients do not themselves depend on the thermodynamic forces. If one would contemplate taking such a dependence into account, one should realize that the assumption of local equilibrium is no longer correct in that case.

It is clear from the above expression for the excess entropy production there is a rather large number of force–flux pairs. In the most general case all Cartesian components of the fluxes couple to all the Cartesian components of the fluxes. This would lead to an immense number of different constitutive coefficients. For a fluid–fluid interface this situation is greatly simplified because of the symmetry of the problem. The fluid–fluid interface is invariant for translation along the interface and for rotation around the normal on the interface. One may therefore distinguish 2×2 tensorial force flux pairs, 2–dimensional vectorial force–flux pairs and scalar force–flux pairs. In view of the Curie symmetry principle, De Groot (1962), forces and fluxes of a different tensorial nature do not couple.

We need one more identity before we can identify all thermodynamic forces and fluxes at the interface. The jump of a product from one side of the interface to the other satisfies

$$(ab)_- = a_- b_+ + a_+ b_- \tag{7.1}$$

Thus one has, for instance,

$$\{\vec{\Pi}_{n,\|} \cdot (\vec{v}_\| - \vec{v}_\|^s)\}_- = \vec{\Pi}_{n,\|,-} \cdot (\vec{v}_\| - \vec{v}_\|^s)_+ + \vec{\Pi}_{n,\|,+} \cdot (\vec{v}_\| - \vec{v}_\|^s)_-$$

$$= \vec{\Pi}_{n,\|,-} \cdot (\vec{v}_{\|,+} - \vec{v}_\|^s) + \vec{\Pi}_{n,\|,+} \cdot \vec{v}_{\|,-} \tag{7.2}$$

The first term contains the slip force when the interface is moved relative to the two bulk phases, and the second term contains the slip force when one bulk phase slides over the other.

In order to get an overview of the forces and fluxes of a different tensorial nature we will tabulate them for each tensorial character:

1. Symmetric traceless tensor

flux: $[\overleftrightarrow{\Pi}^s_{\|,\|} - \frac{1}{2}(\mathrm{tr}\overleftrightarrow{\Pi}^s)(\overleftrightarrow{I} - \vec{n}\vec{n})]^S$ force: $[(\vec{\nabla}_\| \, \vec{v}_\|^s) - \frac{1}{2}(\mathrm{tr}(\vec{\nabla}\, \vec{v}^s))(\overleftrightarrow{I} - \vec{n}\vec{n})]^S$

Here the superindex S indicates the symmetric part of a tensor. The linear constitutive equation which is the result gives the symmetric traceless part of the excess viscous pressure tensor as a constant η^s times the symmetric traceless gradient of the velocity of the excess mass along the interface. The proportionality constant may be interpreted as the interfacial shear viscosity.

2. Antisymmetric traceless tensor

flux: $[\overset{\rightrightarrows}{\Pi}{}^s_{\|,\|}]^A$ force: $(\vec{\nabla}_{\|}\,\vec{v}^s_{\|})^A$

Here the superindex A indicates the antisymmetric part of a tensor. The linear constitutive equation which is the result in this case gives the antisymmetric traceless part of the excess viscous pressure tensor as a constant times the antisymmetric traceless gradient of the velocity of the excess mass along the interface. The proportionality constant may be interpreted as the interfacial rotational viscosity. For a fluid–fluid interface one may use invariance for rotation around the normal on the interface to show that the antisymmetric part of the excess viscous pressure is zero, Waldmann (1967), if the molecules have no interfacial spin, De Groot (1962). As we discussed above the surface is not invariant for rotation around the normal if the principal radii of curvature are different. In fact one may conclude that the rotational surface viscocity will be proportional to the difference of the radii of curvature. In practice this term is expected to be small and therefore unimportant.

3. Vectors

even

fluxes: $\vec{J}^s_{q,\|}$ forces: $\vec{\nabla}\,(1/T^s)$

$\vec{J}^s_{\ell,\|}$ (for $\ell = 1,\ldots,n-1$) $(\vec{F}_{\ell,s} - \vec{F}_{n,s}) - \vec{\nabla}\,[(\mu^s_\ell - \mu^s_n)/T^s]$

odd

fluxes: $\vec{\Pi}_{n,\parallel,-}$	forces: $(\vec{v}_+ - \vec{v}^s)_\parallel$
$\vec{\Pi}_{n,\parallel,+}$	$(\vec{v}_- - \vec{v}^s)_\parallel$
$\vec{\Pi}^s_{\parallel,n}$	$\vec{\nabla}\, v^s_n$

Even and odd gives the behaviour of the corresponding quantities for time reversal. The (n+3) vectorial fluxes can be expressed in terms of the (n+3) vectorial forces using a (n+3)×(n+3) matrix of constitutive coefficients. These coefficients in this matrix are the so-called Onsager coefficients. They are related to the interfacial heat conductivity, interfacial diffusion coefficients, interfacial Dufour coefficients, interfacial thermal diffusion coefficients, coefficients of sliding friction and thermal slip coefficients. The constitutive coefficients for a number of cross effects have no special name. The Onsager symmetry relations state that Onsager coefficients which couple terms with the same symmetry for time reversal will be symmetric while coefficients which couple terms with a different symmetry for time reversal are antisymmetric, De Groot (19620). The number of independent constitutive coefficients thus reduces to $\frac{1}{2}(n+3)(n+4)$.

4. Scalars

even

fluxes: $J_{q,n,-}$	forces: $(1/T)_+ - (1/T^s)$
$J_{q,n,+}$	$(1/T)_-$
$J_{\ell,n,-}$ (for $\ell = 1,\dots,n-1$)	$(\mu_\ell - \mu^s_\ell - \mu_n + \mu^s_n)_-$
$J_{\ell,n,+}$ (for $\ell = 1,\dots,n-1$)	$(\mu_\ell - \mu_n)_+$

odd

fluxes: $\Pi^s \equiv tr(\overleftrightarrow{\Pi}^s)$	forces: $\operatorname{div} \vec{v}^s$
$[\rho(v_n - v_n^s)]_-$	$[(p - p(T^s,\mu_\ell^s) + \Pi_{n,n})/\rho]_+ + \frac{1}{2}\vert\vec{v} - \vec{v}^s\vert_+^2$
$[\rho(v_n - v_n^s)]_+$	$[(p - p(T^s,\mu^s) + \Pi_{n,n})/\rho]_- + \frac{1}{2}\vert\vec{v} - \vec{v}^s\vert_-^2$

The (2n + 3) scalar fluxes can be expressed in terms of the (2n + 3) scalar forces using a (2n + 3)×(2n + 3) matrix of Onsager coefficients. Some of the coefficients can be related to the temperature jump coefficients. The Onsager symmetry relations again state that coefficients which couple terms with the same symmetry for time reversal will be symmetric while coefficients which couple terms with a different symmetry for time reversal are antisymmetric, De Groot (1962). The number of independent constitutive coefficients thus reduces to $\frac{1}{2}(2n + 3)(2n + 4)$.

The total number of constitutive (Onsager) coefficients which gives a full description of all the possible processes at the interface is clearly prohibitively large. For a one component system one needs already 26 constitutive coefficients. In practice most (if not all) these coefficients will be unknown. Furthermore it will almost be impossible to measure all of them. In a practical situation one must decide which of the above fluxes and forces contribute to the dynamical processes in a significant manner and neglect all the others. Though many of these forces and fluxes may thus not play a role in a given situation, it is good to know which ones are neglected in a simplified description. One may otherwise too easily assume that conclusions derived using a simplified description are also valid for a more complicated situation. We will not proceed to discuss these simplifications in any detail. We will here merely mention some of these simplifications. The first is the application to problems where all velocities and gradients are normal to a planar dividing surface .In that case only the scalar fluxes and forces play a role. This situation is usually appropriate for transport through membranes. Furthermore, in view of the fact that fluid–fluid interfaces due to gravity form a horizontal plane, this case is also applicable to problems like evaporation and condensation, Bedeaux (1990). Another typical simplification is that certain densities and currents have no excess at the surface. If one considers for instance a surface between a liquid and its vapor one may neglect the excess density. It is then clear that the interfacial chemical potential is no longer an independent variable. In fact one should in that case take the interfacial chemical potential equal to the extrapolated value in the adjacent liquid, Bedeaux (1990). This not only reduces the number of independent variables describing the interface but also eliminates the corresponding fluxes and forces. A more difficult case is when the heat capacity of the interface may be neglected. Then the interfacial temperature becomes an average of the extrapolated bulk temperatures. The weights in the average depend on the specific heat of the adjacent phases. The most simple case is the interface between a gas and a liquid where one may in good approximation set the interfacial temperature equal to the extrapolated value in the liquid, Bedeaux (1990).

8. EQUILIBRIUM FLUCTUATIONS

The probability of thermal fluctuations around equilibrium of a closed system is given in terms of the fluctuation δS of the total entropy S of the system by

$$P_{eq} \sim \exp\left(\delta S/k_B\right) \tag{8.1}$$

where k_B is Boltzmann's constant. The total entropy is obtained by integration of the entropy density over the volume of the system.

$$S = \int d\vec{r}\, \rho s = \int d\vec{r}\, [\rho^- s^- \theta^- + \rho^s s^s \delta^s + \rho^+ s^+ \theta^+] \tag{8.2}$$

One may write the total entropy as the sum of the contributions from the bulk phases and the contribution from the interface

$$S = S^- + S^s + S^+ \tag{8.3}$$

which are defined for the bulk phases by

$$S^- = \int d\vec{r}\, \rho^- s^- \theta^- = \int_{\xi_1<0} d\vec{r}\, \rho^- s^- \quad \text{and} \quad S^+ = \int d\vec{r}\, \rho^+ s^+ \theta^+ = \int_{\xi_1>0} d\vec{r}\, \rho^+ s^+ \tag{8.4}$$

and for the interface by

$$S^s = \int d\vec{r}\, \rho^s s^s \delta^s = \int d\xi_1 d\xi_2 d\xi_3\, h_1 h_2 h_3\, \rho^s(\xi_2,\xi_3)\, s^s(\xi_2,\xi_3)\, h_1^{-1}\, \delta(\xi_1)$$

$$= \int d\xi_2 d\xi_3\, h_2 h_3\, \rho^s(\xi_2,\xi_3)\, s^s(\xi_2,\xi_3) \tag{8.5}$$

where we used the standard conversion of the integration to curvilinear coordinates, Morse (1953).

The fluctuation δS of the total entropy are on the one hand due to the usual fluctuations of the entropy densities around their equilibrium value and on the other hand to fluctuations in the location of the interface around its equilibrium position, Zielinska (1982). These fluctuations in the location of the interface are of particular interest as they lead to novel behaviour of the correlation functions in the neighbourhood of the interface; behaviour which can not be anticipated on the basis of our understanding of the fluctuations in the bulk regions, Bedeaux (1985; 1986).

The general formula which gives the fluctuation of the total entropy in terms of temperature, composition, density and position of the interface fluctuations is rather complicated. We will therefore restrict ourselves to the more simple special case of the liquid–vapour interface in a one component fluid. This system is placed in the gravitational field $\Psi(\vec{r}) = \rho(\vec{r})gz$ where g is the gravitational acceleration. Furthermore we use that the equilibrium dividing surface is planar. For the equilibrium state of the system the curvilinear coordinates may be choosen such that they coincide with the Cartesian coordinates

$$\xi_{1,eq} = z \;,\quad \xi_{2,eq} = x \;\text{ and }\; \xi_{3,eq} = y \tag{8.6}$$

The equilibrium position of the dividing surface is therefore given by the x–y plane. The nonequilibrium position of the dividing surface is given by $\xi_1(x,y,z) = 0$. In our analysis it is most convenient to use the height of the fluctuating surface

$$h(x,y) \equiv \delta\xi_1(x,y) \equiv \xi_1(x,y) - \xi_{1,eq}(x,y) \tag{8.7}$$

as the variable characterising the location of the interface. The normal on the fluctuating surface is given by

$$\vec{n}(x,y) = (-\frac{\partial h(x,y)}{\partial x}, -\frac{\partial h(x,y)}{\partial y}, 1)/[1 + (-\frac{\partial h(x,y)}{\partial x})^2 + (-\frac{\partial h(x,y)}{\partial y})^2] \tag{8.8}$$

To linear order in the height this gives for the fluctuation of the normal

$$\delta\vec{n}(x,y) = -\vec{\nabla}\, h(x,y) \tag{8.9}$$

A long and rather complicated analysis, Zielinska (1982), leads to the following expression for the fluctuation of the total entropy to quadratic order in the fluctuations of the variables

$$\delta S = (2T_o)^{-1} \int d\vec{r}\, \{[(c_V^-/T_o)(\delta T^-)^2 + (\rho_o^-)^{-2}(\kappa_T^-)^{-1}(\delta\rho^-)^2 + \rho_o^-|\vec{v}^-|^2]\theta(-z)$$

$$[(c_V^+/T_o)(\delta T^+)^2 + (\rho_o^+)^{-2}(\kappa_T^+)^{-1}(\delta\rho^+)^2 + \rho_o^+|\vec{v}^+|^2]\theta(z)$$

$$[(c_V^s/T_o)(\delta T^s)^2 + (\rho_o^s)^{-2}(\kappa_T^s)^{-1}(\delta\rho^s)^2 + \rho_o^s|\vec{v}^s|^2 + \gamma_o|\vec{\nabla}h|^2 - g\rho_{o,-}d^2]\delta(z) \tag{8.10}$$

where the equilibrium values of the various quantities have been indicated by the subindex o. The contributions in the bulk regions have their usual form and contain the specific heat at constant volume $c_V^\pm$ and the isothermal compressibility $\kappa_T^\pm$. Some of the contributions due to the interface are similar to those found in the bulk regions and contain the interfacial specific heat at constant surface area c_V^s and the interfacial isothermal compressibility κ_T^s defined by

$$c_V^s \equiv [\frac{\partial(\rho u)^s}{\partial T^s}]_{\rho^s} \quad \text{and} \quad \kappa_T^s \equiv -(\rho^s)^{-1}[\frac{\partial\rho^s}{\partial\gamma}]_{T^s} \tag{8.11}$$

Two new terms appear, however, both due to the displacement of the surface. The first finds its origin in the increase in surface area due to fluctuations and the second is a gravitational energy contribution.

A very interesting and useful consequence of the above expression for the fluctuation of the entropy is that the equilibrium fluctuations of δT^-, $\delta\rho^-$, $\vec{v}^-$, δT^+, $\delta\rho^+$, $\vec{v}^+$, δT^s, $\delta\rho^s$, $\vec{v}^s$ and h are not coupled. In other words the equilibrium cross correlations of these variables are all zero. One of the reasons to restrict ourselves in this section to the one–component case is the fact that in a multi–component system one has cross correlations of the concentrations of the various components which makes the relevant expressions unpleasantly complicated. For the equilibrium fluctuations of the density we find from the above formulae, eq.(8.1) together with eq.(8.10)

$$< \delta\rho^{\pm}(\vec{r})\, \delta\rho^{\pm}(\vec{r}\,') > = k_B T_o (\rho_o^{\pm})^2 \kappa_T^{\pm}\, \delta(\vec{r} - \vec{r}\,')$$

$$< \delta\rho^{s}(x,y)\, \delta\rho^{s}(x',y') > = k_B T_o (\rho_o^{s})^2 \kappa_T^{s}\, \delta(x - x')\, \delta(y - y') \qquad (8.12)$$

Similar expressions may easily be written down for the temperature and velocity auto–correlation functions.

In our further analysis of the equilibrium fluctuations we shall use the freedom in the choice of the dividing surface to take $\rho^s = 0$. The density is then given by

$$\rho(\vec{r}) = \rho^-(\vec{r})\theta^-(\vec{r}) + \rho^+(\vec{r})\theta^+(\vec{r}) \qquad (8.13)$$

Using the fact that $< \rho^{\pm} > = \rho_o^{\pm}$ and $\delta\theta^- = -\,\delta\theta^+$, a fluctuation of the density around its average value may be written as

$$\delta\rho(\vec{r}) = [\delta\rho^-(\vec{r})]\theta^-(\vec{r}) + [\delta\rho^+(\vec{r})]\theta^+(\vec{r}) - \rho_{o,-}\,\delta\theta^-(\vec{r}) \qquad (8.14)$$

As is clear from this expression the fluctuations of the density are a consequence on the one hand of the fluctuations of the bulk densities $\rho^{\pm}$ and on the other hand of the fluctuations of the (location of the interface dependent) characteristic functions. The density auto–correlation function becomes

$$H(\vec{r},\vec{r}') \equiv < \delta\rho(\vec{r})\delta\rho(\vec{r}') >$$

$$= <\delta\rho^-(\vec{r})\delta\rho^-(\vec{r}')> <\theta^-(\vec{r})\,\theta^-(\vec{r}')> + <\delta\rho^+(\vec{r})\delta\rho^+(\vec{r}')> <\theta^+(\vec{r})\,\theta^+(\vec{r}')>$$

$$+ (\rho_{o,-})^2 <\delta\theta^-(\vec{r})\,\delta\theta^-(\vec{r}')> \qquad (8.15)$$

where we neglected the z-dependence of $\rho_o^\pm$ due to gravity which is small for the system sizes we will consider. Using eq.(8.12) the above equation gives

$$H(\vec{r},\vec{r}') = k_B T_o[(\rho_o^-)^2\kappa_T^- <\theta^-(\vec{r}')> + (\rho_o^+)^2\kappa_T^+ <\theta^+(\vec{r}')>]\,\delta(\vec{r}-\vec{r}')$$

$$+ (\rho_{o,-})^2 <\delta\theta^-(\vec{r})\,\delta\theta^-(\vec{r}')> \qquad (8.16)$$

The first term on the right hand side is the usual short range contribution to the density correlation function with as small modification the appearance of the average of the characteristic functions. The second term is due to the socalled capillary wave fluctuations in the location of the interface and, as will be discussed below, is long range along the interface.

We restrict our explicit discussion below to the average density and to the density auto-correlation function. It is not difficult to extent the analysis to study the behaviour of other variables like the temperature. All these variables will be similarly affected by the capillary wave fluctuations of the interface.

9. EQUILIBRIUM FLUCTUATIONS OF THE HEIGHT

As a first step we will now calculate the height auto-correlation function. From eq.(8.1) together with eq.(8.10) it follows that the equilibrium probability distribution of the height is given by

$$P_{eq}(\{h(x,y)\}) \sim \exp\{-(2k_BT_o)^{-1}\int dxdy[\gamma_o|\nabla h(x,y)|^2 - g\rho_{o,-}h^2(x,y)] \qquad (9.1)$$

As there are no cross correlations with the other fluctuating variables one may use eq.(9.1) to calculate averages of all functions which only depend on the height. This height probability distribution is the usual expression used in the context of the capillary wave theory [Buff 1965].

It is now convenient to introduce the Fourier transform of the height

$$\tilde{h}(k_x,k_y) \equiv \int dxdy\, h(x,y) \exp[-i(k_x x + k_y y)] \tag{9.2}$$

The above probability distribution may then alternatively be written as

$$P_{eq}(\{\tilde{h}(\vec{Q})\}) \sim \exp\{-(8\pi^2 k_B T_o)^{-1} \int d\vec{Q}\, [\gamma_o Q^2 + g\rho_{o,-}]\tilde{h}(\vec{Q})\tilde{h}(-\vec{Q})\} \tag{9.3}$$

where $\vec{Q} \equiv (k_x,k_y)$ and $Q \equiv |\vec{Q}|$. It is good to realise that due to the real nature of $h(x,y)$

$$\tilde{h}(-\vec{Q}) = \tilde{h}^*(\vec{Q}) \tag{9.4}$$

It follows from the above equations that the average height is zero

$$<\tilde{h}(\vec{Q})> = 0 \qquad ======> \qquad <h(x,y)> = 0 \tag{9.5}$$

Furthermore it follows that the height auto-correlation function is given by

$$<\tilde{h}(\vec{Q})\tilde{h}(-\vec{Q}')> = 4\pi^2 k_B T_o [g\rho_{o,-} + \gamma_o Q^2]\, \delta(\tilde{Q} - \tilde{Q}') \equiv 4\pi^2\, \tilde{S}(\tilde{Q})\delta(\tilde{Q} - \tilde{Q}') \tag{9.6}$$

If we introduce the capillary length by

$$L_c \equiv (\gamma_o / g\rho_{o,-})^{1/2} \tag{9.7}$$

one has

$$\tilde{S}(\vec{Q}) = (k_B T_o / g\rho_{o,-})(1 + Q^2 L_c^2)^{-1} \tag{9.8}$$

The inverse Fourier transform of this function is the height auto-correlation function as a function of the position

$$S(\vec{R} - \vec{R}') = \langle h(\vec{R})\, h(\vec{R}') \rangle \tag{9.9}$$

where $\vec{R} \equiv (x,y)$ and similarly for $\vec{R}'$. Inverse Fourier transformation of eq.(9.8) gives

$$S(\vec{R}) = S(R) = (2\pi)^{-2} \int d\vec{Q}\, \exp(i\vec{Q}.\vec{R})\, \tilde{S}(\vec{Q}) = (k_B T_o / 2\pi\gamma_o)\, K_o(R/L_c) \tag{9.10}$$

where K_o is a modified Bessel function of the second kind, Gradshteyn (1965). For small and for large R one has

$$S(R) = \begin{cases} \dfrac{k_B T_o}{2\pi\gamma_o} \ln\left[\dfrac{L_c}{R}\right] & \text{for } R << L_c \\ \dfrac{k_B T_o}{2\gamma_o} \left[\dfrac{L_c}{2\pi R}\right]^{1/2} \exp\left[-\dfrac{R}{L_c}\right] & \text{for } R >> L_c \end{cases} \tag{9.11}$$

It is clear from this expressions that there are long range correlations along the interface. The correlation length is equal to the well known capillary length which for most interfaces is a macroscopic length scale; for water it is for instance in the order of a fraction of a millimeter. A perturbing aspect of the above result is that the correlation length will become infinite under zero gravity conditions. We will return to this point below. Another problem arises for very small distances where the logarithm diverges. This problem finds its origin in the fact that one should not use the description above on a molecular length scale. The usual way out of this dilemna is to simply add a molecular length a to R in the above expression. This only modifies the result if R is comparable to a and there it eliminates the unphysical divergence.

10. THE AVERAGE DENSITY PROFILE

The average density is given by

$$\rho_o(z) \equiv <\rho(\vec{r})> = <\rho^-(\vec{r})\theta^-(\vec{r}) + \rho^+(\vec{r})\theta^+(\vec{r})> = \rho_o^- <\theta^-(\vec{r})> + \rho_o^+ <\theta^+(\vec{r})>$$

$$= \frac{1}{2}(\rho_o^- + \rho_o^+) + \frac{1}{2}(\rho_o^- - \rho_o^+) <\mathrm{sgn}\,(z - h(\vec{R}))> \qquad (10.1)$$

We will now first calculate the probability to find the dividing surface at the heigth z. This probability distribution is defined by

$$P(z) \equiv <\delta(h(\vec{R}) - z)> \qquad (10.2)$$

and is related to the average density profile by

$$\frac{d}{dz}\rho_o(z) = \frac{1}{2}(\rho_o^- - \rho_o^+)P(z) \qquad (10.3)$$

Using a standard identity in the theory of Fourier analysis we have

$$P(z) = (2\pi)^{-1} \int dk_z \exp(ik_z z) <\exp(-ik_z h(\vec{R}))> \qquad (10.4)$$

In view of the fact that the average of h is zero and that further the distribution of h is Gaussian it follows that

$$P(z) = (2\pi)^{-1} \int dk_z \exp(ik_z z - \frac{1}{2} k_z^2 <h^2(\vec{R})>)$$

$$= (2\pi)^{-1} \int dk_z \exp(ik_z z - \frac{1}{2} k_z^2 S(0)) = (2\pi S(0))^{-1/2} \exp\left[-\frac{z^2}{2S(0)}\right] \qquad (10.5)$$

From this expression one may identify the socalled capillary thickness of the interface as

$$W \equiv \sqrt{S(0)} = \left[\frac{k_B T_o}{2\pi\gamma_o} \ln \frac{L_c}{a}\right]^{1/2} \tag{10.6}$$

Using this thickness in eq.(10.5) the height probability distribution function becomes

$$P(z) = (2\pi)^{-1/2}\, W^{-1} \exp\left[-\frac{1}{2}\left(\frac{z}{W}\right)^2\right] \tag{10.7}$$

This is a Gaussian distribution.

Integrating the expression for the height probability distribution gives as average density profile

$$\rho_o(z) = \frac{1}{2}(\rho_o^- + \rho_o^+) + \frac{1}{2}(\rho_o^- - \rho_o^+)\, \mathrm{erf}\,(z/ W\sqrt{2}) \tag{10.8}$$

where erf indicates the error function. A matter which has caused a rather heated discussion, Evans (1981) and Bedeaux (1985), is the fact that the capillary thickness also approaches infinity in the gravity free limit. In the context of a more molecular theory based on the work by van der Waals one obtains a finite value for the thickness of the interface, Evans (1981). The consensus is at present that the capillary wave theory is correct in this matter but a really satisfactory theory combining the two descriptions is as yet lacking.

11. THE EQUILIBRIUM DENSITY AUTO-CORRELATIONFUNCTION

Using methods similar to those used to calculate the average density one may also derive an explicit expression for the contribution to the density autocorrelation function due to capillary waves, Bedeaux (1986). We will not go through this analysis in detail but merely give the result

$$H(\vec{r},\vec{r}\,') = k_B T_o[(\rho_o^-)^2 \kappa_T^- <\theta^-(\vec{r}\,')> + (\rho_o^+)^2 \kappa_T^+ <\theta^+(\vec{r}\,')>]\, \delta(\vec{r} - \vec{r}\,')$$

$$+ (\rho_{o,-})^2 <\delta\theta^-(\vec{r})\, \delta\theta^-(\vec{r}\,')>$$

$$= k_B T_0 \{ \tfrac{1}{2} [(\rho_0^-)^2 \kappa_T^- + (\rho_0^+)^2 \kappa_T^+] + \tfrac{1}{2} [(\rho_0^-)^2 \kappa_T^- - (\rho_0^+)^2 \kappa_T^+] \mathrm{erf}(z/W\sqrt{2}) \} \delta(\vec{r} - \vec{r}')$$

$$+ (\rho_{0,-})^2 \sum_{n=1}^{\infty} \frac{1}{n!} S^n(|\vec{R} - \vec{R}'|) \frac{d^n \rho_0(z)}{dz^n} \frac{d^n \rho_0(z')}{dz'^n} \qquad (11.1)$$

This density auto-correlation function has due to the capillary contribution long range correlations along the interface. That this is the behaviour one should expect was in fact predicted on the basis of the microscopic theory by Wertheim (1976). It is much more difficult to find the z,z' dependence on the basis of the microscopic theory. It is to be expected that the capillary wave result above contains much of the essentials of the behaviour of the correlation function as a function of z,z'. It is possible to express the derivatives of the average density profile in terms of the eigenfunctions of the harmonic oscillator. In this way one may rewrite the capillary wave contribution as an expansion in a complete set of eigenfunctions, Bedeaux (1985; 1986). In the analysis of the properties of the density auto-correlation function this representation in terms of eigenfunctions is very convenient.

12. TIME DEPENDENT FLUCTUATIONS OF THE INTERFACE

In order to calculate the unequal time correlation functions of the variables involved one must give equations of motion for these fluctuating variables. A convenient method to do this is to add random fluxes to the systematic fluxes given by the linear laws. This procedure is analogous to the addition of a random force to the frictional force in the description of the motion of a Brownian particle. The extension of this procedure to a one-component one-phase fluid was given by Landau and Lifshitz (1959). The extension of this procedure to multi-component systems is straightforward. The further extension to two phase fluids was given by Zielinska and Bedeaux (1982). We will again restrict ourselves to a one-component fluid in order to avoid unnecessary complications. In a two-phase system one not only has linear laws for the fluxes in the bulk regions but also one has the linear laws discussed in section 7 for the fluxes along, through and into the interfacial region. To all these fluxes both in the bulk at the surface one must add random fluxes. In the bulk regions this is done in the same way as in the one phase system and we will therefore now first discuss this procedure in this region.

In the bulk regions one writes

$$\overleftrightarrow{\Pi}^{\pm}_{tot} = \overleftrightarrow{\Pi}^{\pm} + \overleftrightarrow{\Pi}^{\pm}_{R} \quad \text{and} \quad \vec{J}^{\pm}_{q,tot} = \vec{J}^{\pm}_{q} + \vec{J}^{\pm}_{q,R} \tag{12.1}$$

The systematic contributions are given by the usual linear phenomenological laws:

$$\Pi^{\pm}_{ij} = -\eta^{\pm}\left(\frac{\partial v^{\pm}_{i}}{\partial x_{j}} + \frac{\partial v^{\pm}_{j}}{\partial x_{i}} - \frac{2}{3}\delta_{ij}\,\text{div}\,\vec{v}^{\pm}\right) - \eta^{\pm}_{v}\,\delta_{ij}\,\text{div}\,\vec{v}^{\pm} \tag{12.2}$$

$$\vec{J}^{\pm}_{q} = -\lambda^{\pm}\,\text{grad}\,T^{\pm} \tag{12.3}$$

where $\eta^{\pm}$ are the shear viscosities, $\eta^{\pm}_{v}$ the bulk viscosities and $\lambda^{\pm}$ the heat conductivities. Furthermore i,j = x,y,z. It should be stressed that $\vec{v}^{\pm}$ and $T^{\pm}$ are now fluctuating fields so that as a consequence also the systematic contribution to the total fluxes is a fluctuating contribution. The source of all these fluctuations are the random fluxes. The average of the random fluxes is zero

$$< \overleftrightarrow{\Pi}_{R} > = 0 \quad \text{and} \quad < \vec{J}_{q} > = 0 \tag{12.4}$$

The random fluxes are furthermore assumed to be Gaussian and white. Finally the socalled fluctuation–dissipation theorem for the random fluxes is

$$< \Pi^{\pm}_{R,ij}(\vec{r},t)\,\Pi^{\pm}_{R,k\ell}(\vec{r},t) >$$

$$= 2k_{B}T_{o}[\eta^{\pm}(\delta_{ik}\delta_{j\ell} + \delta_{i\ell}\delta_{jk} - \frac{2}{3}\delta_{ij}\delta_{k\ell}) + \eta^{\pm}_{v}\,\delta_{ij}\delta_{k\ell}]\,\delta(\vec{r}-\vec{r}')\,\delta(t-t') \tag{12.5}$$

$$< J^{\pm}_{q,R,i}(\vec{r},t)\,J^{\pm}_{q,R,j}(\vec{r},t) > = 2k_{B}T^{2}_{o}\,\lambda^{\pm}\,\delta_{ij}\,\delta(\vec{r}-\vec{r}')\,\delta(t-t') \tag{12.6}$$

There are no correlations of the random pressure tensor with the random heat current. This is a consequence of the fact that they have a different tensorial character. It should be realised that the properties of the random fluxes given above are only consistent with the description of the equilibrium fluctuations if the equal time correlations calculated using the above formulae approach their equilibrium value discussed in the previous sections for large times. If one uses the fully linearised equations of motion it is not difficult to verify this property. In the general non–linear case this is much more difficult to prove and in fact not necessarily always true. In the special case of a one–component one–phase fluid this matter has been analysed in great detail, Van Saarloos (1982), and one finds that it is necessary to replace the socalled Onsager coefficients, $\eta^{\pm}T$,$\eta_v^{\pm}T$ and $\lambda^{\pm}T^2$ by their equilibrium value. As thermal fluctuations are usually small this is not a real limitation of the theory. For the more general case where there are more components or more phases such a proof has not been given. It is rather clear, however, that also in that case one must always use the equilibrium value of the Onsager coefficients.

Also at the interface one should write the fluxes found from the interfacial entropy production as the sum of a systematic and a random contribution. In section 7 a list of these fluxes and the corresponding forces was given. To all these fluxes one must thus add a random flux. One therefore has for the symmetric traceless part of the pressure tensor

$$[\overleftrightarrow{\Pi}^{s}_{\|,\|}]^{S}_{tot} = [\overleftrightarrow{\Pi}^{s}_{\|,\|} - \tfrac{1}{2}(\mathrm{tr}\overleftrightarrow{\Pi}^{s})(\overleftrightarrow{I} - \vec{n}\vec{n})]^{S} + [\overleftrightarrow{\Pi}^{s}_{\|,\|}]^{S}_{R} \qquad (12.7)$$

As discussed in section 7 one may take the antisymmetric part of the pressure tensor zero. For the even vectorial fluxes one has

$$\begin{aligned} [\vec{J}^{s}_{q,\|}]_{tot} &= \vec{J}^{s}_{q,\|} + \vec{J}^{s}_{q,\|,R} \\ [\vec{J}^{s}_{\ell,\|}]_{tot} &= \vec{J}^{s}_{\ell,\|} + \vec{J}^{s}_{\ell,\|,R} \quad (\text{for } \ell = 1, \dots, n-1) \end{aligned} \qquad (12.8)$$

For the odd vectorial fluxes one has

$$[\vec{\Pi}_{n,\parallel,-}]_{tot} = \vec{\Pi}_{n,\parallel,-} + \vec{\Pi}_{n,\parallel,-,R}$$
$$[\vec{\Pi}_{n,\parallel,+}]_{tot} = \vec{\Pi}_{n,\parallel,+} + \vec{\Pi}_{n,\parallel,+,R}$$
$$[\vec{\Pi}^{s}_{\parallel,n}]_{tot} = \vec{\Pi}^{s}_{\parallel,n} + \vec{\Pi}^{s}_{\parallel,n,R} \tag{12.9}$$

For the even scalar fluxes one has

$$[J_{q,n,-}]_{tot} = J_{q,n,-} + J_{q,n,-,R}$$
$$[J_{q,n,+}]_{tot} = J_{q,n,+} + J_{q,n,+,R}$$
$$[J_{\ell,n,-}]_{tot} = J_{\ell,n,-} + J_{\ell,n,-,R} \qquad (\text{for } \ell = 1, \dots, n\text{-}1)$$
$$[J_{\ell,n,+}]_{tot} = J_{\ell,n,+} + J_{\ell,n,+,R} \qquad (\text{for } \ell = 1, \dots, n\text{-}1) \tag{12.10}$$

For the odd fluxes one finally has

$$\Pi^{s}_{tot} \equiv tr(\vec{\vec{\Pi}}^{s}_{tot}) = \Pi^{s} + \Pi^{s}_{R}$$
$$[\rho(v_n - v^{s}_{n})]_{-,tot} = [\rho(v_n - v^{s}_{n})]_{-} + [\rho(v_n - v^{s}_{n})]_{-,R}$$
$$[\rho(v_n - v^{s}_{n})]_{+,tot} = [\rho(v_n - v^{s}_{n})]_{+} + [\rho(v_n - v^{s}_{n})]_{+,R} \tag{12.11}$$

The average of all the random contributions are zero. We will not write this down explicitly as this is easy to do and it saves a considerable amount of space. Writing down the fluctuation-dissipation theorem for the random fluxes is rather more complicated. For the random contribution to the symmetric traceless part of the pressure tensor one has for instance

$$< [\Pi^{s}_{i,j}(x,y,t)]^{S}_{R} \, [\Pi^{s}_{k,\ell}(x',y',t')]^{S}_{R} > =$$

$$2k_B T_o \eta^{s} [\delta_{ik}\delta_{j\ell} + \delta_{i\ell}\delta_{jk} - \delta_{ij}\delta_{k\ell}] \, \delta(x-x')\delta(y-y')\delta(t-t') \tag{12.12}$$

where i, j, k, ℓ = x or y. Notice the fact that as in the discusion of the equilibrium correlation functions we assume that the equilibrium dividing surface is a plane, for which we choose the x,y plane. The random contribution to the symmetric traceless part of the pressure tensor is not correlated to any other random flux as it is the only flux of this tensorial nature. In general random fluxes of a different tensorial nature are not correlated. Furthermore random fluxes with the same tensorial character but with a different sign for time reversal are also not correlated. To use a more positive formulation only random fluxes with the same tensorial character and with the same sign for time reversal are correlated. One may now give the general form of the fluctuation–dissipation theorem for two random fluxes with the same tensorial character and with the same sign for time reversal. The expression is similar to eq.(12.12). On the left hand side one has the product of the two random fluxes at different positions and times. On the right hand side one has k_B times the corresponding Onsager coefficient which couples the corresponding processes times $\delta(x - x')\delta(y - y')\delta(t - t')$ for the scalar random fluxes. For the vectorial fluxes one should also multiply with δ_{ij} where i, j = x or y label the components of these two–dimensional vectors. As is the case in section 7 for the linear relations between the fluxes and forces it is rather laborious to write out all these relations in full detail including the coefficients. As the prescription given above is rather clear, however, and completely general we will not do this.

The dynamic equations describing the fluctuations are now found by substitution of the above total fluxes for currents along and into the interface, including therefore their random contribution, into the various balance equations for flow along the interface. Of course one should also substitute the total fluxes in the bulk regions into the balance equations in these regions. The balance equations for flow along the interface and the expressions for the total fluxes through the interface serve as boundary conditions for the dynamic equations in the bulk regions. It will be clear that the resulting description is excessively complex. As of yet it has not been possible to calculate the time dependence of the various correlation functions in any reasonable approximation. The first such approximation would clearly be to fully linearise the equations of motion around equilibrium. The next step is to decouple the various degrees of freedom until one is left with a simple solvable problem for the variable one wants to consider. It is then also convenient to use the correlation function method where one takes the time dependent correlation function equal to the equilibrium correlation function times the exponential decay found for the variable involved in the absence of fluctuations. In the fully linearised case this method is equivalent to solving the equations including the fluctuating sources.In this way one has for instance analysed fluctuations of the height with some success. Much has to still be done in this context before a reasonable body of knowledge of the behaviour of the time dependence of the various correlation functions will be available, however. Finally it is noted that the fact that boundary conditions must contain a random source term as discussed above is shown rather convincingly in a paper on the derivation of the Langevin equation for the Brownian motion of a spherical particle with a finite slip coefficient through a fluid, Bedeaux (1977).

REFERENCES

Albano, A.M., Bedeaux, D., and Vlieger, J. 1979. Physica 99A: 293

Albano, A.M., Bedeaux, D., and Vlieger, J. 1980. Physica 102A: 105

Albano, A.M., Bedeaux, D. 1987. Physica 147A: 407

Bedeaux, D., Albano, A.M., and Mazur, P. 1976. Physica 82A: 438

Bedeaux, D., Albano, A.M., and Mazur, P. 1977. Physica 88A: 564

Bedeaux, D. and Weeks, J. D. 1985. J. Chem. Phys.82: 972

Bedeaux, D. 1986. Nonequilibrium Thermodynamics and Statistical Physics, in Advances in Chemical Physics, Vol. 64, p.47, Ed. I. Prigogine and Stuart A. Rice. John Wiley & Sons, New York.

Buff, F. P., Lovett, R. A., and Stillinger, F. H. 1965. Phys. Rev. Lett. 15: 621

Defay, R., and Prigigine, I. 1966. Surface Tension and Adsorption. Longmans and Green, London.

De Groot, S.R., and Mazur, P. 1962. Nonequilibrium Thermodynamics. North-Holland, Amsterdam. (reprinted by Dover publications, New York, 1984)

Gibbs, J.W. 1906. The Scientific Papers of J. Willard Gibbs. Vol. I. Longmans and Green, London. (reprinted by Dover publications, New York, 1961)

Gradshteyn, I. S. and Ryzhik, I. M. 1965. Table of Integrals, Series and Products. Academic Press, New York and Londen.

Kovac, J. 1977. Physica 86A: 1

Kovac, J. 1981. Physica 107A: 280

Landau, L. D. and Lifshitz, E. M. 1959. Course of Theoretical Physics, vol. 6, Fluid Mechanics, Pergamon, Londen.
Mikulecky, D.C., and Caplan, S.R. 1966. J. Chem. Phys. 70: 3049
Morse, P.M., and Feshbach, H. 1953. Methods of Theoretical Physics, Vol I, McGraw – Hill, New York.
Ronis, D., Bedeaux, D., and Oppenheim, I. 1978. Physica 90A: 487
Waldmann, L. 1967. Z. Naturforsch. 22A: 1269
Wolff, P.A., and Albano, A.M. 1979. Physica 98A: 491
Zielinska, B. J. A. and Bedeaux, D. 1982. Physica 112A: 265

THE PHYSICS OF ICE-WATER PHASE CHANGE SURFACES

K. Hutter

Technische Hochschule Darmstadt, Darmstadt, Germany

This lecture series outlines a theoretical thermodynamic formulation of phase change processes between ice and water and follows the work of ALTS & HUTTER (1986, 1988a, b, c, 1989).

The phase boundary is modelled in two different ways, firstly as a singular surface bearing mass, momentum, energy and entropy and, secondly, as a boundary layer of finite thickness over which the bulk fields suffer smooth but rapid changes. A complete thermodynamic theory of phase change surfaces is developed using an extension of MÜLLER'S (1972) entropy postulate. Alternatively, the boundary layer description uses known spatially three-dimensional descriptions and achieves a surface formulation by essentially smearing these over the boundary layer thickness.

Comparison of the balance equations and the constitutive relations between the two formulations yields physical interpretations of the surface fields in terms of curvature weighted mean values over bulk field distributions within the layer and a precise understanding of new boundary conditions for the transverse transport of matter and energy across the phase boundary. Kinematic and dynamic consistency conditions between both theories disclose new enlightening results pertaining to the curvature dependence of surface tension and the possibility or impossibility of nuclei formation.

The linear constitutive relations for the flux quantities of a fluid-like viscous phase change surface being prohibitively complex, the postulation of a reversible interface, shows, remarkably, that all constitutive quantities of the surface can be expressed in terms of those of the bulk materials and three additional phenomenological quantities, namely two "viscosities" (which vanish for inviscid phase change surfaces) and a boundary layer thickness. This makes the formulation particularly apt for experimental verification.

Stability considerations for the phase-boundary layer determine the temperature dependence of surface tension and boundary-layer thickness and provide a profound explanation for the phenomena of undercooling and overheating and a deep understanding of the freezing process from free surfaces. Furthermore, the classical theory of surface tension is proven to be inadequate, when applied to phase boundaries, as are its consequences for nucleation processes.

1. Introduction

There are numerous applications where a solid and liquid coexist. We mention magma flow in the transition between the astenosphere and the lithosphere, metallic melts, cloud formations and *temperate ice* in glaciers and ice sheets. Temperate ice is a mixture of ice as a polycrystelline assemblage in which grain boundaries and inclusions are filled with melt water that is generated by the fusion due to shear heating and that is transported through the vein system by percolation and diffusion. The amount of internal surfaces along which phase change processes occur are considerable and energy and mass attributed to these surfaces are most likely not negligible. A prerequisite for a sound theoretical concept of temperate ice is therefore a thermodynamic theory of phase-boundaries between ice and water, needless to mention that such a formulation has its merits also in other applications.

Undoubtedly, impurities (salts) will affect the thermodynamic behaviour of phase boundaries, in particular for small water inclusions where brine concentrations become large. Inclusion of salts requires the application of mixture concepts and complicates the foundations substantially. Being aware of the quantitative significance of the salts and their migration through the phase boundary, but not yet understanding the ice water phase transition in their absence, we decided to study "wet ice" as a pure substance leaving impurities to a further step. The complexity of the deductions in this article seem to prove us right in our approach.

To begin with and by way of motivation, we review the literature on phase boundaries. Our view is that a phase boundary is not a surface of sudden change of mass, energy and entropy; it rather constitutes a thin layer across which mass, energy etc. change smoothly but rapidly between the densities of the adjacent bulk materials. The reason for this is the necessity of molecular adjustment between different molecular arrays of the adjacent bulk phases. Such a boundary layer possesses a thickness of a few molecular distances and therefore carries mass, energy and entropy which in a physically reasonable description should not be neglected. Moreover, due to its different molecular orderings, the boundary layer has constitutive properties which differ from those of the adjacent bulk materials. One manifestation of this is the appearance of surface tension. Another one is the existence of the boundary layer thickness which should be detectably by NMR-techniques of neutron scattering experiments.

More than 100 years ago FARADAY (1842, 1845) and TYNDALL (1858a,b) proposed that the surface of ice in air, that is not too far below the melting point, is covered by a thin liquid-like layer. With this hypothesis many of the peculiar mechanical ice properties such as *regelation* and the *abnormally small coefficient of friction* could be qualitatively understood. Since that time, other explanations followed: *pressure melting* (THOMSON, 1861), *local frictional heating* (BOWDEN & TABOR, 1954), polycrystalline creep (GLEN, 1955; STEINEMANN, 1958). WEYL (1951) opposed the idea that *pressure melting can explain regelation phenomena and small friction*. He claimed instead that due to the asymmetry and the polarizability of the oxygen O_2-ion and because of the non-polarizability of the H^+-proton an electrical double layer must develop on a water surface, the negative part of which forms the outer layer. *Ice is also covered by a liquid-like thin film which itself again is shielded by the electrical double layer.* Its thickness depends on the temperature and the material with which the ice is in contact. With these ideas the peculiarities of regelation, friction on solid supports (other than ice), *penetration*, but also other phenomena like separation of small water droplets in fogs and clouds, *ballo-electricity in* waterfalls and high *chemical reactivity of nascent water and ice surfaces*, as appear in cavitation and crack propagation, find natural explanations (see WEYL, 1951, and the literature given there). FLETCHER (1961, 1966) proved *the stability of smooth transitional surface layers between ice and water* as opposed to sudden discontinuous changes. He showed that the *free energy of the polarized and smoothed surface layer is smaller than the free energy of a layer of the same thickness* but undistorted bulk structure up to the dividing surface. Hence there can be no objection against the existence of smooth boundary layers between different bulk phases. Indirect experimental proofs for the existence of liquid-like films below 0°C are given by NAKAYA & MATSUMOTO (1954), and GUBLER (1982). Direct measurements of film thickness on flat ice by *proton channeling experiments* are due to GOLECKI & JACCARD (1977).

The smooth boundary layers between the bulk phases, ice and water, have a thickness of only a few molecular distances. Compared to the dimensions of the adjacent bulk materials, they are almost infinitely thin and can therefore be described as two-dimensional continua, representing mathematically singular surfaces with their own thermomechanical properties.These surfaces carry mass, momentum, energy and entropy. A thermodynamic theory of phase boundaries can thus be founded on the postulation of conservation laws for surface mass, momentum and energy and a balance law for surface entropy together with constitutive assumptions for surface fields and jump-relations of the bulk fields on and at the phase boundary. We will apply MÜLLER'S (1972) formulation of thermodynamics and extend it to two-dimensional continua. The theory furnishes generalized boundary conditions at phase boundaries between ice and water and field equations for the motion of the phase boundary and its density and temperature.

The description of boundary layers by mass-carrying-singular surfaces is not new. SCRIVEN (1960), GURTIN & MURDOCH (1975), MOECKEL (1974), LINDSAY & STRAUGHAN (1979) and GRAUEL (1980) developed thermodynamic theories on surfaces. This paper follows MOECKEL and LINDSAY & STRAUGHAN but extends their approach to the dynamics and statics of phase transitions. Other authors, mostly from physical chemistry, followed and extended the work of GIBBS (1929, 1948). They correlate the fields on singular surfaces with mean values of excess fields defined over the three-dimensional boundary layer. This approach has the advantage of allowing a physical interpretation of the surface fields. It further provides the possibility of calculating these fields by means of statistical physics. Prominent exponents in this branch of research are BUFF (1956), BUFF & SALTSBURG (1956a, b), SLATTERY (1967) and DEEMER & SLATTERY (1978).

We extend both theories, and perform a comparison of the two. This yields, via a dynamical consistency condition, new results for

- freezing and melting at curved interfaces,
- boundary conditions across phase boundaries,
- curvature dependence of surface tension,
- stability and instability of the boundary layer, and
- existence or non-existence of nucleation phenomena,

all in accord with observations. Moreover, within a first approximation to phase equilibrium all results are explicit in their curvature- and temperature dependence and contain at most one single phenomenological constant, the boundary layer thickness at some reference temperature. The emerging results, deduced from a complex dynamical theory, are therefore surprisingly simple and easily controllable either by experiments or statistical calculations; they provide better insight into the physics of phase transitions at the ice-water interface and the process of nucleation than do previous formulations.

2. Geometry and motion of surfaces

Let the phase boundary between ice and water (or a solid and its liquid phase) be represented by the non-material singular surface ς. This surface is generally curved and divides a region B_t at time t of Euclidean space into the subregions R_t^+ and R_t^- which are occupied by ice and water, respectively. The boundaries of $R_t^{\pm}$, denoted by $S^{\pm}$, join at the common closed curve C_t which is the intersection of the singular surface ς with the surface of B_t (Figure 1). We assume that ς can be covered by a single coordinate system ξ^α ($\alpha = 1, 2$). Any point $\mathbf{x}$ in B_t (and in particular on ς) can be referred to a Cartesian system of coordinates x^k ($k = 1, 2, 3$) relative to a fixed reference point in Euclidean space. At a reference time $t = t_R$, the ice-water material body B occupies the reference configuration R_{t_R} in Euclidean space, with singular surface Σ covered by the independent coordinates Ξ^Γ ($\Gamma = 1, 2$). The corresponding point $\mathbf{X}$ in R_{t_R} is referred to the coordinates X^K ($K = 1, 2, 3$), again relative to a fixed reference point in Euclidean space. Any smooth motion of ς in Euclidean space can then be described by the relations

$$\begin{aligned} \xi^\alpha &= \xi^\alpha_\varsigma(\Xi^\Gamma, t) \\ \mathbf{x} &= \chi(\xi^\alpha, t) = \chi(\xi^\alpha_\varsigma(\Xi^\Gamma, t), t) = \chi_R(\Xi^\Gamma, t), \end{aligned} \tag{2.1}$$

with the index R denoting the description relative to the reference configuration.

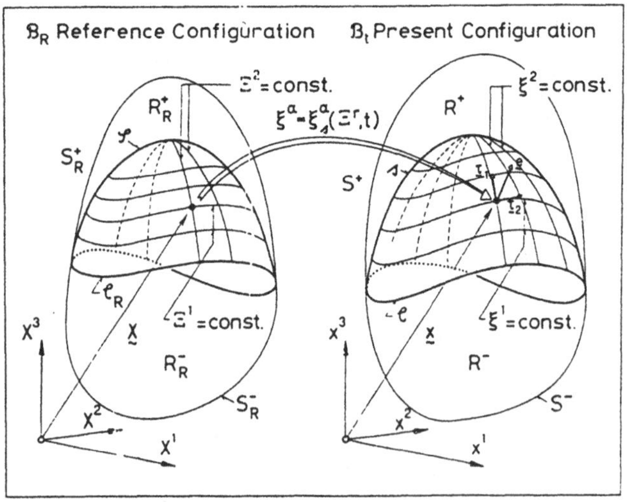

Fig. 1: Body with singular surface in its reference and present configurations, respectively.

The tangent vectors $\boldsymbol{\tau}_\alpha$ with components

$$\tau_\alpha^k := \frac{\partial x^k(\xi^\beta, t)}{\partial \xi^\alpha} \tag{2.2}$$

and the unit normal vector e with components

$$e^k := \left[\frac{(\boldsymbol{\tau}_1 \times \boldsymbol{\tau}_2)}{|\boldsymbol{\tau}_1 \times \boldsymbol{\tau}_2|}\right]^k = \frac{\varepsilon^k_{\ lm}\tau_1^l\tau_2^m}{\sqrt{\varepsilon^r_{\ st}\tau_1^s\tau_2^t\ \varepsilon_{rpq}\tau_1^p\tau_2^q}} \tag{2.3}$$

will be used as basis vectors. They define the metric tensor

$$g_{\alpha\beta} = \boldsymbol{\tau}_\alpha \cdot \boldsymbol{\tau}_\beta = \boldsymbol{\tau}_\beta \cdot \boldsymbol{\tau}_\alpha = g_{\beta\alpha}, \tag{2.4}$$

$$g^{\alpha\beta} g_{\beta\gamma} = \delta^\alpha_\gamma \tag{2.5}$$

and its inverse $g^{\alpha\beta}$, both of which are used (as usual) to raise and lower surface indices of vectors and tensors from co- to contra-variant components and vice versa. We use the summation convention according to which summation has to be performed over diagonally repeated greek ($\alpha = 1, 2$) indices.

The spatial variation of the basis vectors within ς are given by

$$\begin{aligned} \boldsymbol{\tau}_{\alpha;\beta} &= \boldsymbol{\tau}_{\alpha,\beta} - \Gamma^\mu_{\alpha\beta}\boldsymbol{\tau}_\mu = b_{\alpha\beta}\boldsymbol{e}, \\ \boldsymbol{e}_{;\beta} &= \boldsymbol{e}_{,\beta} = b^\alpha_\beta \boldsymbol{\tau}_\alpha, \end{aligned} \tag{2.6}$$

where $\Gamma^\mu_{\alpha\beta}$ is CHRISTOFFEL's symbol for the surface coordinates and $b_{\alpha\beta}$ the covariant components of the curvature tensor, *i.e.*

$$\begin{aligned} \Gamma^\mu_{\alpha\beta} &= g^{\mu\gamma}\boldsymbol{\tau}_\gamma \cdot \boldsymbol{\tau}_{\alpha,\beta} = \Gamma^\mu_{\beta\alpha}, \\ b_{\alpha\beta} &= \boldsymbol{e} \cdot \boldsymbol{\tau}_{\alpha,\beta} = b_{\beta\alpha} = -\boldsymbol{\tau}_\alpha \cdot \boldsymbol{e}_{,\beta}. \end{aligned} \tag{2.7}$$

Furthermore, commas denote partial derivatives and semicolons covariant differentiations with respect to surface coordinates. For covariant and contravariant components V_α, V^α of a surface vector and covariant, contravariant and mixed components $T_{\alpha\beta}, T^{\alpha\beta}, T^\alpha_{\cdot\beta}$ of a surface tensor these are defined by

$$\begin{aligned} V_{\alpha;\beta} &:= V_{\alpha,\beta} - \Gamma^\mu_{\alpha\beta}V_\mu, \\ V^\alpha_{\ ;\beta} &:= V^\alpha_{\ ,\beta} + \Gamma^\alpha_{\beta\mu}V^\mu, \\ T_{\alpha\beta;\gamma} &:= T_{\alpha\beta,\gamma} - \Gamma^\mu_{\alpha\gamma}T_{\mu\beta} - \Gamma^\mu_{\beta\gamma}T_{\alpha\mu}, \\ T^{\alpha\beta}_{\ \ ;\gamma} &:= T^{\alpha\beta}_{\ \ ,\gamma} + \Gamma^\alpha_{\gamma\mu}T^{\mu\beta} + \Gamma^\beta_{\gamma\mu}T^{\alpha\mu}, \\ T^\alpha_{\ \beta;\gamma} &:= T^\alpha_{\ \beta,\gamma} + \Gamma^\alpha_{\gamma\mu}T^\mu_{\ \beta} - \Gamma^\mu_{\beta\gamma}T^\alpha_{\ \mu}. \end{aligned} \tag{2.8}$$

Note that $g_{\alpha\beta;\gamma} = 0$ and $g^{\alpha\beta}{}_{;\gamma} = 0$.

The velocity $\boldsymbol{w}$ of the non-material surface is the time derivative of the motion at constant reference surface coordinates

$$\boldsymbol{w} = \partial_t \boldsymbol{\chi}_R = \boldsymbol{\tau}_\alpha \partial_t \xi^\alpha_\varsigma + \partial_t \boldsymbol{\chi} = \boldsymbol{\tau}_\alpha w^\alpha + w_n \boldsymbol{e}, \tag{2.9}$$

with $\partial_t := \partial/\partial t$,

$$w^\alpha = g^{\alpha\beta} \boldsymbol{\tau}_\beta \cdot \boldsymbol{w} = \partial_t \xi^\alpha_\varsigma \quad \text{and} \quad w_n = \boldsymbol{e} \cdot \boldsymbol{w}. \tag{2.10}$$

Having defined $\boldsymbol{w}$, we can calculate the rate of change of the basis vectors, and so the metric and curvature tensors. For later use we list these here without proof:

$$\partial_t \boldsymbol{\tau}_\alpha = w_{n,\alpha} \boldsymbol{e} - w_n b^\beta_\alpha \boldsymbol{\tau}_\beta, \qquad \partial_t \boldsymbol{e} = -g^{\alpha\beta} w_{n,\beta} \boldsymbol{\tau}_\alpha, \tag{2.11}$$

$$\partial_t g_{\alpha\beta} = -2 w_n b_{\alpha\beta}, \quad \partial_t g^{\alpha\beta} = 2 w_n b^{\alpha\beta}, \quad \partial_t b_{\alpha\beta} = w_{n;\alpha\beta} - w_n b^\gamma_\alpha b_{\gamma\beta}, \tag{2.12}$$

$$\begin{aligned} \dot{\boldsymbol{\tau}}_\alpha &= \partial_t \boldsymbol{\tau}_\alpha + \boldsymbol{\tau}_{\alpha,\beta} w^\beta = (\Gamma^\beta_{\alpha\gamma} w^\gamma - w_n b^\beta_\alpha) \boldsymbol{\tau}_\beta + (w_{n,\alpha} + b_{\alpha\beta} w^\beta) \boldsymbol{e}, \\ \dot{\boldsymbol{e}} &= \partial_t \boldsymbol{e} + \boldsymbol{e}_{,\gamma} w^\gamma = -(w_{n,\alpha} + b_{\alpha\gamma} w^\gamma) g^{\alpha\beta} \boldsymbol{\tau}_\beta, \end{aligned} \tag{2.13}$$

$$\begin{aligned} \dot{g}_{\alpha\beta} &= \partial_t g_{\alpha\beta} + g_{\alpha\beta,\gamma} w^\gamma = (\Gamma^\mu_{\alpha\gamma} g_{\mu\beta} + \Gamma^\mu_{\beta\gamma} g_{\mu\alpha}) w^\gamma - 2 w_n b_{\alpha\beta}, \\ \dot{g}^{\alpha\beta} &= \partial_t g^{\alpha\beta} + g^{\alpha\beta}{}_{,\gamma} w^\gamma = -(\Gamma^\alpha_{\mu\gamma} g^{\mu\beta} + \Gamma^\beta_{\mu\gamma} g^{\mu\alpha}) w^\gamma + 2 w_n b^{\alpha\beta}, \\ \dot{b}_{\alpha\beta} &= \partial_t b_{\alpha\beta} + b_{\alpha\beta,\gamma} w^\gamma = w_{n;\alpha\beta} - w_n b^\gamma_\alpha b_{\gamma\beta} + b_{\alpha\beta,\gamma} w^\gamma. \end{aligned} \tag{2.14}$$

In what follows we also use the two invariants of the curvature tensor

$$K_M = \frac{1}{2} b^\alpha_\alpha = \frac{1}{2}(K_1 + K_2), \qquad K_G = \det(b^\beta_\alpha) = K_1 K_2. \tag{2.15}$$

These are the mean and the gaussian curvatures of ς, respectively, where K_1, K_2 are the principal curvatures, the eigenvalues of b^β_α.

Using (2.2), the surface elements da and da_R in the present and reference configuration are, respectively

$$da = \sqrt{g}\, d\xi^1 d\xi^2, \quad da_R = \sqrt{G}\, d\Xi^1 d\Xi^2, \tag{2.16}$$

with

$$g = \det(g_{\alpha\beta}) \quad \text{and} \quad G = \det(G_{\Gamma\Delta}) \quad \text{and} \quad G_{\Gamma\Delta} = \frac{\partial \boldsymbol{\chi}_R}{\partial \Xi^\Gamma} \cdot \frac{\partial \boldsymbol{\chi}_R}{\partial \Xi^\Delta}.$$

Combining $(2.16)_{1,2}$ with

$$d\xi^1 d\xi^2 = det(\xi^\alpha_{\varsigma,\Gamma})\, d\Xi^1 d\Xi^2 = J_\varsigma\, d\Xi^1 d\Xi^2$$

yields

$$da = \sqrt{g/G}\, J_\varsigma\, da_R\,. \tag{2.17}$$

(2.17) relates the surface element in the present configuration with that in the reference configuration. Its time rate of change, following the surface ς, is expressible as

$$\frac{d}{dt}da = (w^{\alpha}{}_{;\alpha} - 2w_n K_M)da. \tag{2.18}$$

This formula is needed to establish the following transport relation for the surface. Let ψ_ς be a surface scalar. Then for smooth processes on ς

$$\frac{d}{dt}\int_{\varsigma_t} \psi_\varsigma \, da = \int_{\varsigma_t} \left[\frac{\partial \psi_\varsigma}{\partial t} + (\psi_\varsigma w^\alpha)_{;\alpha} - 2w_n K_M \psi_\varsigma\right] da. \tag{2.19}$$

Finally, we quote the divergence theorem for the surface. It reads

$$\int_{C_t} \phi^\alpha_\varsigma h_\alpha \, ds = \int_{\varsigma_t} \phi^\alpha_{\varsigma;\alpha} \, da, \tag{2.20}$$

where

$$h_\alpha = \boldsymbol{\tau}_\alpha \cdot (\boldsymbol{\tau}_\beta \times \boldsymbol{e}) \frac{\partial \xi^\beta_C}{ds} \tag{2.21}$$

is the unit normal vector in the tangent plane on ς along C (see Fig. 1), and s is the arc length parameter on C .

Clearly in order that all the above transformations be valid, the fields involved and the motion need to be continuously differentiable. In other words no discontinuities or singularities are assumed to arise on ς.

3. Balance laws on phase boundaries

The phase boundary between ice and water is represented by a moving non-material singular orientable surface. This is essentially GIBBS' (1929) idea. The general form of a balance relation on such a surface (Fig. 2) is given by:

$$\frac{d}{dt}\int_{\varsigma(t)} \psi_\varsigma \, da = -\int_{C(t)} \phi^\alpha_\varsigma h_\alpha \, ds - \int_{\varsigma(t)} [\![\boldsymbol{\phi} + \psi_v \otimes \boldsymbol{W}]\!] \cdot \boldsymbol{e} \, da + \int_{\varsigma(t)} (\pi_\varsigma + \sigma_\varsigma) \, da \tag{3.1}$$

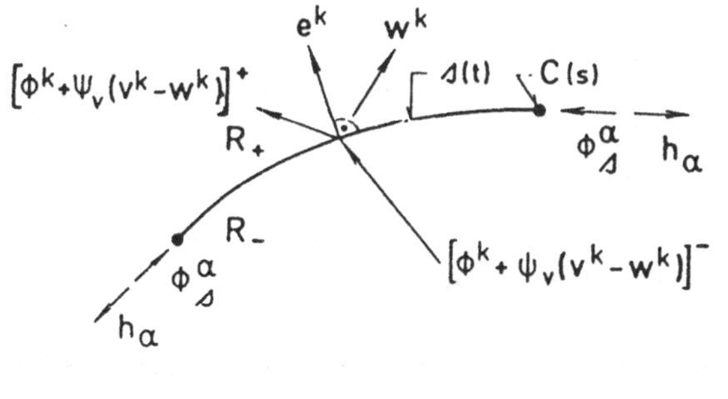

Fig. 2: Flux contribution on a non-material singular surface.

Here ds denotes the differential arc length along $C(t)$, which is oriented counterclockwise around the surface normal $\boldsymbol{e}$. This curve is chosen to be material in the sense that it possesses the same tangential velocity w^α as do material particles sitting instantaneously on the surface. The singular surface as a whole, however, is non-material, because matter may cross it when it represents a phase boundary. Moreover, da is the area element on $\varsigma(t)$, and d/dt denotes total time derivative following the velocity $\boldsymbol{w} = w_n \boldsymbol{e} + w^\alpha \boldsymbol{\tau}_\alpha$ of $\varsigma(t)$ in Euclidean space. We have also defined $\boldsymbol{W} := \boldsymbol{v} - \boldsymbol{w}$. The field quantities have the meaning:

$\pi_\varsigma, \sigma_\varsigma$ — surface production and supply rate densities of $\Psi(t)$,

$\boldsymbol{\phi}(\boldsymbol{x}, t)$ — nonconvective bulk influx of $\Psi(t)$ through the material surface of the bulk phase, which moves with velocity $\boldsymbol{v}$ in Euclidean space,

$\psi_\varsigma(\xi^\beta, t)$ — surface density of $\Psi(t)$,

$\phi_\varsigma^\alpha(\xi^\beta, t)$ — nonconvective influx rate of $\Psi(t)$ through the boundary curve $C(t)$ which moves with the surface velocity whose components are w^α,

$\psi_v(\boldsymbol{x}, t)$ — bulk density of $\Psi(t)$.

Lastly, we have

$$[\![\boldsymbol{\phi} + \psi_v \otimes \boldsymbol{W}]\!] \cdot \boldsymbol{e} := [\boldsymbol{\phi}_+ + \psi_v^+ \otimes \boldsymbol{W}_+] \cdot \boldsymbol{e} - [\boldsymbol{\phi}_- + \psi_v^- \otimes \boldsymbol{W}_-] \cdot \boldsymbol{e}$$

as the contribution of the bulk flux of $\Psi(t)$ through $\varsigma(t)$ to the balance of $\Psi(t)$ on $\varsigma(t)$, with $\boldsymbol{W}_\pm := \boldsymbol{v}_\pm - \boldsymbol{w}$. The quantities $\boldsymbol{\phi}_+$, $\psi_v^\pm$ and $\boldsymbol{v}_\pm$ represent the limiting values of the bulk fields $\boldsymbol{\phi}$, ψ and $\boldsymbol{v}$, respectively, as the singular surface is approached from R_+ and R_-, respectively. This "jump" contribution to the surface balance (3.1) relation is derivable from the balance equation for the entire region $(R_+ \cup \varsigma \cup R_-)(t)$. Our aim is the derivation of a local version of (3.1) when sufficient differentiability assumptions are satisfied. To this end, we first interchange in (3.1) the time derivative and the integration. This is done with the aid of the transport theorem (2.19). Secondly, the divergence theorem is applied to the surface integral term. This process yields

$$\int_{\varsigma(t)} [(\partial_t \psi_\varsigma) - 2\psi_\varsigma K_M w_n + (\phi_\varsigma^\alpha + \psi_\varsigma w^\alpha)_{;\alpha} + [\![\boldsymbol{\phi} + \psi_v \otimes \boldsymbol{W}]\!] \cdot \boldsymbol{e} - (\pi_\varsigma + \sigma_\varsigma)]\, da = 0. \qquad (3.2)$$

Since, beyond the differentiability assumptions, we assume that $\varsigma(t)$ is arbitrary, the integrand in (3.2) must also vanish, *i.e.*,

$$(\partial_t \psi_\varsigma) - 2\psi_\varsigma K_M w_n + (\phi_\varsigma^\alpha + \psi_\varsigma w^\alpha)_{;\alpha} = (\pi_\varsigma + \sigma_\varsigma) - [\![\boldsymbol{\phi} + \psi_v \otimes \boldsymbol{W}]\!] \cdot \boldsymbol{e}. \qquad (3.3)$$

This is the local form of a balance equation for any additive quantity ψ_ς defined on a moving non-material singular surface.

With the identifications of Table 1 the special balance equations for mass, momentum, energy and entropy are obtained:

quantity	ψ_ς	ϕ_ς^α	π_ς	σ_ς	ψ_v	ϕ
mass	ϱ_ς	0	0	0	ϱ	0
momentum	$\varrho_\varsigma \boldsymbol{w}$	$-\boldsymbol{t}_\varsigma^\alpha$	0	$\varrho_\varsigma \boldsymbol{g}$	$\varrho \boldsymbol{v}$	$-\boldsymbol{t}$
energy	$\varrho_\varsigma \varepsilon_\varsigma$	$q_\varsigma^\alpha - \boldsymbol{w} \cdot \boldsymbol{t}_\varsigma^\alpha$	0	$\varrho_\varsigma \boldsymbol{w} \cdot \boldsymbol{g} + \varrho_\varsigma r_\varsigma$	$\varrho \varepsilon$	$\boldsymbol{q} - \boldsymbol{t}^{\mathrm{T}} \boldsymbol{v}$
entropy	$\varrho_\varsigma s_\varsigma$	ϕ_ς^α	π_ς	σ_ς	ϱs	$\boldsymbol{\phi}$

Table 1: Thermodynamic fields for surface balance relations.

The specific quantities in the first column are surface mean density ϱ_ς, surface momentum density $\varrho_\varsigma \boldsymbol{w}$, surface energy density $\varrho_\varsigma \varepsilon_\varsigma := \varrho_\varsigma(u_\varsigma + \frac{1}{2}\boldsymbol{w}\cdot\boldsymbol{w})$, and and surface entropy density $\varrho_\varsigma s_\varsigma$, all per unit area. u_ς and s_ς are the surface internal energy and surface entropy per unit mass, respectively. The second column contains the non-convective surface fluxes per unit time and unit length relative to particles moving with velocity w^α along moving material lines. $\boldsymbol{t}_\varsigma^\alpha$ is the surface stress, $q_\varsigma^\alpha - \boldsymbol{w}\cdot\boldsymbol{t}_\varsigma^\alpha$ the energy flux, where q_ς^α is the surface heat flux, and ϕ_ς^α is the surface entropy flux. The corresponding mass flux vanishes, because $C(t)$ is referred to moving material lines. The third column lists the production densities on the surface; those of mass, momentum and energy vanish since these quantities are conserved. The entropy production density in the surface is assumed to satisfy the inequality $\pi_\varsigma \geq 0$, such that entropy is produced within the phase boundary whenever phase-transition processes are irreversible. In the fourth column the supply densities due to external fields such as gravity and electromagnetism are collected. $\boldsymbol{g}$ is the gravitational force per unit mass acting on the surface particles, r_ς is the heat supply due to radiative absorption per unit mass and unit time and σ_ς is the entropy supply density, which will be specified later. The bulk densities in the fifth column are mass density ϱ, momentum density $\varrho\boldsymbol{v}$, where $\boldsymbol{v}$ is the material velocity, energy density $\varrho\varepsilon := \varrho(u + \frac{1}{2}\boldsymbol{v}\cdot\boldsymbol{v})$ and entropy density ϱs in the adjacent bulk materials. u and s denote the specific internal energy and the specific entropy, respectively. $\boldsymbol{t}$ is CAUCHY's stress tensor, $\boldsymbol{q}$ the heat flux vector and $\boldsymbol{\phi}$ the entropy flux vector. Later we shall also briefly discuss the balance law of angular momentum, but because we will eventually regard the phase change surface as a non-polar continuum we do not include it in Table 1.

The surface stress $\boldsymbol{t}_\varsigma^\alpha$ can be decomposed according to

$$\boldsymbol{t}_\varsigma^\alpha := \boldsymbol{\tau}_\beta S^{\beta\alpha} + \boldsymbol{e} S^\alpha \ . \tag{3.4}$$

$S^{\beta\alpha}$ are the contravariant components of the tensor of surface tension, traction being positive; β indicates the direction of force and α the coordinate line to whose unit length the force is referred; S^α is a surface shear stress in the direction normal to the surface and refers to the unit length of the coordinate lines. As we will make plausible below, it is a consequence of the conservation of moment of momentum that

$$S^{\beta\alpha} = S^{\alpha\beta} \qquad \text{and} \qquad S^\alpha = 0. \tag{3.5}$$

Substituting then from the above Table 1 into equation (3.3), using the decomposition (3.4), the conservation of angular momentum in the form (3.5), and the convective time derivative

$$\dot{\psi}_\varsigma = \partial_t \psi_\varsigma + w^\alpha \psi_{\varsigma,\alpha} \, , \tag{3.6}$$

we obtain specific balance relations for surface mass, linear surface momentum, energy and entropy:

$$\begin{aligned}
\dot{\varrho}_\varsigma &= -\varrho_\varsigma (w^\alpha{}_{;\alpha} - 2K_M w_n) - \mathcal{R} \\
\varrho_\varsigma \dot{w}^\alpha &= -\varrho_\varsigma \Gamma^\alpha_{\beta\gamma} w^\beta w^\gamma + \varrho_\varsigma w_n (g^{\alpha\beta} w_{n,\beta} + 2b^\alpha_\beta w^\beta) + S^{\alpha\beta}{}_{;\beta} + \varrho_\varsigma \boldsymbol{\tau}^\alpha \cdot \boldsymbol{g} - C^\alpha \\
\varrho_\varsigma \dot{w}_n &= -\varrho_\varsigma w^\alpha (w_{n,\alpha} + b_{\alpha\beta} w^\beta) + S^{\alpha\beta} b_{\beta\alpha} + \varrho_\varsigma \boldsymbol{e} \cdot \boldsymbol{g} - C_n \\
\varrho_\varsigma \dot{u}_\varsigma &= -q^\alpha_{\varsigma;\alpha} + S^{\alpha\beta}(w_{(\alpha;\beta)} - w_n b_{\alpha\beta}) + \varrho_\varsigma r_\varsigma - \mathcal{U} \\
\varrho_\varsigma \dot{s}_\varsigma &= -\phi^\alpha_{\varsigma;\alpha} + \pi_\varsigma + \sigma_\varsigma - \mathcal{S}
\end{aligned} \tag{3.7}$$

and the jump quantities

$$\begin{aligned}
\mathcal{R} &:= [\![\varrho \boldsymbol{W}]\!] \cdot \boldsymbol{e} \\
C^\alpha &:= \boldsymbol{\tau}^\alpha \cdot [\![\varrho \boldsymbol{W} \otimes \boldsymbol{W} - \boldsymbol{t}]\!] \cdot \boldsymbol{e} \\
C_n &:= \boldsymbol{e} \cdot [\![\varrho \boldsymbol{W} \otimes \boldsymbol{W} - \boldsymbol{t}]\!] \cdot \boldsymbol{e} \\
\mathcal{U} &:= [\![\boldsymbol{q} - \boldsymbol{W} \cdot \boldsymbol{t} + \varrho((u - u_\varsigma) + \tfrac{1}{2} W^2) \boldsymbol{W}]\!] \cdot \boldsymbol{e} \\
\mathcal{S} &:= [\![\boldsymbol{\phi} + \varrho (s - s_\varsigma) \boldsymbol{W}]\!] \cdot \boldsymbol{e},
\end{aligned} \tag{3.8}$$

where $W^2 := \boldsymbol{W} \cdot \boldsymbol{W} = (\boldsymbol{v} - \boldsymbol{w}) \cdot (\boldsymbol{v} - \boldsymbol{w})$. Equation $(3.7)_1$, represents the surface mass balance, $(3.7)_{2,3}$ are the momentum equations within the phase change surface and perpendicular to it; furthermore, $(3.7)_4$ is the balance of internal surface energy and $(3.7)_5$ constitutes the balance of surface entropy. As usual, semicolons denote covariant differentiations, and conservation of moment of momentum has been invoked, that is $S^\alpha = 0$, and $S^{\alpha\beta} = S^{\beta\alpha}$. In the following we shall also employ the *surface stretching*

$$D_{\alpha\beta} := w_{(\alpha;\beta)} - w_n b_{\alpha\beta} = (g_{\alpha\gamma} w^\gamma)_{;\beta} - w_n b_{\alpha\beta}. \tag{3.9}$$

Finally, for irreversible phase change processes the entropy production density must be non-negative

$$\pi_\varsigma \geq 0 \qquad \text{(second law)}. \tag{3.10}$$

Remark on angular momentum. Assuming a non-polar concept for the bulk solid and fluid does not imply that the phase change surface be also a classical non-polar continuum. Indeed the boundary layer concept that will be introduced in the next section shows that, strictly, the phase change surface ought to be treated as a two dimensional COSSERAT surface, like a shell. Thus if one introduces the quantities

$$\begin{array}{ll} \varrho_\varsigma \boldsymbol{a}_\varsigma & \text{surface spin density,} \\ -\boldsymbol{M}_\varsigma^\alpha & \text{surface spin non-convective flux,} \\ \varrho_\varsigma \boldsymbol{m}_\varsigma & \text{surface spin supply density,} \end{array} \tag{3.11}$$

$$\begin{array}{ll} \boldsymbol{x} \times \varrho_\varsigma \boldsymbol{w} & \text{surface moment of momentum density,} \\ -\boldsymbol{x} \times \boldsymbol{t}_\varsigma^\alpha & \text{non-convective surface moment of momentum flux} \\ & \text{(negative torque of surface force density),} \\ \boldsymbol{x} \times \varrho_\varsigma \boldsymbol{g} & \text{surface moment of momentum supply density} \\ & \text{(torque per unit area of gravitational force),} \end{array} \tag{3.12}$$

and postulates a conservation law for the respective sums of the spin- and moment of momentum quantities (angular momentum) an analogous procedure as above shows that the following *spin balance relation* must hold:

$$\partial_t(\varrho_\varsigma \boldsymbol{a}_\varsigma) - 2\varrho_\varsigma \boldsymbol{a}_\varsigma K_M w_n + (\varrho_\varsigma \boldsymbol{a}_\varsigma w^\alpha - \boldsymbol{M}_\varsigma^\alpha)_{;\alpha} = \varrho_\varsigma \boldsymbol{m}_\varsigma + S^{[\beta\alpha]} \boldsymbol{\tau}_\alpha \times \boldsymbol{\tau}_\beta + S^\alpha \boldsymbol{\tau}_\alpha \times \boldsymbol{e} \, . \tag{3.13}$$

It follows from this equation that moment of momentum is conserved if and only if all spin quantities (3.11) vanish and

$$S^{[\alpha\beta]} = 0, \qquad S^\alpha = 0. \tag{3.14}$$

It is a consequence of 'micro'-physics that the requirements (3.14) cannot strictly hold. For, according to WEYL (1951) and FLETCHER (1961, 1966) the ice-water interface is electrically polarized. Consequently, the interface is a polar material which in general possesses internal angular momentum and, if it is modelled by a singular surface, also surface spin. In equilibrium and close to equilibrium, however, the direction of polarization is fixed or nearly fixed to the normal direction of the interface. In this case the macroscopic spin $\boldsymbol{a}_\varsigma$ vanishes. Spin supply $\boldsymbol{m}_\varsigma$ due to heat radiation requires coupling between angular momentum of the electromagnetic field of heat radiation and molecular spin of molecules via the gyromagnetic ratio and leads only to a macroscopic effect for macroscopically polarized radiation and optically active molecules (helix-structure of molecules). Heat radiation is macroscopically unpolarized and water molecules are optically inactive molecules. Hence, $\boldsymbol{m}_\varsigma = \boldsymbol{0}$ and (3.13) reduces to

$$- \boldsymbol{M}_{\varsigma;\alpha}^\alpha = S^{[\beta\alpha]} \boldsymbol{\tau}_\alpha \times \boldsymbol{\tau}_\beta + S^\alpha \boldsymbol{\tau}_\alpha \times \boldsymbol{e} \, . \tag{3.15}$$

If we nonetheless use (3.14) in ensuing developments, we should be aware that interphase layer thicknesses are thick enough that electrical polarization is overshadowed by "statistical averaging" over the thickness of the phase change region.

4. Physical interpretation of surface fields

So far the phase boundary was considered as a mathematical singular surface which is equipped with its own constitutive properties. Physically, however, the phase boundary is actually a thin layer across which bulk fields change "smoothly," although this change may be considerable and takes place only over a few Å, see Fig. 3. Also, because of the polarizability of ice and water, foreign molecules may be absorbed to the phase boundary. The mass density across the layer may thus have a distribution as shown by the dotted line of Fig. 4.

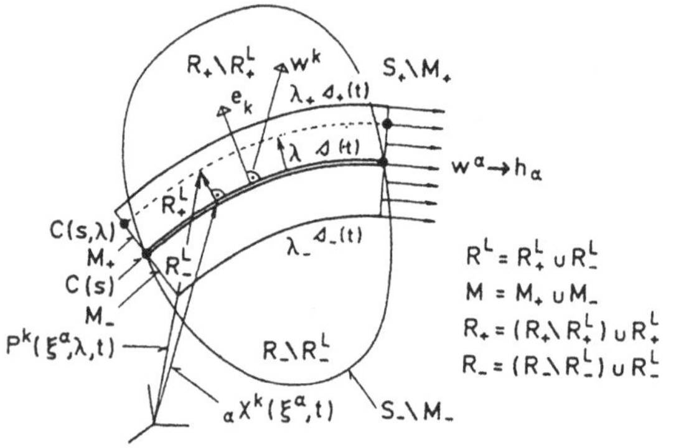

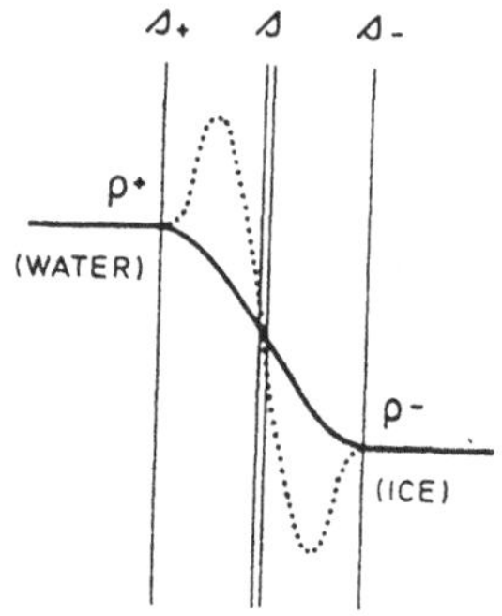

Fig. 3: Two-dimensional schematic of a thin boundary layer between two phases (for notation see Fig. 2).

Fig. 4: Possible density distributions across the phase boundary between ice and water. ······ with absorption, ——— without absorption.

So far, the concept of a surface field was employed. Here, we express these quantities in terms of mean values of bulk fields over a boundary layer, which, a posteriori, justifies the mathematical approach of the previous chapter. The mathematical model we use is based on the work of Buff (1956) and Deemer & Slattery (1978), which we have had to modify to account properly for the effect of curvature at the interface boundary.

4.1 Equivalence of the boundary layer and the singularity surface viewpoints

We start with the general form of a local (*i.e.*, differential) balance relation

$$(\partial_t \psi_v) + \operatorname{div}(\psi_v \boldsymbol{v} + \boldsymbol{\phi}) = \pi_v + \sigma_v \tag{4.1}$$

for an additive thermodynamic quantity Ψ_v in a three-dimensional ("bulk") region R with

$$\Psi_v(t) := \int_{R(t)} \psi_v(\boldsymbol{x}, t)\, dv \ ,$$

where $\psi_v, \boldsymbol{v}$ and $\boldsymbol{\phi}$ have the same meaning as before, and π_v and σ_v are the production rate and supply rate densities of Ψ_v in R at time t. Integrating (4.1) over the non-material boundary layer region $R^L = R^L_+ \cup R^L_-$ with surface $\partial R^L = M \cup \varsigma_+ \cup \varsigma_-$, which move with the velocity $\boldsymbol{w}$ of the dividing singular surface ς that intersects R^L (see Fig. 3), we obtain

$$\frac{d}{dt}\int_{R^L(t)} \psi_v \, dv + \int_{\partial R^L(t)} (\psi_v \boldsymbol{W} + \boldsymbol{\phi}) \cdot \boldsymbol{n} \, da = \int_{R^L(t)} (\pi_v + \sigma_v) \, dv, \tag{4.2}$$

where $\boldsymbol{n}$ is the outer unit normal to the surface R^L. The reduction of (4.2) to an expression involving integrals over the dividing surface ς is largely technical and requires arguments of differential geometry. Physically one must recognize that the surface fields are formed as the excess fields over the boundary layer of the true fields and the extensions of the field values at the boundary layer edge. Denoting these extended fields by subscripts $\pm$ (depending upon on which side they apply) the balances

$$\frac{d}{dt}\int_{R^L_\pm(t)} \psi_v^\pm \, dv + \int_{\partial R^L_\pm(t)} (\psi_v^\pm \boldsymbol{W}^\pm + \boldsymbol{\phi}^\pm) \cdot \boldsymbol{n} \, da = \int_{R^L_\pm(t)} (\pi_v^\pm + \sigma_v^\pm) \, dv \tag{4.3}$$

can be formed and subtracted from the respective portions of (4.2). What then remains is the division of the volume integration into successive integrations perpendicular to the singular surface ς (coordinate λ, see Fig. 3) and a second integration within the surface. For the difference of (4.2) and (4.3) this process yields

$$\frac{d}{dt}\int_{\varsigma(t)} \psi_\varsigma \, da = \int_{\varsigma(t)} \{\pi_\varsigma + \sigma_\varsigma - [\![\psi_v \boldsymbol{W} + \boldsymbol{\phi}]\!] \cdot \boldsymbol{e}\} \, da - \int_{C(t)} \phi_\varsigma^\alpha h_\alpha \, ds \tag{4.4}$$

in which

$$\begin{aligned}
\psi_\varsigma &:= \int_{\lambda_-}^{\lambda_+} (\psi_v - \psi_v^\pm) \sqrt{\frac{g(\lambda)}{g}} \, d\lambda, \\
\pi_\varsigma &:= \int_{\lambda_-}^{\lambda_+} (\pi_v - \pi_v^\pm) \sqrt{\frac{g(\lambda)}{g}} \, d\lambda, \\
\sigma_\varsigma &:= \int_{\lambda_-}^{\lambda_+} (\sigma_v - \sigma_v^\pm) \sqrt{\frac{g(\lambda)}{g}} \, d\lambda, \\
\phi_\varsigma^\alpha &:= \int_{\lambda_-}^{\lambda_+} \boldsymbol{\tau}^\alpha(\lambda) \left[(\psi_v \boldsymbol{W} - \psi_v^\pm \boldsymbol{W}^\pm) + (\boldsymbol{\phi} - \boldsymbol{\phi}^\pm)\right] \sqrt{\frac{g(\lambda)}{g}} \, d\lambda,
\end{aligned} \tag{4.5}$$

$$\psi_v^\pm = \begin{cases} \psi_v^+ & \text{for } \lambda > 0, \\ \psi_v^- & \text{for } \lambda < 0, \end{cases} \quad \textit{etc.} \tag{4.6}$$

and

$$\sqrt{\frac{g(\lambda)}{g}} = (1 - 2\lambda K_M + \lambda^2 K_G), \tag{4.7}$$

$$\tau^\alpha(\lambda) = \frac{1}{1 - 2\lambda K_M + \lambda^2 K_G}[\delta^\alpha_\beta - \lambda(2K_M\delta^\alpha_\beta - b^\alpha_\beta)]\tau^\alpha, \tag{4.8}$$

$$h_\alpha = \tau_\alpha \cdot (\tau_\beta \times e)\frac{d\xi^\beta_C}{ds}. \tag{4.9}$$

The factor $\sqrt{g(\lambda)/g}$ accounts for the change in surface area parallel to ς and expresses it in terms of λ and the mean and gaussian curvatures of ς, while $\tau^\alpha(\lambda) = g^{\alpha\beta}(\lambda)\tau_\beta(\lambda)$ is the dual of the tangent vector $\tau_\alpha(\lambda)$ of the surface parallel to ς at the distance λ. Finally, h_α is the outer unit normal vector to the bounding curve $C(t)$ in the local plane of the surface $\varsigma(t)$, where s is the arc length along $C(t)$ (see Figs. 2, 3).

Equations (4.4) have the form of the general balance law (3.1) on non-material singular surfaces, and the definitions (4.5) establish the correspondence. Accordingly, the surface fields are weighted mean values of bulk excess fields over the boundary layer of thickness $\lambda_+ - \lambda_-$. For flat surfaces with $g_{\alpha\beta}(\lambda) = g_{\alpha\beta}$ and $\tau^\alpha(\lambda) = \tau^\alpha$, the curvature-dependent weighting disappears.

It is worthwile to pause here and to emphasize why the surface fields must be expressed as the weighted integrals over the thickness of the phase change surface of the *excess fields*, i. e. the difference $\psi_v - \psi_v^\pm$. When treating the phase change region as a singular surface, the bulk regions extend to the very edge of this surface, and thus in a respective boundary value problem ψ_v will be determined in the entire regions R^+ and R^-. Would, therefore, in the definitions of the surface fields the extended bulk fields $\psi_v^\pm$, etc. be omitted (see (4.5)), they would, in the phase change boundary layer be counted twice, so that an incorrect surface theory would emerge. In the literature this point seems to have caused some confusion.

We now decompose the bulk velocity, heat flux, entropy flux and CAUCHY stress within the layer into tangential and normal components with respect to the singular surfaces:

$$\begin{aligned}
v(\lambda) &= v^\alpha(\lambda)\tau_\alpha + v_n(\lambda)e, \\
q(\lambda) &= q^\alpha(\lambda)\tau_\alpha + q_n(\lambda)e, \\
\phi(\lambda) &= \phi^\alpha(\lambda)\tau_\alpha + \phi_n(\lambda)e, \\
t(\lambda) &= t^{\alpha\beta}(\lambda)\tau_\alpha \otimes \tau_\beta + t^\alpha_{\cdot n}(\lambda)\tau_\alpha \otimes e + t^{\cdot\alpha}_n(\lambda)e \otimes \tau_\alpha + t_n(\lambda)e \otimes e.
\end{aligned} \tag{4.10}$$

If we substitute the specific fields in Table 1 into (4.5), using (3.4), (3.14) and (4.6) - (4.8), then the following expressions for the surface fields are obtained:

(i) *for the balance law of mass:*

$$\varrho_\varsigma = \int_{\lambda_-}^{\lambda_+} (\varrho - \varrho^\pm)(1 - 2\lambda K_M + \lambda^2 K_G)\,d\lambda,$$

surface mass density

$$
\begin{aligned}
\phi_\varsigma^\alpha &= \int_{\lambda_-}^{\lambda_+} (\varrho\boldsymbol{v} - \varrho^\pm\boldsymbol{v}^\pm)\cdot\boldsymbol{\tau}^\beta[\delta_\beta^\alpha - \lambda(2K_M\delta_\beta^\alpha - b_\beta^\alpha)]\,d\lambda \\
&\quad -(\boldsymbol{w}\cdot\boldsymbol{\tau}^\beta)\int_{\lambda_-}^{\lambda_+} (\varrho - \varrho^\pm)[\delta_\beta^\alpha - \lambda(2K_M\delta_\beta^\alpha - b_\beta^\alpha)]\,d\lambda, \qquad (4.11)
\end{aligned}
$$

surface mass flux,

$$
\pi_\varsigma = 0, \quad \sigma_\varsigma = 0,
$$

surface mass production rate and supply rate densities.

Recall now that we chose the surface coordinates as lines that are frozen to the motion of the surface particles. This choice requires that $\phi_\varsigma^\alpha = 0$. Consequently

$$
w^\beta \int_{\lambda_-}^{\lambda_+} (\varrho - \varrho^\pm)[\delta_\beta^\alpha - \lambda(2K_M\delta_\beta^\alpha - b_\beta^\alpha)]\,d\lambda = \int_{\lambda_-}^{\lambda_+} (\varrho v^\beta - \varrho^\pm v_\pm^\beta)[\delta_\beta^\alpha - \lambda(2K_M\delta_\beta^\alpha - b_\beta^\alpha)]\,d\lambda\,. \qquad (4.12)
$$

The tangential velocity $w^\beta = \boldsymbol{w}\cdot\boldsymbol{\tau}^\beta$ of the surface is determined by the mass density and the tangential components $v^\beta = \boldsymbol{v}\cdot\boldsymbol{\tau}^\beta$ of the material velocity field within the boundary layer.
(ii) *for the surface momentum balance:*

$$
\begin{aligned}
\varrho_\varsigma w_n &= \int_{\lambda_-}^{\lambda_+} (\varrho v_n - \varrho^\pm v_n^\pm)(1 - 2\lambda K_M + \lambda^2 K_G)\,d\lambda, \\
\varrho_\varsigma w^\alpha &= \int_{\lambda_-}^{\lambda_+} (\varrho v^\alpha - \varrho^\pm v^\alpha)(1 - 2\lambda K_M + \lambda^2 K_G)\,d\lambda,
\end{aligned}
$$

surface momentum density,

$$
\begin{aligned}
S^{\alpha\beta} &= \int_{\lambda_-}^{\lambda_+} [(t^{\alpha\gamma} - t_\pm^{\alpha\gamma}) - (\varrho W^\alpha W^\gamma - \varrho^\pm W_\pm^\alpha W_\pm^\gamma)][\delta_\gamma^\beta - \lambda(2K_M\delta_\gamma^\beta - b_\gamma^\beta)]\,d\lambda, \qquad (4.13) \\
S^\alpha &= \int_{\lambda_-}^{\lambda_+} [(t_n^\gamma - t_{n\pm}^\gamma) - (\varrho W_n W^\gamma - \varrho^\pm W_n^\pm W_\pm^\gamma)][\delta_\gamma^\beta - \lambda(2K_M\delta_\gamma^\beta - b_\gamma^\beta)]\,d\lambda,
\end{aligned}
$$

surface stress,

$$
\boldsymbol{\pi}_\varsigma = \mathbf{0}, \quad \boldsymbol{\sigma}_\varsigma = \varrho_\varsigma \boldsymbol{g},
$$

surface momentum production rate and supply rate densities.

Notice that $(4.13)_2$ defines the tangential velocity w^α of the singular surface in terms of the excess bulk momentum density. However, (4.12) must also hold, and this yields restrictions that will be discussed below.

(iii) *for surface energy balance:*

$$\varrho_\varsigma u_\varsigma = \int_{\lambda_-}^{\lambda_+} [\varrho(u + \frac{1}{2}W^2) - \varrho^\pm(u^\pm + \frac{1}{2}W_\pm^2)]\,(1 - 2\lambda K_M + \lambda^2 K_G)\, d\lambda,$$

surface internal energy density,

$$\begin{aligned} q_\varsigma^\alpha = \int_{\lambda_-}^{\lambda_+} & [(q^\gamma - q_\pm^\gamma) - (W_\beta t^{\beta\gamma} - W_\beta^\pm t_\pm^{\beta\gamma}) - (W_n t_n^\gamma - W_n^\pm t_{n\pm}^\gamma) \\ & +(\varrho(u + \tfrac{1}{2}W^2)W^\gamma - \varrho^\pm(u^\pm + \tfrac{1}{2}W_\pm^2)W_\pm^\gamma][\delta_\gamma^\beta - \lambda(2K_M\delta_\gamma^\beta - b_\gamma^\beta)]\, d\lambda, \end{aligned} \tag{4.14}$$

surface heat flux,

$$\varrho_\varsigma \pi_\varsigma = 0, \quad \varrho_\varsigma r_\varsigma = \int_{\lambda_-}^{\lambda_+} (\varrho r - \varrho^\pm r^\pm)\,(1 - 2\lambda K_M + \lambda^2 K_G)\, d\lambda,$$

surface energy production rate and supply rate densities.

These relations show that the surface density of the internal energy $\varrho_\varsigma u_\varsigma$ is the curvature weighted mean value across the phase boundary of the excess-internal energy plus the diffusive kinetic energy density within the layer. Further, the surface heat flux is a (differently weighted) mean value of the excess-bulk heat flux, relative stress power and relative convective energy flux inside the layer. The mathematical constructions of these expressions have a formal similarity to corresponding quantities in thermodynamic mixture theory, Müller (1972).

(iv) *for surface entropy balance:*

$$\varrho_\varsigma s_\varsigma := \int_{\lambda_-}^{\lambda_+} (\varrho s - \varrho^\pm s^\pm)(1 - 2\lambda K_M + \lambda^2 K_G)\, d\lambda,$$

surface entropy density,

$$\pi_\varsigma := \int_{\lambda_-}^{\lambda_+} (\pi_v - \pi_v^\pm)(1 - 2\lambda K_M + \lambda^2 K_G)\, d\lambda, \tag{4.15}$$

surface entropy production rate density,

$$\sigma_\varsigma := \int_{\lambda_-}^{\lambda_+} (\sigma_v - \sigma_v^\pm)(1 - 2\lambda K_M + \lambda^2 K_G)\, d\lambda,$$

surface entropy supply rate density

$$\phi_\varsigma^\alpha \quad := \quad \int_{\lambda_-}^{\lambda_+} [(\phi^\beta - \phi_\pm^\beta) + (\varrho s W^\beta - \varrho^\pm s^\pm W_\pm^\beta)][\delta_\gamma^\beta - \lambda(2K_M\delta_\gamma^\beta - b_\gamma^\beta)]\, d\lambda,$$

surface entropy flux.

In the interest of emphasizing the physics, we have omitted the detailed derivation of these expressions, which can be found in Alts & Hutter (1986).

4.2 Consistency conditions

The requirement of vanishing transverse shear stress $S^\alpha = 0$, the first necessary condition that angular momentum be conserved, corresponds to the membrane approximation, the second being the symmetry requirement $S^{\alpha\beta} = S^{\beta\alpha}$ of the surface tension. According to $(4.13)_{3,4}$ imposition of these requirements corresponds to three constraints on the possible profiles of the Cauchy stress; these are called the *dynamical consistency conditions*. However, even when these conditions are fulfilled, the emerging theory is more general than the classical surface tension models which require isotropy of $S^{\alpha\beta}$. A quantity corresponding to it is the trace $\frac{1}{2}S_\alpha^\alpha$, or

$$\sigma := \frac{1}{2}S_\alpha^\alpha = \frac{1}{2}S^{\alpha\beta}g_{\beta\alpha} = \int_{\lambda_-}^{\lambda_+} [(t^{\alpha\gamma} - t_\pm^{\alpha\gamma}) - (\varrho W^\alpha W^\gamma - \varrho^\pm W_\pm^\alpha W_\pm^\gamma)][g_{\gamma\alpha} - \lambda(2K_M g_{\gamma\alpha} - b_{\gamma\alpha})]\, d\lambda. \tag{4.16}$$

The other constraints have already been mentioned in connection with relations (4.12) and $(4.13)_2$. Indeed, eliminating w^α between these quantities results in the relation

$$b_\alpha^\beta I^\alpha - K_G\delta_\alpha^\beta II^\alpha + K_G(2K_M\delta_\alpha^\beta - b_\alpha^\beta)III^\alpha = 0 \quad (\beta = 1,2), \tag{4.17}$$

where I^α, II^α and III^α are defined by

$$\begin{aligned} I^\alpha &:= \int_{\lambda_-}^{\lambda_+} \Upsilon^\alpha\, d\lambda \int_{\lambda_-}^{\lambda_+} R\lambda\, d\lambda - \int_{\lambda_-}^{\lambda_+} \Upsilon^\alpha\lambda\, d\lambda \int_{\lambda_-}^{\lambda_+} R\, d\lambda\,, \\ II^\alpha &:= \int_{\lambda_-}^{\lambda_+} \Upsilon^\alpha\, d\lambda \int_{\lambda_-}^{\lambda_+} R\lambda^2\, d\lambda - \int_{\lambda_-}^{\lambda_+} \Upsilon^\alpha\lambda^2\, d\lambda \int_{\lambda_-}^{\lambda_+} R\, d\lambda\,, \\ III^\alpha &:= \int_{\lambda_-}^{\lambda_+} \Upsilon^\alpha\lambda\, d\lambda \int_{\lambda_-}^{\lambda_+} R\lambda^2\, d\lambda - \int_{\lambda_-}^{\lambda_+} \Upsilon^\alpha\lambda^2\, d\lambda \int_{\lambda_-}^{\lambda_+} R\lambda\, d\lambda, \end{aligned} \tag{4.18}$$

respectively, with

$$\Upsilon^\alpha(\lambda) := \varrho(\lambda)v^\alpha(\lambda) - \varrho^\pm v_\pm^\alpha, \qquad R(\lambda) := \varrho(\lambda) - \varrho^\pm\,. \tag{4.19}$$

For flat phase boundaries, relations (4.17) are trivially satisfied. For curved phase boundaries, however, they represent two conditions on λ_+ and λ_- and fix the positions of ς_+ and ς_-

relative to the singular surface $\varsigma(t)$ at $\lambda = 0$. Equivalent parameters are the thickness $d_\varsigma := \lambda_+ - \lambda_-$ of the boundary layer and $a_\varsigma = \lambda_+ + \lambda_-$, a measure of "eccentricity." Given the position of the singular surface ς (at $\lambda = 0$), its curvature, the distributions $\varrho(\lambda)$ and $v^\alpha(\lambda)$ of the density and the tangential velocity within the layer and their values $\varrho^\pm$ and $v^\alpha_\pm$ in the adjacent bulk phases, the thickness d_ς and the eccentricity a_ς emerge as functions of the curvature of ς and functionals of the distributions $\varrho(\lambda)$ and $v^\alpha(\lambda)$. Because the actual distributions of ϱ and v^α across the phase boundary layer are not known, they can be chosen such as to make d_ς independent of the curvature and a_ς dependent on the curvature only. This guarantees that the bounding surfaces ς_+ and ς_- are parallel to the singular surface ς and that the tangential velocity of the singular surface satisfies (4.12) and $(4.13)_2$. Only under these conditions does the singular surface represent a valid continuum physical representation of the layer of phase transition.

5. Calculation of surface fields from boundary-layer field distributions

From a thermodynamics point of view the next step to follow would now be the postulation of constitutive relations and their reduction by thermodynamic principles. We postpone this part and proceed by presenting results about surface fields as they emerge, when the structure of the boundary layer fields is prescribed. We believe that in this way a better feeling for the physics is obtained.

We now assume that the membrane approximation is valid ($S^{[\alpha\beta]} = 0 = S^\alpha$) and postulate the distribution of the bulk fields over the thickness coordinate λ. Cubic fields are the simplest possible representations, as it can be demonstrated (ALTS & HUTTER, 1988a) that linear distributions result in internal inconsistencies when trying to satisfy the relation (4.17). Thus higher order representations of the boundary layer profiles must be postulated. Thus, let f be any field quantity satisfying the postulated boundary layer profile (Fig. 5)

$$f(\xi^\alpha,\lambda,t) = f_0(\xi^\alpha,t) + f_1(\xi^\alpha,t)\,\lambda + f_3(\xi^\alpha,t)\,\lambda^3\,, \quad \text{with} \quad f_1 := \left(\frac{\partial f}{\partial \lambda}\right)_{\xi^\alpha,t}|_{\lambda=0}. \tag{5.1}$$

The coefficient functions $f_0(\xi^\alpha,t)$ and $f_3(\xi^\alpha,t)$ are determined by the boundary conditions $f^\pm(\xi^\alpha,t) := f(\xi^\alpha,\lambda_\pm,t)$, and thus take the forms

$$\begin{aligned} f_0(\xi^\alpha,t) &= \frac{1}{2}(f^+ + f^-) - \frac{a_\varsigma}{2d_\varsigma(3a_\varsigma^2 + d_\varsigma^2)}\{(a_\varsigma^2 + 3d_\varsigma^2)[\![f]\!] + 2d_\varsigma(a_\varsigma^2 - d_\varsigma^2)f_1\}\,, \\ f_3 &= \frac{8}{2d_\varsigma(3a_\varsigma^2 + d_\varsigma^2)}\{[\![f]\!] - d_\varsigma f_1\}\,, \end{aligned} \tag{5.2}$$

where $d_\varsigma = \lambda_+ - \lambda_-$ and $a_\varsigma = \lambda_+ + \lambda_-$ are the *thickness* and *eccentricity* of the layer, respectively (as defined in the last section). These families of profiles with the adjustable normal slopes $f_1(\xi^\alpha,t)$ contain sufficient freedom to satisfy the constraints (4.17) and to allow a membrane approximation for the phase interface. This is also the reason why quadratic terms are ignored in (5.1). Those interested in the computational details may consult ALTS & HUTTER (1986).

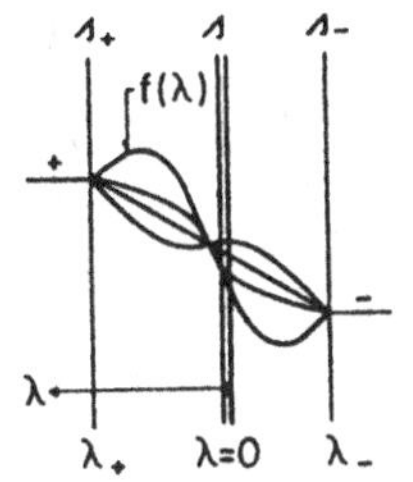

Fig. 5: Family of cubic profiles for the bulk fields within the boundary layer.

The constraints (4.17) can be satisfied if the slopes ϱ_1 and v_1^α of the density and the tangential velocity profiles are proportional to the corresponding jumps

$$\varrho_1 = \frac{X}{d_\varsigma}[\![\varrho]\!] , \qquad v_1^\alpha = \frac{X}{d_\varsigma}[\![v^\alpha]\!] \tag{5.3}$$

with the same factor of proportionality, X/d_ς, the identities (4.17) reduce then to two non-linear equations for the dimensionless factor X and the dimensionless eccentricity $\alpha := a_\varsigma/d_\varsigma$ which require that

$$\alpha = \alpha(k_M, k_G) , \qquad X = X(k_M, k_G) \tag{5.4}$$

are functions of the dimensionless curvatures $k_M = K_M d_\varsigma$ and $k_G = K_G d_\varsigma$. Numerical solutions of these equations show that, within the physical permissible interval $-1 \leq \alpha \leq 1$ exactly two families, $\alpha_1(k_M, k_G)$ and $\alpha_2(k_M, k_G)$, exist; Fig. 6. Note that it is crucial for the reference surface ($\lambda = 0$) not to lie outside the boundary layer defined by λ^+ and λ^-. This condition corresponds exactly to $-1 \leq \alpha \leq 1$.

Using these results, the surface density and tangential velocity follow from $(4.11)_1$ and $(4.13)_2$ and are expressible as

$$\varrho_\varsigma = [\![\varrho]\!]\frac{d_\varsigma}{2}F_1(k_M, k_G) , \tag{5.5}$$

$$\varrho_\varsigma w^\alpha = [\![\varrho v^\alpha]\!]\frac{d_\varsigma}{2}F_1(k_M, k_G) + [\![\varrho]\!][\![v^\alpha]\!]\frac{d_\varsigma}{6}F_2(k_M, k_G) . \tag{5.6}$$

The dimensionless functions F_1 and F_2 are functions of $k_M, k_G, \alpha(k_M, k_G)$ and $X(k_M, k_G)$, and can explicitly be written down in terms of these variables, but α and X are expressible in the form (5.4), as indicated in (5.5) and (5.6). Plots for spheres, are presented in Fig. 6.

Let us interpret (5.5) first. F_1 can be positive or negative. Its sign must be chosen such that $\varrho_\varsigma > 0$. Hence, two cases may be distinguished, which, for spherical geometries, are explained in Fig. 7. (Recall that the +side of the interface is always that side into which the surface normal vector points).

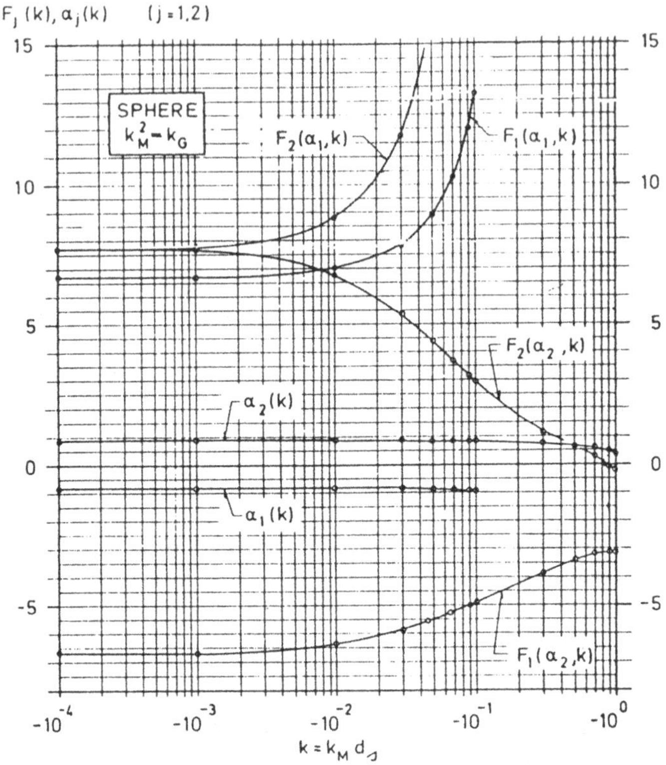

Fig. 6: Solutions of equation (4.17) for spherical geometry, $K_M^2 = K_G$. Plotted on semi-logarithmic scale is $\alpha_1 < 0$ and $\alpha_2 > 0$ as a function of $k_M = K_M d_\varsigma$. For $|k_M| > 0.1$, one has $|\alpha_1| > 1$, which is unphysical and therefore not shown. Since the boundary layer formulation breaks down when K_M is of order d_ς, results for $|k_M| > 1$ are also not shown. Also displayed are the values of the functions F_1 and F_2 appearing in (5.5) and (5.6). Dependence on curvature is significant only for $|k_M| > 10^{-3}$. We use the notation $k = k_M$

CASE A   I $\begin{matrix}\uparrow e_k \\ \left(\underline{\underset{-}{W}}\right) \\ + \end{matrix}$ $\left.\begin{matrix}\varrho_+ = \varrho_I \\ \varrho_- = \varrho_w\end{matrix}\right\} [\varrho] = \varrho_I - \varrho_w < 0$

$\varrho_\varDelta > 0 \quad\longrightarrow\quad F_1 < 0 \; : \; \underline{\alpha_2 > 0}$

CASE B   W $\begin{matrix}\uparrow e_k \\ \left(\underline{\overline{I}}\right) \\ + \end{matrix}$ $\left.\begin{matrix}\varrho_+ = \varrho_w \\ \varrho_- = \varrho_I\end{matrix}\right\} [\varrho] = \varrho_w - \varrho_I > 0$

$\varrho_\varDelta > 0 \quad\longrightarrow\quad F_1 > 0 \; : \; \underline{\alpha_1 < 0}$

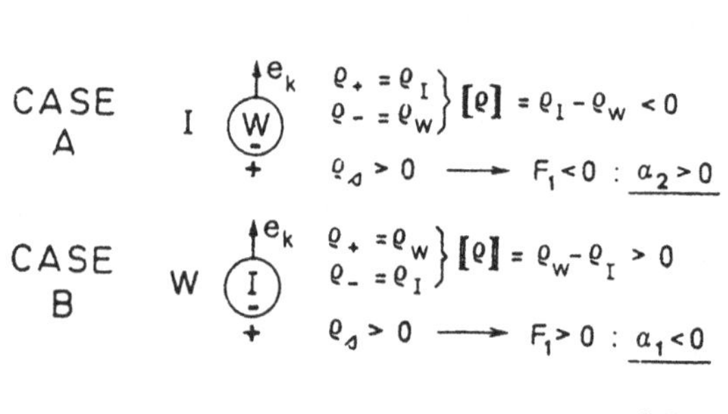

Fig. 7: Relative positions of ice (I) and water (W) and choice of sign of dimensionless eccentricity.

For a water inclusion in ice (case A) the positive family $1 > \alpha_2(k_M, k_G) > 0$ of eccentricities must be chosen. Since for this family (4.17) possesses solutions for all curvatures, water inclusions in ice exist down to bubble sizes of nucleation dimensions. (Note that this is a physical limitation and not a mathematical one, as indicated by the above analysis). The situation is different, when ice inclusions in water are considered (case B). In this case the negative family $-1 < \alpha_1(k_M, k_G) < 0$ must be selected. Since for this family (4.17) possesses no physically meaningful ($|\alpha_1| < 1$) solution when $|k_M| > 0.1$, we conclude that very small ice inclusions in water (of nucleation dimensions) cannot exist. Ice nucleation in steam, however, is possible because this corresponds to (case A). From this we infer, that nucleation of ice from pure water must start from the water surface. This, in fact, is observed in nature. The non-existence of ice-nucleation in water may also explain the phenomenon of undercooling in pure water. Ice formation in water, however, is possible, when nucleation kernels are present with dimensions that are larger than ten times the boundary layer thickness, or $|k_M| \leq 0.1$. Clearly impurities such as salts play this role.

Let us discuss the order of magnitude of ϱ_ς. Assume a thickness $d_\varsigma = 100$ Å $= 10^{-6}$ cm at the normal freezing point. The densities of water and ice at 0°C and 1 bar are $\varrho_w = 0.9992$ g/cm^3, (HUTTER & TRÖSCH, 1976, p. 58) and $\varrho_I = 0.9164$ g/cm^3 (HOBBS, 1974, p. 348). For a flat phase boundary we have $F_1 = 6.689$; thus

$$\varrho_\varsigma = (\varrho_w - \varrho_I)\frac{d_\varsigma}{2}|F_1(0,0)| = 2.81 \cdot 10^{-7}\,\frac{\text{g}}{\text{cm}^2} \qquad \text{(ice/water)}.$$

Alternatively, for a flat phase boundary between water and its vapour at normal pressure and room temperature one obtains with $d_\varsigma = 100$ Å

$$\varrho_\varsigma = (\varrho_w - \varrho_I)\frac{d_\varsigma}{2}|F_1(0,0)| = 3.35 \cdot 10^{-6}\,\frac{\text{g}}{\text{cm}^2} \qquad \text{(water/vapour)}\,.$$

These are very small numbers; consequently, if there is only a single boundary between large masses of ice and water, its mass contribution to the total mass of ice and water is negligible. However, in an ice-water mixture as in ice slush, where the area of internal phase boundaries per unit volume of the mixture is considerable, the mass contribution of the phase boundaries to the total mass of the mixture can no longer be ignored.

From (5.5) no inferences can be drawn about the curvature dependence of ϱ_ς, unless the curvature dependence of the jump $[\![\varrho]\!]$ is known. For compressible bulk materials, ice and water, there must be a curvature dependence of $[\![\varrho]\!]$, since the pressure in the inclusion is curvature dependent and heavily increasing for decreasing size of the inclusion. However, this dependence is not known. Therefore, let us consider two extreme cases.

1. Bulk material incompressible; then $[\![\varrho]\!] = [\![\varrho]\!]_0 =$ constant and equal to the value for the flat interface and ϱ_ς is curvature dependent through F_1.
2. $\varrho_\varsigma = \varrho_\varsigma^0 =$ constant and equal to the value of the flat interface; then $[\![\varrho]\!]$ is curvature dependent.

In Fig. 8 both cases are displayed for spherical geometry. Case 1 is shown by curves (1A) and (1B), case 2 by the straight line (2). Physically plausible cases lie between these two.

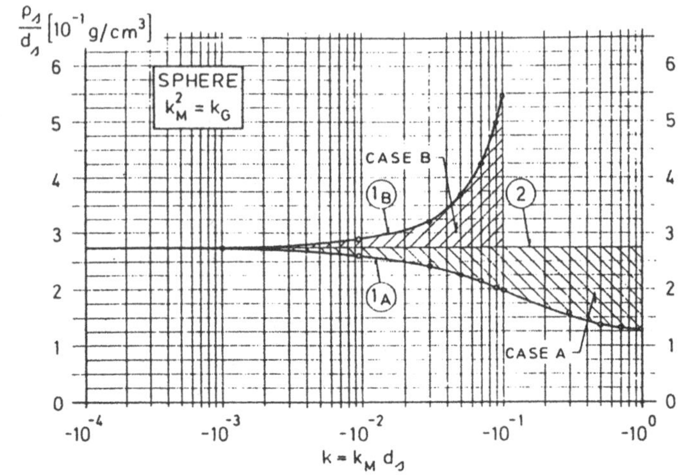

Fig. 8: Possible domains for the curvature dependence of surface density ϱ_s. Curves (1A) and (1B) display formula (5.6), for which $[\![\varrho]\!] = [\![\varrho_0]\!]$ = constant, line ② is the curve $\varrho_s = \varrho_s^0$ = constant.

Case A: Spherical water inclusion in ice,
Case B: Spherical ice inclusion in water.
Realistic curvature dependence of ϱ_s/d_s must lie in the shaded area.

The boundary layer model is also valid for the fluid-vapour phase transition. In this case the critical point can be reached, at which the phase boundary must disappear, since the phases are no longer distinguishable. This property is included in the foregoing results, for with $[\![\varrho]\!] \to 0$ also $\varrho_\varsigma \to 0$ for all curvatures: the surface field ϱ_ς vanishes when the phase boundary layer disappears. All other surface fields must have similar behaviour at the critical point.

A brief discussion of formula (5.6) for surface momentum may be added. For adherence of the bulk material, viz $v_+^\alpha = v_-^\alpha$ we deduce $w^\alpha = v_+^\alpha = v_-^\alpha$: the tangential velocity of the interface equals the material velocities of the adjacent bulk materials. For $v_+^\alpha \neq v_-^\alpha$ the surface velocity w^α differs from both.

Next, consider the normal motion of the interface. Inserting the cubic profile for $v_n(\lambda)$ into $(4.13)_1$ and choosing the normal slope

$$v_{n1} = \frac{X}{d_\varsigma}[\![v_n]\!] \tag{5.7}$$

we may show that

$$\varrho_\varsigma w_n = [\![\varrho v_n]\!]\frac{d_\varsigma}{2}F_1(k_M, k_G) + [\![\varrho]\!][\![v_n]\!]\frac{d_\varsigma}{6}F_2(k_M, k_G)\,. \tag{5.8}$$

For a material surface, $w_n = v_n^+ = v_n^-$ and (5.8) reduces to (5.5), as it should. Similarly, at the critical point, where $[\![\rho]\!] = 0$, (5.8) reduces to triviality. For a real phase boundary with $w_n \neq 0$, (5.8) requires $v_n^+ = v_n^-$.

More complex are the results for surface tension and transverse shear. Inserting cubic profiles for the boundary layer fields $t^{\alpha\beta}(\lambda) \neq t^{\beta\alpha}(\lambda)$ and $t_n^\alpha(\lambda)$ into $(4.13)_{2,3}$ yields relations for the tensor $S^{\alpha\beta}$ of surface tension with, in general, non-vanishing antisymmetric contribution $S^{[\alpha\beta]}$ and normal stress S^{α}. The normal slopes $t_1^{[\alpha\beta]}$ and t_{n1}^{α} of the stress distributions within the layer can be chosen such that the requirements of the membrane and non-polarity approximations are satisfied. The resulting expression for $t_1^{[\alpha\beta]}$ and t_{n1}^{α} that are obtained by requiring $S^{[\alpha\beta]} = 0$ and $S^{\alpha} = 0$ can be substituted into the expression that was obtained for $S^{(\alpha\beta)}$. What obtains for $S^{(\alpha\beta)}$ contains only the symmetric slopes $t_1^{(\alpha\beta)}$ as adjustable parameters. It will be determined such that the mechanical equilibrium condition

$$S^{\alpha\beta}|_E b_{\beta\alpha} = [\![p_E]\!] - (\varrho_\varsigma e)|_E \cdot g\,, \tag{5.9}$$

following from $(3.7)_3$, is identically satisfied. (Equilibrium is defined as a state with uniform temperature, vanishing velocities, vanishing tangential and normal shear stresses within the layer and hydrostatic stress states in the adjacent bulk materials, thus assigning the phase change region essentially fluid properties). $p_E^{\pm}$ denote the hydrostatic pressures in the bulk materials. Using the constitutive assumption

$$t_1^{(\alpha\beta)} = t_1^{(\alpha\beta)}|_E = t_1^E g^{\alpha\beta} + t_2^E (b^{\alpha\beta} - K_M g^{\alpha\beta})\,, \tag{5.10}$$

in which t_1^E and t_2^E are two scalar coefficients, the equilibrium and non-equilibrium tensor surface tensions, as well as the equilibrium scalar surface tension, can be shown to take the forms

$$\begin{aligned}
S^{(\alpha\beta)}|_E &= \sigma_E g^{\alpha\beta} + \sigma_2^E (b^{\alpha\beta} - K_M g^{\alpha\beta}) \\
S^{(\alpha\beta)} - S^{(\alpha\beta)}|_E &= -[\![t^{\gamma\delta} + p_E g^{\gamma\delta}]\!]\frac{d_\varsigma}{2}\left\{G_1 \delta_\gamma^\alpha \delta_\delta^\beta + \frac{d_\varsigma}{4} G_2 \delta_\gamma^{(\alpha}\delta_\delta^{\beta)} + \frac{d_\varsigma^2}{16} G_3 b_\gamma^\alpha b_\delta^\beta\right\} \\
\sigma_E &= [\![p_E]\!]\frac{d_\varsigma}{2} S_1 + t_1^E \frac{d_\varsigma^2}{3} S_3 + t_2^E \frac{d_\varsigma}{3}(k_M^2 - k_G) S_4,
\end{aligned} \tag{5.11}$$

respectively, where the functions $G_i(k_M, k_G)$ $(i = 1,2,3)$ and $S_j(k_M, k_G)$ $(j = 1,3,4)$ are given by ALTS & HUTTER (1986). Later we will also show that $\sigma_2^E = 0$. It should be emphasized that in the derivation of (5.11) the convective contributions in $(4.13)_3$ have been ignored, because they are quadratic in the velocity. We regard this as reasonable as most water in wet ice is trapped, and velocities must be small in this case.

Non-equilibrium surface tension is caused by the departure of the tangential stresses in the adjacent bulk materials from their corresponding hydrostatic pressures. Fluid-like phase interfaces cannot sustain surface shear stresses in equilibrium. Consequently neglecting the gravitational contribution ($g = 0$) we deduce from (5.9) and (2.15)

$$\sigma_E k_M = [\![p_E]\!]\frac{d_\varsigma}{2} \tag{5.12}$$

which is Laplace's formula. This relation is akin to (5.10); it relates the pressure jump $[\![p_E]\!]$ to the surface tension σ_E, but no information about a curvature dependence of σ_E is obtained. Common practice in the classical theory of surface tension is to assume σ_E to be independent of the curvatures, and this seems to be corroborated by measurements. But all measurements of surface tension known to us cover a regime where the dimensionless curvatures k_M, k_G are practically zero. So, the statement that σ_E is independent of the curvatures is nothing other than an assumption, whose validity is doubtful in the vicinity of nucleation dimensions for which curvatures are large.

To obtain some information in limiting cases, we combine $(5.11)_1$ and (5.12) via the pressure jump $[\![p_E]\!]$; this yields

$$\sigma_E = \frac{1}{(1-k_M S_1)}\left\{t_1^E \frac{d_\varsigma^2}{3} S_3 + t_2^E \frac{d_\varsigma}{3}(k_M^2 - k_G)S_4\right\} . \tag{5.13}$$

This is a relation for the curvature dependence of σ_E, if the curvature dependences of t_1^E and t_2^E are known. Since these dependences are presently unknown, (5.13) requires experimental investigation. There are two alternatives that we may consider instead. First, $\sigma_E = \sigma_E^0$ is independent of curvature. Second, $t_1^E = t_1^0$ and $t_2^E = t_2^0$ are assumed to be independent of curvature and given by their flat-surface-limit values $t_1^0 = (3/d_\varsigma)(\sigma_E^0/S_3^0)$ and $t_2^0 = -(3/d_\varsigma)(\sigma_E^0/S_3^0)\cdot(S_4^0/S_3^0)$. The real curvature dependence of σ_E should lie between these two extremes. Computations were done for spherical geometry, Fig. 9.

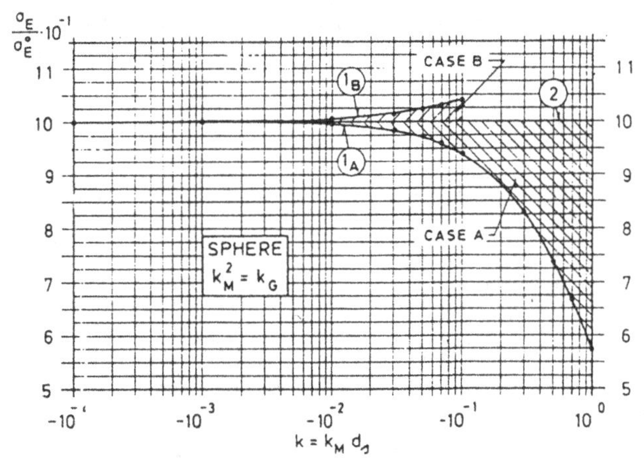

Fig. 9: Possible domains for the curvature dependence of the surface tension σ_E. Curves (1A) and (1B) display results obtained with formula (5.15) for which $t_1^E = t_1^0$, $t_2^E = t_2^0$.

Case A: Spherical water inclusion in ice,
Case B: Spherical ice inclusion in water.

Curve (2) is for $\sigma_E = \sigma_E^0 =$ constant. Realistic curvature dependences of σ_E must lie in the shaded areas.

A stability analysis of the boundary layer in equilibrium may determine the proper curvature dependence of t_1^E and t_2^E.

The results for the surface internal energy, entropy, heat flux and entropy flux are simply collected, since they are unconstrained. Substituting cubic boundary-layer distributions for

$u(\lambda), q^\alpha(\lambda)$ and $\phi^\alpha(\lambda)$ into $(4.14)_{1,2}$, $(4.15)_{1,4}$, thereby neglecting all convective contributions and choosing

$$u_1 = \frac{X}{d_\varsigma}[\![u]\!], \quad s_1 = \frac{X}{d_\varsigma}[\![s]\!], \quad q_1^\alpha = \frac{X}{d_\varsigma}[\![\phi^\alpha]\!], \quad q_1^\alpha = \frac{X}{d_\varsigma}[\![\phi^\alpha]\!] \tag{5.14}$$

for the normal slopes at $\lambda = 0$, the following results are obtained:

$$\varrho_\varsigma u_\varsigma = [\![\varrho u]\!]\frac{d_\varsigma}{2}F_1(k_M, k_G) + [\![\varrho]\!][\![u]\!]\frac{d_\varsigma}{6}F_2(k_M, k_G), \tag{5.15}$$

$$\varrho_\varsigma s_\varsigma = [\![\varrho s]\!]\frac{d_\varsigma}{2}F_1(k_M, k_G) + [\![\varrho]\!][\![s]\!]\frac{d_\varsigma}{6}F_2(k_M, k_G), \tag{5.16}$$

and

$$q_\varsigma^\alpha = -[\![q^\beta]\!]\frac{d_\varsigma}{2}\left\{J_1\delta_\beta^\alpha + \frac{d_\varsigma}{4}J_2 b_\beta^\alpha\right\}, \tag{5.17}$$

$$\phi_\varsigma^\alpha = -[\![\phi^\beta]\!]\frac{d_\varsigma}{2}\left\{J_1\delta_\beta^\alpha + \frac{d_\varsigma}{4}J_2 b_\beta^\alpha\right\}. \tag{5.18}$$

The curvature dependent functions $J_j(k_M, k_G)$ $(j = 1, 2)$ are given in Alts & Hutter (1986). The choices (5.14) are akin to the slopes (5.3) of the density and tangential velocity profiles, which are necessary consequences of the constraints (4.15). Hence, (5.14) represents the assumption that all boundary-layer fields possess the same relative normal slope at $\lambda = 0$ (similarity!). This makes (5.15)-(5.18) the simplest possible connections between the jumps of bulk fields and the corresponding surface fields. The relations also guarantee that the surface fields vanish when the critical point is approached (a necessary requirement for disappearance of the phase boundary).

Finally we mention that the common structure in all above representations is that the surface fields depend upon the jumps of the corresponding bulk fields, the boundary layer thickness and the mean and Gaussian curvatures of the surface. The dependence on the jumps of the bulk fields is linear. These properties will guide us in the selection of the constitutive relations.

6. Constitutive Theory

The basic objective of thermodynamics on non-material singular surfaces representing phase change surfaces is the determination of the fields

$$\varrho_\varsigma(\xi^\alpha, t), \quad \chi_\varsigma(\xi^\alpha, t), \quad T_\varsigma(\xi^\alpha, t) \tag{6.1}$$

of density, motion and surface temperature. These can be determined from the balance equations $(3.17)_{1-4}$ for mass, momentum and energy, provided for given supplies g, r_ς the surface fields

$$S^{\alpha\beta}(\xi^\gamma, t), \quad q_\varsigma^\alpha(\xi^\gamma, t), \quad u_\varsigma(\xi^\gamma, t) \tag{6.2}$$

and the jumps

$$\mathcal{R}(\xi^\gamma, t), \quad \mathcal{C}^\alpha(\xi^\gamma, t), \quad \mathcal{C}_n(\xi^\gamma, t), \quad \mathcal{U}(\xi^\gamma, t) \tag{6.3}$$

are related in a "materially dependent manner" to the basic fields (6.1). Such relations are called "constitutive equation" for phase boundaries. They must reflect the peculiarities of the phase-change interface in non-equilibrium, which embrace dynamics and thermodynamics of freezing and melting or evaporation and condensation, effects of surface tension, internal friction etc. Moreover, the layer between a crystal and its melt, for instance, must somehow represent a transition between the anisotropic structure on the solid side and the isotropic structure on the fluid side. The transition layer between a solid and its fluid is neither a solid nor a fluid, it is something in between. Corresponding statements hold for the fluid-gas transition layer. Furthermore, the transition layer contains extremely large density gradients, velocity gradients and (possibly) temperature gradients in the normal direction. Thermodynamic constitutive theories or non-equilibrium statistical theories, that would embrace all these peculiarities of the phase-change layer do presently not exist, even though first attempts of such formulations do exist (see Murdoch, this volume). In this light we therefore prefer to take a deterministic view and must then be aware that thicknesses of phase change surfaces be so large that, for example, effects of Brownian motion are smeared out.

6.1 Constitutive equations for surface fields

Constitutive equations for the surface and the jump fields will be treated separately. The surface fields

$$C_\varsigma := \{S^{(\alpha\beta)}, q_\varsigma^\alpha, u_\varsigma, \phi_\varsigma^\alpha, s_\varsigma\} \tag{6.4}$$

must be related to the independent fields (6.1) by constitutive equations. For this we model the interface between the bulk material as a fluid-like, viscous and heat conducting (two dimensional) continuum, which is mathematically defined by the constitutive equations

$$C_\varsigma(\xi^\gamma, t) := \Im_\varsigma[\varrho_\varsigma, T_\varsigma, w^\alpha, w_n, T_{\varsigma,\beta}, w^\alpha{}_{;\beta}, w_{n,\beta}, \boldsymbol{\tau}_\alpha, \boldsymbol{e}, \boldsymbol{\tau}_{\alpha;\beta}, \boldsymbol{e}_{;\beta}](\xi^\gamma, t). \tag{6.5}$$

T_ς is the surface temperature which may be different from the temperatures T_+ and T_- in the adjacent bulk materials. The surface gradients $T_{\varsigma,\alpha}$, $w^\alpha{}_{;\beta}$ and $w_{n,\beta}$ account for heat conduction along, and internal viscous friction within the phase boundary, respectively. The constitutive fields may also depend on the surface geometry, as expressed in (6.5) by the dependence on $\boldsymbol{\tau}_\alpha, \boldsymbol{e}, \boldsymbol{\tau}_{\alpha;\beta}$ and $\boldsymbol{e}_{;\beta}$.

6.2 Constitutive equations for jump fields

For vanishing surface fields the jump contributions

$$\mathcal{R}, \quad \mathcal{C}^\alpha, \quad \mathcal{C}_n, \quad \mathcal{U}, \quad \mathcal{S} \tag{6.6}$$

in the balance laws (3.7) necessarily vanish and the surface balance laws degenerate in this case to the classical boundary conditions at non-material moving surfaces. Real phase boundaries that bear mass, momentum, energy, etc., however, require constitutive assumptions for these jump quantities.

If the Cauchy stress in the adjacent bulk materials is decomposed into the equilibrium stress and the extra stress, $\boldsymbol{t}_\pm = -p_E^\pm \mathbf{1} + \hat{\boldsymbol{t}}_\pm$ and if the bulk material entropy flux $\boldsymbol{\phi}_\pm$ is identified with $q_\pm/T_\pm$ (where T denotes the absolute temperature) then direct inspection of (3.8), shows that the quantities (6.6) are explicit functions of the limiting values

$$\begin{aligned} W_+^\alpha &:= \boldsymbol{W}_+ \cdot \boldsymbol{\tau}^\alpha, & \hat{t}_{n+}^\alpha &:= \boldsymbol{\tau}^\alpha \cdot \hat{\boldsymbol{t}}_+ \cdot \boldsymbol{e}, \\ W_n^+ &:= \boldsymbol{W}_+ \cdot \boldsymbol{e}, & \hat{t}_n^+ &:= \boldsymbol{e} \cdot \hat{\boldsymbol{t}}_+ \cdot \boldsymbol{e}, \\ \theta_+ &:= T_+ - T_-, & q_n^+ &:= \boldsymbol{q}_+ \cdot \boldsymbol{e} \end{aligned} \tag{6.7}$$

of the relative tangential velocity W_+^α, shear stress t_{n+}^α relative normal velocity W_n^+, normal stress t_n^+, temperature difference θ_+ and heat flux q_n^+. The jump contributions (3.7) contain then explicitly $[\![\varrho]\!], [\![u]\!], [\![p_E]\!]$ and $[\![s]\!]$, and the following jump quantities

$$\mathcal{J}_\varsigma : \begin{cases} \mathcal{V}^\alpha := \boldsymbol{\tau}^\alpha \cdot [\![\boldsymbol{v}]\!], & \mathcal{G}^\alpha := \boldsymbol{\tau}^\alpha \cdot [\![\hat{\boldsymbol{t}}]\!] \cdot \boldsymbol{e}, \\ \mathcal{V}_n := \boldsymbol{e} \cdot [\![\boldsymbol{v}]\!], & \mathcal{G}_n := \boldsymbol{e} \cdot [\![\hat{\boldsymbol{t}}]\!] \cdot \boldsymbol{e}, \\ \mathcal{K} := [\![\frac{1}{T}]\!], & \mathcal{Q} := \boldsymbol{e} \cdot [\![\boldsymbol{q}]\!] \end{cases} \tag{6.8}$$

and can be rewritten as

$$\begin{aligned} \mathcal{R} &= [\![\varrho]\!] W_n^+ + (\varrho^+ - [\![\varrho]\!]) \mathcal{V}_n, \\ \mathcal{C}^\alpha &= -\mathcal{G}^\alpha + \mathcal{R} W_+^\alpha + (\varrho^+ - [\![\varrho]\!]) \mathcal{V}^\alpha (W_n^+ - \mathcal{V}_n), \\ \mathcal{C}_n &= -\mathcal{G}_n + [\![p_E]\!] + [\![\varrho]\!] W_{n+}^2 + (\varrho^+ - [\![\varrho]\!]) \mathcal{V}_n (2W_n^+ - \mathcal{V}_n), \\ \mathcal{U} &= \mathcal{Q} + \Big\{ [\![\varrho(u - u_\varsigma + \tfrac{1}{2}W^2)]\!] [\![p_E]\!] - \mathcal{G}_n + \varrho^+ (u^+ - u_\varsigma + \tfrac{1}{2}W_+^2) + p_E^+ - \hat{t}_n^+ \Big\} (W_n^+ - \mathcal{V}_n) \\ &\quad - \mathcal{G}^\alpha (W_+^\alpha - \mathcal{V}^\alpha) - \mathcal{V}^\alpha \hat{t}_{n+}^\alpha, \\ \mathcal{S} &= \mathcal{K} q_n^+ + \left\{ \frac{1}{T_+} - \mathcal{K} \right\} \mathcal{Q} + [\![\varrho(s - s_\varsigma)]\!] (W_n^+ - \mathcal{V}_n) + \mathcal{V}_n \varrho^+ (s^+ - s_\varsigma) \, . \end{aligned} \tag{6.9}$$

The fields (6.7) may be independently prescribed on one side of the phase boundary. The outcome on the other side, however, depends on this input and the constitutive properties of the phase interface. Hence, the jump contributions (6.8) may depend on the variables listed in (6.5) and (6.7):

$$\mathcal{J}_\varsigma := \Re_\varsigma [\varrho_\varsigma, T_\varsigma, w^\alpha, w_n, T_{\varsigma,\beta}, w^\alpha{}_{;\beta}, w_{n,\beta}; \boldsymbol{\tau}_\alpha, \boldsymbol{e}, \boldsymbol{\tau}_{\alpha;\beta}, \boldsymbol{e}_{;\beta}; \varrho^+, T_+; W_+^\alpha, W_n^+, \hat{t}_{n+}^\alpha, \hat{t}_n^+, \theta_+, q_n^+]. \tag{6.10}$$

Inserting the constitutive equations for the surface fields and the jump contributions into $(3.7)_{1-4}$ yields a set of field equations for the determination of $\varrho_\varsigma(\xi^\alpha, t), \boldsymbol{x}(\xi^\alpha, t)$ and $T_\varsigma(\xi^\alpha, t)$. Every solution for given supplies $g(\xi^\alpha, t)$ and $r_\varsigma(\xi^\alpha, t)$, and given bulk fields $\varrho^+(\xi^\alpha, t), T^+(\xi^\alpha, t)$, $v_+^\alpha(\xi^\beta, t)$, $v_n^+(\xi^\beta, t)$, $t_{n+}^\alpha(\xi^\beta, t)$, $t_n^+(\xi^\beta, t)$ and $q_n^+(\xi^\beta, t)$ on one side of the phase boundary is called a *thermodynamic process* upon phase interfaces between ice and water. In practice such thermodynamic processes are constructed by solving the CAUCHY initial value problem to the emerging system of partial differential equations; existence of solutions is guaranteed by the CAUCHY-KOWALEVSKY theorem.

6.3 Connection to the boundary layer theory

The constitutive postulates of the form (6.10) for the quantities defined in (6.8) have in parts found explicit expressions already earlier in the boundary layer theory of section 5, where cubic boundary layer profiles with specific choices for the normal slopes were assumed. The results are given in (5.5), (5.6), (5.8), (5.9), (5.15) and (5.16) and can be rewritten in the form

$$
\begin{aligned}
&[\![\varrho]\!] = \frac{\varrho_\varsigma}{d_\varsigma}\frac{2}{F_1}\,, \qquad [\![u]\!] = (u^+ - u_\varsigma)\frac{1}{\mathcal{F}}\,, \qquad [\![s]\!] = (s^+ - s_\varsigma)\frac{1}{\mathcal{F}}\,,\\
&[\![p_E]\!] = S^{\alpha\beta}|_E b_{\beta\alpha} + (\varrho_\varsigma e)|_E \cdot g\,, \qquad [\![v^\alpha]\!] = \mathcal{V}^\alpha = W_+^\alpha\frac{1}{\mathcal{F}}\,, \qquad [\![v_n]\!] = \mathcal{V}_n = W_n^+\frac{1}{\mathcal{F}}\,,\\
&\mathcal{F} := 1 - \frac{1}{3}\frac{F_2}{F_1} - \frac{\varrho^+}{\varrho_\varsigma}\frac{d_\varsigma}{2}F_1\,.
\end{aligned}
\tag{6.11}
$$

It is the equilibrium condition $(6.11)_4$ that prompted us to separate the hydrostatic pressure from the argument functions $\hat{t}^\alpha_{n+}$ and $\hat{t}^+_n$. Thus, given the boundary layer results (6.11), constitutive relations need be postulated only for the quantities (6.4) and $\mathcal{G}^\alpha, \mathcal{G}_n, \mathcal{K}, \mathcal{Q}$ in (6.8).

The boundary layer formulation has also led to explicit representations of tangential flux quantities, such as stress in (5.11), heat flux in (5.17) and entropy flux in (5.18). These may be written in the form

$$
\begin{aligned}
[\![\hat{t}^{\alpha\beta}]\!] &= -\frac{2}{d_\varsigma}\frac{1}{\Delta}(S^{(\alpha\beta)} - S^{(\alpha\beta)}|_E)\left\{\Delta_1\delta^\alpha_\gamma\delta^\beta_\delta - \frac{d_\varsigma}{4}\Delta_2\delta^{(\alpha}_\gamma b^{\beta)}_\delta + \frac{d_\varsigma^2}{32}\Delta_3 b^\alpha_\gamma b^\beta_\delta\right\}\,,\\
[\![q^\beta]\!] &= -\frac{2}{d_\varsigma}\frac{1}{\Delta_J}q^\alpha_\varsigma\left\{(J_1 + \frac{1}{2}k_M J_2)\delta^\alpha_\beta - J_2\frac{d_\varsigma}{4}b^\alpha_\beta\right\}\,,\\
[\![\phi^\beta]\!] &= -\frac{2}{d_\varsigma}\frac{1}{\Delta_J}\phi^\alpha_\varsigma\left\{(J_1 + \frac{1}{2}k_M J_2)\delta^\alpha_\beta - J_2\frac{d_\varsigma}{4}b^\alpha_\beta\right\}\,,
\end{aligned}
\tag{6.12}
$$

where $[\![\hat{t}^{\alpha\beta}]\!] = [\![t^{\gamma\delta} + p_E g^{\gamma\delta}]\!]$ and

$$
\Delta_J := J_1^2 + \frac{1}{2}k_M J_1 J_2 + \frac{1}{16}k_G J_2^2\,. \tag{6.13}
$$

The functions Δ_j ($j = 1,2,3$) and Δ are given in Appendix C of ALTS & HUTTER (1986), as are the dimensionless functions J_1 and J_2, which depend on the dimensionless curvatures k_M and k_G only. Relations (6.12) also have the structure (6.10) for the general constitutive equations of jump quantities. Expressions (6.12) will be used later to relate the adjacent bulk temperatures T_+, T_- to the surface temperature temperature T_ς.

7. Restrictions upon constitutive relations

The functional form of the constitutive equations may be further restricted via the principles of *material frame indifference*, *material symmetry* and *entropy*. The application of these principles to the basic constitutive assumptions (6.5) and (6.10) results in a thermodynamic constitutive theory of phase boundaries.

7.1 Principle of material frame indifference

This principle requires invariance of the constitutive equations under the transformations of the form

$$\mathrm{x}^* = \boldsymbol{Q}_t(\mathrm{x} - \boldsymbol{c}_t)\,, \tag{7.1}$$

representing a change of Euclidean observer $\Sigma \rightarrow \Sigma^*$ in which $\boldsymbol{Q}_t$ is a time-dependent proper orthogonal tensor and $\boldsymbol{c}_t$ a time-dependent translation of the origin of Σ^* relative to Σ, see Fig. 10. Scalars S, vectors $\boldsymbol{v}$ and second-order tensors $\boldsymbol{T}$ which transform under change of Euclidean observer $\Sigma \rightarrow \Sigma^*$ via the relations

$$S^* = S, \qquad \boldsymbol{v}^* = \boldsymbol{Q}_t\boldsymbol{v}, \quad \text{and} \quad \boldsymbol{T}^* = \boldsymbol{Q}_t\boldsymbol{T}\boldsymbol{Q}_t^{\mathrm{T}} \tag{7.2}$$

respectively, are called Euclidean (observer-independent) scalars, vectors and tensors, respectively. In particular, it is usually assumed that the following bulk (material) fields are required by the principle to be independent of Euclidean observer:

$$\begin{array}{ll} \varrho, u, s, T & \text{Euclidean scalars} \\ \boldsymbol{q}, \boldsymbol{\phi} & \text{Euclidean vectors} \\ \boldsymbol{t} & \text{Euclidean second-order tensor.} \end{array} \tag{7.3}$$

The transformations of certain kinematic fields, such as tangent vectors, velocity and so on under change of Euclidean observer can be deduced from the following two representations of the motion of a surface element on $\varsigma(t)$ relative to Σ^* and Σ:

$$\begin{aligned} \mathrm{x}^* &= \chi^*(\xi^\alpha, t) = \chi_R^*(\Xi^\Gamma, t), \\ \mathrm{x} &= \chi(\xi^\alpha, t) = \chi_R(\Xi^\Gamma, t). \end{aligned} \tag{7.4}$$

From these and the Euclidean observer transformation (7.1), one can show that

$$\left.\begin{array}{r} g_{\alpha\beta}, b_{\alpha\beta}, \\ v^\alpha - w^\alpha, v_n - w_n, \\ D_{\alpha\beta} = w_{(\alpha;\beta)} - b_{\alpha\beta}w_n \end{array}\right\} \quad \text{transform as Euclidean scalars}$$

$$\left.\begin{array}{r} \boldsymbol{\tau}_\alpha, \boldsymbol{e}, \\ \boldsymbol{W} = \boldsymbol{v} - \boldsymbol{w} \end{array}\right\} \quad \text{transform as Euclidean vectors} \tag{7.5}$$

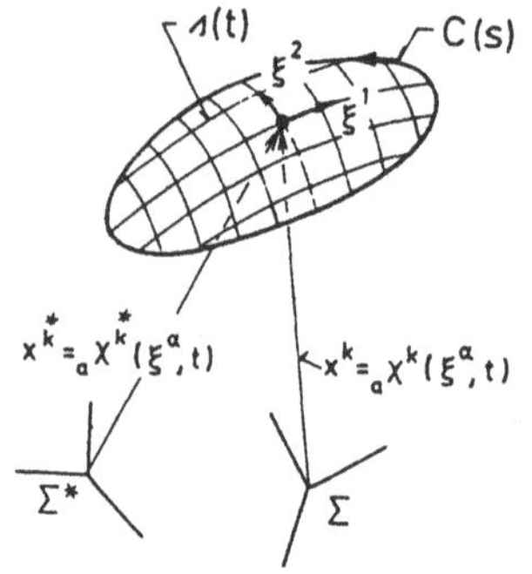

Fig. 10: Singular surface $\varsigma(t)$ with respect to observer frames Σ and Σ^*.

With the transformation properties (7.5) being established, it follows from the boundary layer representations (4.11)-(4.15) (these are the integral representations of the surface fields) that the surface fields

$$\varrho_\varsigma, u_\varsigma, s_\varsigma, q_\varsigma^\alpha, \phi_\varsigma^\alpha, S^{\alpha\beta} \tag{7.6}$$

transform as Euclidean scalars. In addition, we assume that the surface temperature T_ς is a Euclidean scalar. With these results, all quantities listed in (6.4) and (6.8) are objective scalars. It may then be shown that material frame indifference allows the quantities listed in (6.4) and (6.8) to depend only on scalar combinations of the arguments; this yields

$$\begin{aligned} C_\varsigma(\xi^\gamma, t) &= \Im_\varsigma[\varrho_\varsigma, T_\varsigma, T_{\varsigma,\beta}, D_{\alpha\beta}, g_{\alpha\beta}, b_{\alpha\beta}](\xi^\gamma, t) \\ \mathcal{J}_\varsigma(\xi^\gamma, t) &= \Re_\varsigma[\varrho_\varsigma, T_\varsigma, T_{\varsigma,\beta}, D_{\alpha\beta}, g_{\alpha\beta}, b_{\alpha\beta}; \varrho^+, T_+; W_+^\alpha, W_n^+, \hat{t}_{n+}^\alpha, \hat{t}_n^+, \theta_+, q_n^+](\xi^\gamma, t). \end{aligned} \tag{7.7}$$

7.2 Principle of material symmetry

The fluid-like boundary layer between ice and water should provide a transition from ice crystal of hexagonal symmetry on one side of the layer to isotropic water on the other side. The local symmetry group of the singular surface representing this layer, is thus a compromise of the symmetry groups of the ice grains and the water at locally opposite sides of the layer. Compact glacier ice consists of an entity of randomly oriented crystals which form grains. In cold glacier ice most of the crystal boundaries are solid. They are not phase interfaces, but act as slip surfaces and are primarily responsible for the macroscopically viscous behaviour. In temperate glacier ice, however, melting occurs at some of these crystalline boundaries, NYE & FRANK (1969). Grains are formed of which the interfaces to the water phase are covered by the fluid-like boundary layer, WEYL (1951). These act as slip surfaces, and connect differently oriented ice grains with water, see Fig. 11, but slip is only marginally increased because of the different orientations of the grains and their interconnections. Passing along $\varsigma(t)$ over distances which are large compared to the sizes of the adjacent ice crystallites, the local symmetry group of the phase interface is likely to change, however more conspicuously along the surface than orthogonally to it. Over such distances the mean symmetry group cannot be far from transverse isotropy with respect to the local normal vector $\boldsymbol{e}$. This means that the constitutive properties of the phase boundary are unchanged under arbitrary time independent local rotations and (in plane) reflections. More specifically, they are isotropic functions under orthogonal changes of the actual surface coordinates. With the coordinate change $\xi^\alpha = \xi^\alpha(\xi^{\bar\alpha})$, these transformations are defined by

$$H^\alpha_{\bar\alpha} := \frac{\partial \xi^\alpha}{\partial \xi^{\bar\alpha}} \quad \text{and} \quad H^{\bar\alpha}_\alpha := \frac{\partial \xi^{\bar\alpha}}{\partial \xi^\alpha} \tag{7.8}$$

such that

$$H^{\bar\alpha}_\gamma H^\gamma_{\bar\beta} = \delta^{\bar\alpha}_{\bar\beta}, \qquad H^{\bar\gamma}_\beta H^\alpha_{\bar\gamma} = \delta^\alpha_\beta \quad \text{and} \quad \det H^\alpha_{\bar\alpha} = \frac{1}{\det H^{\bar\alpha}_\alpha} = \pm 1,$$

where ξ^α and $\xi^{\bar\alpha}$ are two sets of actual surface coordinates and $H^\alpha_{\bar\alpha}$ belongs to the full orthogonal group $O(2)$ in two dimensions.

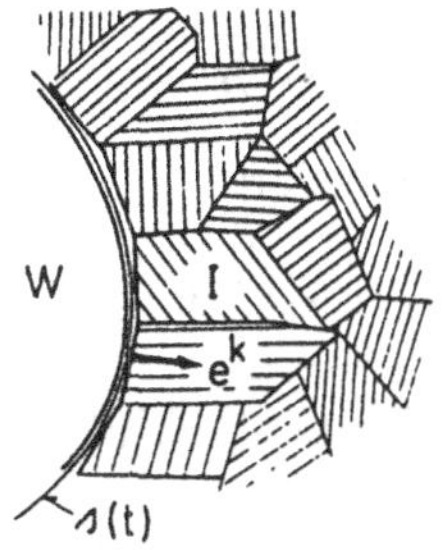

Fig. 11: Schematic view of polycrystalline glacier ice, phase-boundary layer and water inclusion or vein.

In passing it should also be remarked that the principle of material symmetry is a statement invoking *material* properties and should therefore be formulated in terms of referential coordinates $\Xi^\alpha = \Xi^\alpha(\Xi^{\bar\alpha})$. If this is done in (7.8) in terms of the actual coordinates, then it is tacitly assumed that every actual configuration is a possible reference configuration. In three dimensions such a concept has been introduced by Noll to be typical of a fluid for which the unimodular group is the symmetry group. Because of this analogy the full orthogonal group for $H^{\bar\alpha}_\alpha$ is therefore the smallest symmetry group for which (7.8) is meaningful.

On this basis it can then be shown that the constitutive equations (7.7) are isotropic functions of their arguments, see Alts & Hutter (1986, 1988 b). With the *irreducible isotropic invariants* of the basic surface fields

$$\begin{aligned}
K_M &:= \frac{1}{2} b^\alpha_\alpha, \qquad K_G := \frac{1}{2}(b^\alpha_\alpha b^\beta_\beta - b^\alpha_\beta b^\beta_\alpha), \\
I_D &:= D^\alpha_\alpha, \qquad II_D := b^\alpha_\beta D^\beta_\alpha \qquad III_D := \frac{1}{2}(D^\alpha_\alpha D^\beta_\beta - D^\alpha_\beta D^\beta_\alpha), \\
H_1 &:= g^{\alpha\beta} T_{\varsigma,\alpha} T_{\varsigma,\beta}, \\
H_2 &:= b^{\alpha\beta} T_{\varsigma,\alpha} T_{\varsigma,\beta}, \\
H_3 &:= D^{\alpha\beta} T_{\varsigma,\alpha} T_{\varsigma,\beta},
\end{aligned} \tag{7.9}$$

$$H_4 \quad := \quad g^{\mu\nu}(b^\alpha_\mu D^\beta_\nu - D^\alpha_\mu b^\beta_\nu)T_{\varsigma,\alpha}T_{\varsigma,\beta} \qquad ,$$

the "linearized" versions of the constitutive functions of the surface fields read

$$\begin{aligned}
u_\varsigma &= u_\varsigma(\varrho_\varsigma, T_\varsigma, K_M, K_G, I_D, II_D, III_D, H_1, H_2, H_3, H_4),\\
s_\varsigma &= s_\varsigma(\varrho_\varsigma, T_\varsigma, K_M, K_G, I_D, II_D, III_D, H_1, H_2, H_3, H_4),\\
q^\alpha_\varsigma &= -(\kappa g^{\alpha\beta} + \kappa_2(b^{\alpha\beta} - K_M g^{\alpha\beta})T_{\varsigma,\beta},\\
\phi^\alpha_\varsigma &= -(\phi g^{\alpha\beta} + \phi_2(b^{\alpha\beta} - K_M g^{\alpha\beta}))T_{\varsigma,\beta},\\
S^{\alpha\beta} &= (\sigma + \mu_1 I_D + \mu_2 II_D)g^{\alpha\beta} + (\sigma_2 + \mu_3 I_D + \mu_4 II_D)(b^{\alpha\beta} - K_M g^{\alpha\beta}))\\
&+ \mu_2 D^{\alpha\beta} + \mu_6 g^{\mu\nu}(b^{(\alpha}_\mu D^{\beta)}_\nu - D^{(\alpha}_\mu b^{\beta)}_\nu),
\end{aligned} \tag{7.10}$$

and the corresponding ones for the jump fields are

$$\begin{aligned}
\mathcal{V}_n &= \mathcal{V}_1 I_D + \mathcal{V}_2 II_D + \mathcal{V}_3 W^+_n + \mathcal{V}_4 \hat{t}^+_n + \mathcal{V}_5 q^+_n + \mathcal{V}_6 \theta_+,\\
\mathcal{G}_n &= \mathcal{G}_1 I_D + \mathcal{G}_2 II_D + \mathcal{G}_3 W^+_n + \mathcal{G}_4 \hat{t}^+_n + \mathcal{G}_5 q^+_n + \mathcal{G}_6 \theta_+,\\
\mathcal{Q} &= \mathcal{Q}_1 I_D + \mathcal{Q}_2 II_D + \mathcal{Q}_3 W^+_n + \mathcal{Q}_4 \hat{t}^+_n + \mathcal{Q}_5 q^+_n + \mathcal{Q}_6 \theta_+,\\
\mathcal{K} &= \mathcal{K}_1 I_D + \mathcal{K}_2 II_D + \mathcal{K}_3 W^+_n + \mathcal{K}_4 \hat{t}^+_n + \mathcal{K}_5 q^+_n + \mathcal{K}_6 \theta_+,\\
\mathcal{V}^\alpha &= (\mathcal{D}_1 g^{\alpha\beta} + \mathcal{D}_2(b^{\alpha\beta} - K_M g^{\alpha\beta}))T_{\varsigma,\beta} + (\mathcal{D}_3 g^{\alpha\beta} + \mathcal{D}_4(b^{\alpha\beta} - K_M g^{\alpha\beta}))W^+_\beta\\
&+ (\mathcal{D}_5 g^{\alpha\beta} + \mathcal{D}_6(b^{\alpha\beta} - K_M g^{\alpha\beta}))\hat{t}_{\beta+},\\
\mathcal{G}^\alpha &= (\mathcal{C}_1 g^{\alpha\beta} + \mathcal{C}_2(b^{\alpha\beta} - K_M g^{\alpha\beta}))T_{\varsigma,\beta} + (\mathcal{C}_3 g^{\alpha\beta} + \mathcal{C}_4(b^{\alpha\beta} - K_M g^{\alpha\beta}))W^+_\beta\\
&+ (\mathcal{C}_5 g^{\alpha\beta} + \mathcal{C}_6(b^{\alpha\beta} - K_M g^{\alpha\beta}))\hat{t}_{\beta+},
\end{aligned} \tag{7.11}$$

where

$$W^+_\alpha := g_{\alpha\beta}W^\beta_+ \qquad \text{and} \qquad \hat{t}^+_\alpha := g_{\alpha\beta}\hat{t}^\beta_+. \tag{7.12}$$

In these, the representations for the transport quantities $q^\alpha_\varsigma, \phi^\alpha_\varsigma$ and $S^{\alpha\beta}$ within the phase interface are linearized in the deformation rate $D_{\alpha\beta}$ and the temperature gradient $T_{\varsigma,\alpha}$. The coefficients $\kappa, \kappa_2, \phi, \phi_2, \sigma, \sigma_2, \mu_1, \ldots, \mu_6$ may however still be functions of $\varrho_\varsigma, T_\varsigma, K_M$ and K_G. The representations for the jump quantities are also linearized in the indicated variables $I_D, II_D, W^+_n, \hat{t}^+_n, q^+_n, \theta_+, T_{\varsigma,\alpha}, W^+_\beta$ and $t^n_{\beta+}$. Here the jump coefficients $\mathcal{V}^1, \ldots$ may be functions of $\varrho_\varsigma, T_\varsigma, K_M, K_G, \varrho^+$ and T^+. We regard these linearizations to be sufficiently general to characterize the behaviour of phase boundaries between ice and water. The non-linear irreducible representations could easily be written down, however the formulae are prohibitively lengthy and seem to allow only minor improvements compared with the leading order terms in (7.10) and (7.11). The representations (7.10) and (7.11) could also have been written in a somewhat shorter form, using the tensors $g_{\alpha\beta}$ and $b_{\alpha\beta}$ instead of $g^{\alpha\beta}$ and $b^{\alpha\beta} - K_M g^{\alpha\beta}$. The chosen form, however, has special advantages for spherical surfaces, for which $b^{\alpha\beta} - K_M g^{\alpha\beta} = 0$.

The linear representations (7.11) for non-spherical geometry contain 36 unknown transport coefficients, obviously too many for any reasonable theory of the dynamics of phase transitions. We shall reduce their number later by physically motivated additional requirements.

8. Entropy principle

The entropy principle is the most restrictive among the general constraints imposed on the constitutive functions. It guarantees positive entropy production and is the expression of the *second law of thermodynamics*. We use the following generalization of Müller's (1972) form of this principle:

(i) There exists an additive surface entropy. It satisfies the balance law (3.7) for every surface point and all times.

(ii) The specific surface entropy s_ς, its non-convective flux ϕ^α_ς and its input rate density $\mathcal{S}$ from the adjacent bulk materials are given by constitutive equations, which must satisfy material frame indifference and material symmetry requirements. For heat conducting, viscous interfaces between ice and water in the membrane approximation they are given by (7.7) with the specifications (6.7), (6.8) and $(6.9)_5$ and if the phase boundary is isotropic by $(7.10)_{2,4}$, (7.11) and $(6.9)_5$.

(iii) The entropy-supply rate density σ_ς, is a linear combination of the momentum and internal energy supply rate densities:

$$\sigma_\varsigma = \lambda_\varsigma \varrho_\varsigma r_\varsigma + \varrho_\varsigma(\lambda^\varsigma_n \boldsymbol{e} + \lambda^\alpha_\varsigma \boldsymbol{\tau}_\alpha) \cdot \boldsymbol{g} \tag{8.1}$$

where the parameters λ^ς, λ^ς_α and λ^ς_n are assumed independent of the source terms r_ς and $\boldsymbol{g}$.

(iv) There exist ideal heat conducting singular quasi-material lines ℓ, which move with velocity w^α and divide different surface materials (Fig. 12). Let h_α be the unit normal vector to the singular line inside the local tangential planes of the phase interface. Assume that the surface temperature T_ς is continuous across this line (definition of a dia-thermal line). A dia-thermal line is called ideal if the normal jump of the entropy flux vanishes across the line whenever the normal component of the heat flux is continuous. Thus,

$$\ell \text{ is ideal} \Longleftrightarrow [\![T_\varsigma]\!] = 0, \quad [\![q^\alpha_\varsigma h_\alpha]\!] = 0 \Longrightarrow [\![\phi^\alpha_\varsigma h_\alpha]\!] = 0\,. \tag{8.2}$$

(v) The entropy production rate density π_ς is non-negative for every thermodynamics process:

$$\pi_\varsigma \geq 0 \quad \text{for all thermodynamic processes.} \tag{8.3}$$

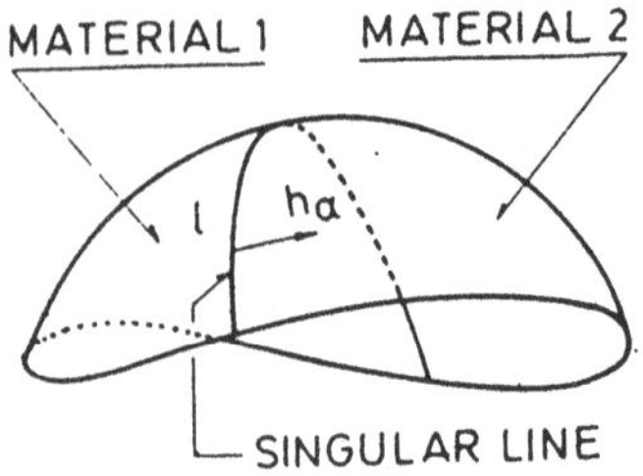

Fig. 12: Singular line ℓ dividing two materials on $\varsigma(t)$, representing the phase boundary.

The entropy inequality (8.3) is the key to placing further restrictions on the constitutive equations. Since (8.3) is required to hold for *all* thermodynamic processes, any acceptable solution of the local mass, momentum, energy balances via the constitutive assumptions must also be a solution of the local entropy balance such that (8.3) holds. Conversely, any acceptable solution of the local entropy balance via constitutive relations satisfying (8.3) must also satisfy the local mass, momentum and energy balances. The balance relations $(3.7)_{1-4}$ can therefore be viewed as constraints on the acceptable solutions of the local entropy balance such that (8.3) holds. Liu (1972) has developed this view of the other balance relations as constraints on the entropy balance, leading to the following form of the entropy inequality

$$\begin{aligned}\pi_\varsigma \;=\;& \varrho_\varsigma \dot{s}_\varsigma + \phi^\alpha_{\varsigma;\alpha} - \varrho_\varsigma[(\lambda^\varsigma_n e + \lambda^\alpha_\varsigma \tau_\alpha)\cdot g - \lambda_\varsigma r_\varsigma] + S\\ &-\Lambda^\varsigma_\varrho\{\dot{\varrho}_\varsigma + \varrho_\varsigma D^\alpha_\alpha + \mathcal{R}\}\\ &-\Lambda^\varsigma_\alpha\left\{\varrho_\varsigma \dot{w}^\alpha + \varrho_\varsigma \Gamma^\alpha_{\beta\gamma} w^\beta w^\gamma - \varrho_\varsigma w_n(g^{\alpha\beta} w_{n,\beta} + 2b^\alpha_\beta w^\beta) - S^{\alpha\beta}{}_{;\beta} - \varrho_\varsigma \tau^\alpha\cdot g + C^\alpha\right\}\\ &-\Lambda^\varsigma_n\left\{\varrho_\varsigma \dot{w}_n + \varrho_\varsigma w^\alpha(w_{n,\alpha} + b_{\alpha\beta} w^\beta) - S^{\alpha\beta} b_{\beta\alpha} - \varrho_\varsigma e\cdot g + C_n\right\}\\ &-\Lambda^\varsigma_\varepsilon\left\{\varrho_\varsigma \dot{u}_\varsigma + q^\alpha_{\varsigma;\alpha} - S^{\alpha\beta} D_{\alpha\beta} - \varrho_\varsigma r_\varsigma + \mathcal{U}\right\} \ge 0\,,\end{aligned} \tag{8.4}$$

where the quantities $\Lambda^\varsigma_\varrho, \Lambda^\varsigma_\alpha, \Lambda^\varsigma_n, \Lambda^\varsigma_\varepsilon$ are the (Lagrange) parameters associated with the constraints placed on the entropy balance by the other balance relations.

At this point, the inequality holds for *arbitrary* surface fields $\varrho_\varsigma(\xi^\alpha,t)$, $\chi(\xi^\alpha,t)$, $T_\varsigma(\xi^\alpha,t)$ and bulk fields $W^\alpha_+(\xi^\beta,t)$, $W^+_n(\xi^\beta,t)$, $\hat{t}^\alpha_{n+}(\xi^\beta,t)$, $\hat{t}^+_n(\xi^\beta,t)$, $\theta_+(\xi^\beta,t)$ and and $q^+_n(\xi^\beta,t)$ on one side of the phase boundary. Inserting the constitutive relations (7.7) for the surface fields into (8.4), performing the differentiations and using $g_{\alpha\beta;\gamma} = 0$, we obtain a form of (8.4) containing explicitly the following derivatives:

$$\begin{gathered}\dot{\varrho}_\varsigma, \dot{T}_\varsigma, \dot{T}_{\varsigma,\alpha}, \dot{D}_{\alpha\beta}, \dot{g}_{\alpha\beta}, \dot{b}_{\alpha\beta}, \dot{w}^\alpha, \dot{w}_n,\\ \varrho_{\varsigma,\alpha}, T_{\varsigma,\alpha}, w_{n,\alpha}, T_{\varsigma;\alpha\beta}, D_{\alpha\beta;\gamma}, b_{\alpha\beta;\gamma}.\end{gathered} \tag{8.5}$$

The parameters $\lambda^\varsigma_n, \lambda^\alpha_\varsigma, \lambda_\varsigma, \Lambda^\varsigma_\varrho, \Lambda^\varsigma_\alpha, \Lambda^\varsigma_n$ and $\Lambda^\varsigma_\varepsilon$ may in general depend on the independent constitutive surface fields $\varrho_\varsigma, T_\varsigma, T_{\varsigma,\beta}, D_{\alpha\beta}, g_{\alpha\beta}$ and $b_{\alpha\beta}$ in (7.7) and their derivatives (8.5).

However, if we assume that *none* of the parameters depend on $\dot{w}^\alpha$ and $\dot{w}_n$, then (8.4) will be linear in $\dot{w}^\alpha$ and $\dot{w}_n$, and it could be violated unless

$$\Lambda^\varsigma_\alpha = 0, \quad \Lambda^\varsigma_n = 0 \tag{8.6}$$

Likewise, if we assume that none of the parameters depend on $\dot{\varrho}_\varsigma$ and $\dot{T}_\varsigma$, (8.4) will be linear in these derivatives, and again could be violated unless

$$\frac{\partial s_\varsigma}{\partial \varrho_\varsigma} = \Lambda^\varsigma_\varepsilon \frac{\partial u_\varsigma}{\partial \varrho_\varsigma} - \frac{\Lambda^\varsigma_\varrho}{\varrho_\varsigma}, \qquad \frac{\partial s_\varsigma}{\partial T_\varsigma} = \Lambda^\varsigma_\varepsilon \frac{\partial u_\varsigma}{\partial T_\varsigma}. \tag{8.7}$$

Assuming the partial derivatives of s_ς and u_ς in (8.7) are regular, and noting that these partial derivatives are by assumption functions only of $\varrho_\varsigma, T_\varsigma, T_{\varsigma,\beta}, D_{\alpha\beta}, g_{\alpha\beta}$ and $b_{\alpha\beta}$, we see that the conditions (8.7) actually require the parameters $\Lambda^\varsigma_\varrho$ and $\Lambda^\varsigma_\varepsilon$ to be at most functions of these surface fields as well. In this case, the inequality (8.4) will be linear in all the other derivatives appearing in (8.5) except the surface temperature gradient $T_{\varsigma,\beta}$, and so could be violated unless we have

$$\begin{aligned}
\frac{\partial s_\varsigma}{\partial g_{\alpha\beta}} &= \Lambda^\varsigma_\varepsilon \frac{\partial u_\varsigma}{\partial g_{\alpha\beta}}, & \frac{\partial s_\varsigma}{\partial b_{\alpha\beta}} &= \Lambda^\varsigma_\varepsilon \frac{\partial u_\varsigma}{\partial b_{\alpha\beta}},\\
\frac{\partial s_\varsigma}{\partial T_{\varsigma,\alpha}} &= \Lambda^\varsigma_\varepsilon \frac{\partial u_\varsigma}{\partial T_{\varsigma,\alpha}}, & \frac{\partial s_\varsigma}{\partial D_{\alpha\beta}} &= \Lambda^\varsigma_\varepsilon \frac{\partial u_\varsigma}{\partial D_{\alpha\beta}},\\
\frac{\partial \phi^\gamma_\varsigma}{\partial \varrho_\varsigma} &= \Lambda^\varsigma_\varepsilon \frac{\partial q^\gamma_\varsigma}{\partial \varrho_\varsigma}, & \frac{\partial \phi^\gamma_\varsigma}{\partial b_{\alpha\beta}} &= \Lambda^\varsigma_\varepsilon \frac{\partial q^\gamma_\varsigma}{\partial b_{\alpha\beta}},\\
\frac{\partial \phi^{(\gamma}_\varsigma}{\partial T_{\varsigma,\alpha)}} &= \Lambda^\varsigma_\varepsilon \frac{\partial q^{(\gamma}_\varsigma}{\partial T_{\varsigma,\alpha)}}, & \frac{\partial \phi^\gamma_\varsigma}{\partial D_{\alpha\beta}} &= \Lambda^\varsigma_\varepsilon \frac{\partial q^\gamma_\varsigma}{\partial D_{\alpha\beta}}.
\end{aligned} \tag{8.8}$$

The inequality (8.4) is also explicitly linear in the external source terms, $\boldsymbol{\tau}_\alpha \cdot \boldsymbol{g}$, $\boldsymbol{e} \cdot \boldsymbol{g}$ and r_ς, which by assumption may be arbitrarily varied. For instance, relative to the direction of gravity the phase boundary can be arbitrarily oriented and the specific supply of internal energy can be varied by varying the external sources of heat radiation. Consequently, the inequality could again be violated, unless

$$\lambda^\varsigma = \Lambda^\varsigma_\varepsilon, \qquad \lambda^\varsigma_\alpha = 0, \qquad \lambda^\varsigma_n = 0. \tag{8.9}$$

Using these results, there remains the residual inequality

$$\pi_\varsigma = \left\{ \frac{\partial \phi^\alpha_\varsigma}{\partial T_\varsigma} - \Lambda^\varsigma_\varepsilon \frac{\partial q^\alpha_\varsigma}{\partial T_\varsigma} \right\} T_{\varsigma,\alpha} + (\Lambda^\varsigma_\varepsilon S^{\alpha\beta} - \Lambda^\varsigma_\varrho \varrho_\varsigma g^{\alpha\beta}) D_{\alpha\beta} + (\mathcal{S} - \Lambda^\varsigma_\varrho \mathcal{R} - \Lambda^\varsigma_\varepsilon \mathcal{U}) \geq 0 \;, \tag{8.10}$$

which holds for arbitrary surface fields, and arbitrary bulk fields on one side of the phase interface.

9. Thermodynamics of phase interfaces

In this section we discuss the consequences of the restrictions (8.7) - (8.10) for the constitutive relations. These can be summarized as follows:

(i) Entropy supply and entropy flux are proportional to heat supply and heat flux,

$$\sigma_\varsigma = \frac{1}{T_\varsigma}\varrho_\varsigma r_\varsigma , \qquad \phi^\alpha_\varsigma = \frac{1}{T_\varsigma} q^\alpha_\varsigma , \tag{9.1}$$

where T_ς is the absolute surface temperature.

(ii) The Lagrange-parameter $\Lambda^\varsigma_\varepsilon$ is a *universal* function of the surface temperature and can be identified with $1/T_\varsigma$, viz.

$$\Lambda^\varsigma_\varepsilon(T_\varsigma) = \frac{1}{T_\varsigma}. \tag{9.2}$$

(iii) The Lagrange-parameter $\Lambda^\varsigma_\varrho$ is given by the surface tension σ (i.e., the isotropic inviscid part of the membrane stress $S^{\alpha\beta}$, see $(7.10)_3$)

$$\Lambda^\varsigma_\varrho = \frac{\sigma(T_\varsigma, \varrho_\varsigma)}{\varrho_\varsigma T_\varsigma}. \tag{9.3}$$

(iv) Surface entropy, energy and tension are only functions of T_ς and ϱ_ς:

$$\begin{aligned} s_\varsigma &= s_\varsigma(T_\varsigma, \varrho_\varsigma) , \qquad & u_\varsigma &= u_\varsigma(T_\varsigma, \varrho_\varsigma), \\ \sigma &= \sigma(T_\varsigma, \varrho_\varsigma) , \qquad & \sigma_2 &= 0 \end{aligned} \tag{9.4}$$

and are related by the Gibbs equation

$$ds_\varsigma = \frac{1}{T_\varsigma}\left(du_\varsigma + \frac{\sigma}{\varrho^2_\varsigma}\, d\varrho_\varsigma \right) \tag{9.5}$$

Since $\sigma_2 = 0$, the equilibrium surface tension does not depend on the differential geometric properties of the surface.

To show (i) and (ii) notice that (8.1) and (8.9) imply

$$\sigma_\varsigma = \Lambda^\varsigma_\varepsilon \varrho_\varsigma r_\varsigma \ . \tag{9.6}$$

To satisfy the relations $(8.8)_{4-8}$, assume that

$$\phi^\alpha_\varsigma = \Lambda^\varsigma_\varepsilon q^\alpha_\varsigma . \tag{9.7}$$

Insertion shows, that $(8.8)_{4-8}$ can be satisfied for $q^\alpha_\varsigma \neq 0$ (which we assume) provided that $\Lambda^\varsigma_\varepsilon$ is independent of ϱ_ς, $T_{\varsigma,\alpha}$, $b_{\alpha\beta}$, and $D_{\alpha\beta}$. Hence we have $\Lambda^\varsigma_\varepsilon(T_\varsigma, g_{\alpha\beta})$. Now $(8.7)_1$ implies that $\Lambda^\varsigma_\varepsilon$ is a surface scalar. Because the only scalars which can be constructed from the metric alone are $\mathrm{tr}\mathbf{g} = g^{\alpha\beta} g_{\alpha\beta} = 2$, $\mathrm{tr}\mathbf{g}^2 = g^\alpha_\beta g^\beta_\alpha = \delta^\alpha_\beta \delta^\beta_\alpha = 2$, which are simply constants, we deduce that $\Lambda^\varsigma_\varepsilon$ is only a function of the surface temperature

$$\Lambda^\varsigma_\varepsilon = \Lambda^\varsigma_\varepsilon(T_\varsigma) \tag{9.8}$$

Relations (9.7) and (9.8) are sufficient conditions for $(8.8)_{4-8}$ to be satisfied. For isotropic interfaces they can also be shown to be necessary using the representations $(7.10)_{3,4}$. For non-isotropic phase boundaries, necessity of (9.7) has not been shown, but will be assumed here, because the result (9.8) is appealing and physically reasonable. On this basis, the results (9.1) and (9.2) are stringent.

The Lagrange parameter $\Lambda_\varepsilon^\varsigma$ may in general depend on the particular material of the phase-boundary layer, since via $(8.7)_1$ it is expressible in terms of constitutive functions. Using postulate (iv) of the entropy principle, however, it can be shown that $\Lambda_\varepsilon^\varsigma$ is independent of the materials in contact at the phase boundary, making $\Lambda_\varepsilon^\varsigma$ in this context a universal function of the surface temperature. We do this as follows. Consider four bulk materials, S, M, I, W. a solid S that does not mix with ice I, its melt M which does not mix with water W (for instance wax and its melt or oil and its solid). Each bulk material is connected to its neighbour either by a phase-boundary layer or a discontinuity surface. The singular line in statement (iv) connects then all points of intersection of the four materials, see Fig. 13. The intersecting line is assumed to be an ideal heat conductor with T_ς satisfying (8.2). Denoting the surface fluxes at the (W, I)-Interface by $q^\alpha_{\varsigma(W,I)}, \phi^\alpha_{\varsigma(W,I)}$, and those at the (M, S)-interface by $q^\alpha_{\varsigma(M,S)}, \phi^\alpha_{\varsigma(M,S)}$, relations (8.2) can be written explicitly as

$$(\phi^\alpha_{\varsigma(W,I)} - \phi^\alpha_{\varsigma(M,S)})h_\alpha = 0 \quad \text{for} \quad (q^\alpha_{\varsigma(W,I)} - q^\alpha_{\varsigma(M,S)})h_\alpha = 0 \, . \tag{9.9}$$

Relations (9.7) and (9.8) hold for all heat conducting, viscous phase interfaces, hence

$$\phi^\alpha_{\varsigma(W,I)} = \Lambda^\varsigma_{\varepsilon(W,I)}(T_\varsigma)q^\alpha_{\varsigma(W,I)}, \qquad \phi^\alpha_{\varsigma(M,S)} = \Lambda^\varsigma_{\varepsilon(M,S)}(T_\varsigma)q^\alpha_{\varsigma(M,S)} \, . \tag{9.10}$$

Combining (9.9) and (9.10) yields

$$\Lambda^\varsigma_{\varepsilon(W,I)}(T_\varsigma) = \Lambda^\varsigma_{\varepsilon(M,S)}(T_\varsigma) := \Lambda^\varsigma_\varepsilon(T_\varsigma). \tag{9.11}$$

Thus, $\Lambda_\varepsilon^\varsigma$ is independent of the particular phase-boundary layer material, a universal function of the surface temperature.

The identification (9.2) defines the surface temperature in an analogous fashion as for bulk materials. More precisely, (9.2) should be written as $\Lambda_\varepsilon^\varsigma = 1/\theta_\varsigma(T_\varsigma)$, where T_ς is the *empirical* surface temperature and θ_ς the absolute surface temperature. The identification $\theta_\varsigma = T_\varsigma$ requires that $\theta_\varsigma(T_\varsigma)$ be monotone and that appropriate scales for temperature can be chosen, see Carathéodory (1955).

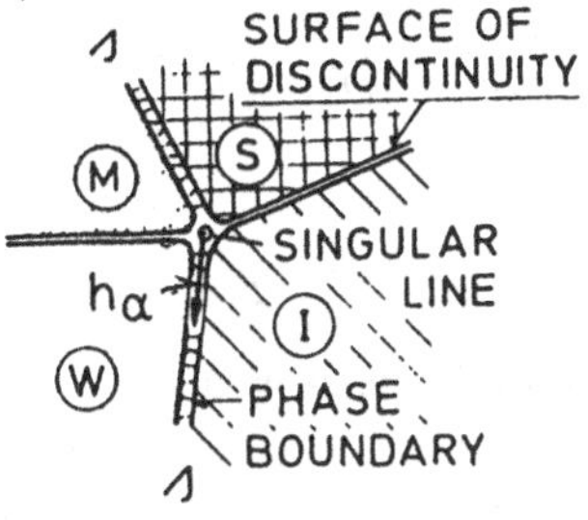

Fig. 13: Intersection of four bulk materials S,M,I,W and singular line.

To show (iii) and (iv) assume that the entropy is smooth (twice continuously differentiable in all fields) and integrable. The mixed second derivatives of (8.7) and $(8.8)_{1-4}$ are then interchangeable, yielding the following integrability conditions for the entropy:

$$\frac{\partial u_\varsigma}{\partial \varrho_\varsigma} = \frac{T_\varsigma^2}{\varrho_\varsigma}\frac{\partial \Lambda_\varrho^\varsigma}{\partial T_\varsigma},$$

$$\frac{\partial u_\varsigma}{\partial T_{\varsigma,\alpha}} = 0, \quad \frac{\partial u_\varsigma}{\partial g_{\alpha\beta}} = 0, \quad \frac{\partial u_\varsigma}{\partial b_{\alpha\beta}} = 0, \quad \frac{\partial u_\varsigma}{\partial D_{\alpha\beta}} = 0, \tag{9.12}$$

$$\frac{\partial \Lambda_\varrho^\varsigma}{\partial T_{\varsigma,\alpha}} = 0, \quad \frac{\partial \Lambda_\varrho^\varsigma}{\partial g_{\alpha\beta}} = 0, \quad \frac{\partial \Lambda_\varrho^\varsigma}{\partial b_{\alpha\beta}} = 0, \quad \frac{\partial \Lambda_\varrho^\varsigma}{\partial D_{\alpha\beta}} = 0.$$

Thus, the internal energy and the Lagrange parameter $\Lambda_\varrho^\varsigma$ are only functions of ϱ_ς and T_ς. By (8.7) and $(8.8)_{1-4}$, the same holds also for the entropy. Summarizing,

$$u_\varsigma = u_\varsigma(\varrho_\varsigma, T_\varsigma), \qquad s_\varsigma = s_\varsigma(\varrho_\varsigma, T_\varsigma), \qquad \Lambda_\varrho^\varsigma = \Lambda_\varrho^\varsigma(\varrho_\varsigma, T_\varsigma)\ . \tag{9.13}$$

The remaining equations (8.7) and $(8.8)_{1-4}$ can thus be condensed to the differential form[1]

$$ds_\varsigma = \frac{1}{T_\varsigma}\left(du_\varsigma + \frac{T_\varsigma}{\varrho_\varsigma}\Lambda_\varrho^\varsigma\, d\varrho_\varsigma\right)\ . \tag{9.14}$$

This is GIBBS' *differential* for the surface entropy. We emphasize, that all these relations hold in non-equilibrium and for phase interfaces with no specified symmetry group.

To identify $\Lambda_\varrho^\varsigma$ we must rely on the representation $(7.10)_5$ for the tensor of surface tension which holds only for isotropic phase interfaces (to which the following conclusions are restricted). We split $(7.10)_5$ into an inviscid and a viscous contribution:

$$\begin{aligned} S_N^{\alpha\beta} &:= \sigma g^{\alpha\beta} + \sigma_2(b^{\alpha\beta} - K_M g^{\alpha\beta}) \\ S_V^{\alpha\beta} &:= (\mu_1 I_D + \mu_2 II_D)g^{\alpha\beta} + (\mu_3 I_D + \mu_4 II_D)(b^{\alpha\beta} - K_M g^{\alpha\beta}) \\ &+ \ \mu_5 D^{\alpha\beta} + \mu_6 g^{\gamma\nu}(b_\gamma^{(\alpha} D_\nu^{\beta)} - D_\gamma^{(\alpha} b_\nu^{\beta)}) \end{aligned} \tag{9.15}$$

(The σ in this formula should not be confused with σ_ς in (9.6)). Using (9.7), (9.2) and $T_\varsigma > 0$, inequality (8.10) can be rewritten as

$$T_\varsigma \pi_\varsigma = (S_N^{\alpha\beta} - T_\varsigma \Lambda_\varrho^\varsigma \varrho_\varsigma g^{\alpha\beta})D_{\alpha\beta} + S_V^{\alpha\beta} D_{\alpha\beta} - \frac{1}{T_\varsigma} q_\varsigma^\alpha T_{\varsigma,\alpha} + \mathcal{P} \geq 0\ , \tag{9.16}$$

where

$$\mathcal{P} := T_\varsigma \mathcal{S} - (T_\varsigma \Lambda_\varrho^\varsigma \mathcal{R} + \mathcal{U}). \tag{9.17}$$

The coefficients σ and σ_2 in $(9.15)_1$ are, in general, functions of $\varrho_\varsigma, T_\varsigma, K_M$ and K_G. $\Lambda_\varrho^\varsigma$, however, is only a function of ϱ_ς and T_ς. Consequently, the first term in (9.16) is linear in

[1]It was pointed out by Professor BÉDEAUX that, according to GIBBS, the relations (9.13) and (9.14) ought to have a curvature dependence in addition to the dependence on ϱ_ς and T_ς. Such a dependence, however, cannot arise via the constitutive assumption (7.7). In work in progress, SVENDSEN & HUTTER show that a curvature dependence arises in GIBBS' relation only if the time-derivative $\dot{b}_{\alpha\beta}$ of the surface curvature is assumed as an independent constitutive field.

$D_{\alpha\beta}$, and does not contain $T_{\varsigma,\alpha}$; thus the inequality (9.16) is valid for arbitrary $D_{\alpha\beta}$, and it could be violated unless

$$S_N^{\alpha\beta} - T_\varsigma \Lambda_\varrho^\varsigma \varrho_\varsigma g^{\alpha\beta} = (\sigma - T_\varsigma \Lambda_\varrho^\varsigma \varrho_\varsigma) g^{\alpha\beta} + \sigma_2 (b^{\alpha\beta} - K_M g^{\alpha\beta}) = 0. \tag{9.18}$$

The surface tensors $g_{\alpha\beta}$ and $g^{\alpha\beta} - K_M\, g^{\alpha\beta}$ are independent in non-equilibrium and may thus have any value, implying

$$\sigma = \sigma(\varrho_\varsigma, T_\varsigma) = T_\varsigma \Lambda_\varrho^\varsigma \varrho_\varsigma, \qquad \sigma_2 = 0. \tag{9.19}$$

$\Lambda_\varrho^\varsigma(\varrho_\varsigma, T_\varsigma)$ is determined by the scalar surface tension σ, density ϱ_ς and temperature T_ς. GIBBS' differential for the surface entropy thus becomes

$$ds_\varsigma = \frac{1}{T_\varsigma}\left[du_\varsigma + \frac{\sigma}{\varrho_\varsigma^2}\, d\varrho_\varsigma\right]. \tag{9.20}$$

Remark 1: Note that in classical theories of surface tension (9.20) is not used in this form. For the special case that the temperature T_ς and $\varrho_\varsigma = m_\varsigma / A_\varsigma$ are uniform over a phase boundary with mass m_ς and area A_ς, the total surface entropy $S_\varsigma = m_\varsigma\, s_\varsigma$, internal energy $U_\varsigma = m_\varsigma\, u_\varsigma$ and surface tension are connected by the relation

$$dS_\varsigma = \frac{1}{T_\varsigma}(dU_\varsigma - \sigma dA_\varsigma) - \frac{1}{T_\varsigma} G_\varsigma \frac{1}{m_\varsigma} dm_\varsigma \tag{9.21}$$

where

$$G_\varsigma := U_\varsigma - T_\varsigma S_\varsigma - \sigma A_\varsigma = m_\varsigma \left\{ u_\varsigma - T_\varsigma s_\varsigma - \frac{\sigma}{\varrho_\varsigma} \right\} =: m_\varsigma g_\varsigma \tag{9.22}$$

is the total free enthalpy. The differential (9.21) is an immediate consequence of (9.20). For surfaces with constant mass ($dm_\varsigma = 0$) it implies the well known result for the surface entropy. However, growing phase boundaries do not preserve their mass, and since $G_\varsigma \neq 0$ (as we shall show later) the application of (9.21) without the last term to problems of nucleation growth (as is done) is questionable. ▲

Finally, let us deduce the integrated forms for the constitutive equations of surface internal energy and surface entropy. From (9.12) and $(9.19)_1$ and the definition of the specific heat capacity at constant surface density we conclude that

$$c_\varrho^\varsigma := \frac{\partial u_\varsigma}{\partial T_\varsigma}, \qquad \frac{\partial u_\varsigma}{\partial \varrho_\varsigma} = -\frac{1}{\varrho_\varsigma^2}\left[\sigma - T_\varsigma \frac{\partial \sigma}{\partial T_\varsigma}\right], \tag{9.23}$$

relations which imply the integrability condition

$$\frac{\partial c_\varrho^\varsigma}{\partial \varrho_\varsigma} = \frac{T_\varsigma}{\varrho_\varsigma^2} \frac{\partial^2 \sigma}{\partial T_\varsigma^2}. \tag{9.24}$$

The entropy derivatives follow then from (9.20) or (8.7), $(9.19)_1$ and (9.23):

$$\frac{\partial s_\varrho^\varsigma}{\partial \varrho_\varsigma} = \frac{1}{\varrho_\varsigma^2}\frac{\partial \sigma}{\partial T_\varsigma}, \qquad \frac{\partial s_\varrho^\varsigma}{\partial T_\varsigma} = \frac{c_\varrho^\varsigma}{T_\varsigma}. \tag{9.25}$$

Integrating (9.24) between a reference state $(\varrho_\varsigma^0, T_\varsigma^0)$ and an arbitrary state $(\varrho_\varsigma, T_\varsigma^0)$ yields

$$c_\varrho^\varsigma(T_\varsigma, \varrho_\varsigma) = c_\varrho^\varsigma(T_\varsigma, \varrho_\varsigma^0) + T_\varsigma \frac{\partial^2}{\partial T_\varsigma^2} \int\limits_{\varrho_\varsigma^0}^{\varrho_\varsigma} \frac{\sigma(T_\varsigma, \varrho_\varsigma')}{\varrho_\varsigma'^2}\, d\varrho_\varsigma' . \tag{9.26}$$

Using this result (9.23) and (9.25) can be integrated and yields

$$\begin{aligned} u_\varsigma(T_\varsigma, \varrho_\varsigma) &= u_\varsigma^0 + \int\limits_{T_\varsigma^0}^{T_\varsigma} c_\varrho^\varsigma(T_\varsigma', \varrho_\varsigma^0)\, dT_\varsigma' - \left(1 - T_\varsigma \frac{\partial}{\partial T_\varsigma}\right) \int\limits_{\varrho_\varsigma^0}^{\varrho_\varsigma} \frac{\sigma(T_\varsigma, \varrho_\varsigma')}{\varrho_\varsigma'^2}\, d\varrho_\varsigma' , \\ s_\varsigma(T_\varsigma, \varrho_\varsigma) &= s_\varsigma^0 + \int\limits_{T_\varsigma^0}^{T_\varsigma} \frac{c_\varrho^\varsigma(T_\varsigma', \varrho_\varsigma^0)}{T_\varsigma'}\, dT_\varsigma' + \frac{\partial}{\partial T_\varsigma} \int\limits_{\varrho_\varsigma^0}^{\varrho_\varsigma} \frac{\sigma(T_\varsigma, \varrho_\varsigma')}{\varrho_\varsigma'^2}\, d\varrho_\varsigma' . \end{aligned} \tag{9.27}$$

Except for the constants of integration $u_\varsigma^0 = u_\varsigma^0(\varrho_\varsigma^0, T_\varsigma^0)$ and $s_\varsigma^0 = s_\varsigma^0(\varrho_\varsigma^0, T_\varsigma^0)$, surface entropy and internal energy are known, whenever the specific heat capacity $c_\varrho^\varsigma(\varrho_\varsigma^0, T_\varsigma)$ is known as a function of temperature and density.

We emphasize that all these results are valid in non-equilibrium. Further conclusion can be obtained from the inequality (9.16); these, however, refer to equilibrium.

10. Thermostatic equilibrium

10.1 Equilibrium properties

In view of $(9.16)_2$, $(7.10)_3$, (7.11) and $(6.9)_{4,5}$ the entropy production density is a function of the variables

$$X_A = \{D_{\alpha\beta}, T_{\varsigma,\alpha}; W_+^\alpha, W_n^+, \hat{t}_{n+}^\alpha, \hat{t}_n^+, q_n^+, \theta_+\} \tag{10.1}$$

with the property that π_ς vanishes whenever $X_A = 0$. The condition $X_A = 0$ is called thermostatic equilibrium and will be denoted by the index E. From (9.16) it follows that π_ς assumes its minimum value zero in equilibrium. Of necessity then

$$\left.\frac{\partial \pi_\varsigma}{\partial X_A}\right|_E = 0, \qquad \left.\frac{\partial \pi_\varsigma}{\partial X_A \partial X_B}\right|_E \quad \text{is positive semi-definite.} \tag{10.2}$$

Evaluation of (10.2) furnishes the equilibrium conditions for phase interfaces. We omit the detailed steps in exploiting conditions $(10.2)_1$ and only list the results which are

$$\begin{aligned} T_\varsigma \left.\frac{\partial \pi_\varsigma}{\partial W_+^\alpha}\right|_E &= \mathcal{G}_\alpha|_E = 0, \\ T_\varsigma \left.\frac{\partial \pi_\varsigma}{\partial \hat{t}_{n+}^\alpha}\right|_E &= \mathcal{V}_\alpha|_E = 0, \\ T_\varsigma \left.\frac{\partial \pi_\varsigma}{\partial W_n^+}\right|_E &= \mathcal{G}_n|_E - [\![\varrho(g - g_\varsigma)]\!]|_E = 0, \end{aligned}$$

$$
\begin{aligned}
T_\varsigma \frac{\partial \pi_\varsigma}{\partial \hat{t}_n^+}\bigg|_E &= \mathcal{V}_n|_E = 0, \\
T_\varsigma \frac{\partial \pi_\varsigma}{\partial \theta_+}\bigg|_E &= -\frac{1}{T_\varsigma}\mathcal{Q}|_E - [\varrho^+(g^+ - g_\varsigma)]|_E \frac{\partial \mathcal{V}_n}{\partial \theta_+} = 0, \\
T_\varsigma \frac{\partial \pi_\varsigma}{\partial q_n^+}\bigg|_E &= T_\varsigma \mathcal{K}|_E = 0, \qquad (10.3)\\
T_\varsigma \frac{\partial \pi_\varsigma}{\partial T_{\varsigma,\alpha}}\bigg|_E &= -\frac{1}{T_\varsigma} q_\varsigma^\alpha|_E = 0, \\
T_\varsigma \frac{\partial \pi_\varsigma}{\partial D_{\alpha\beta}}\bigg|_E &= S_V^{\alpha\beta}|_E - T_\varsigma \mathcal{Q}|_E \frac{\partial \mathcal{K}}{\partial D_{\alpha\beta}} - [\varrho^+(g^+ - g_\varsigma)]|_E \frac{\partial \mathcal{V}_n}{\partial D_{\alpha\beta}} = 0,
\end{aligned}
$$

where differentiation with respect to $\theta_+ = T_+ - T_-$ is understood as differentiation with respect to T_+ at fixed T_ς. In addition use of the representations (7.11) and $(9.15)_2$ yields

$$
\begin{aligned}
&\mathcal{G}_n|_E = 0 \qquad \mathcal{Q}|_E = 0 \qquad S_V^{\alpha\beta}|_E = 0, \\
&\frac{\partial \mathcal{V}_n}{\partial \theta_+}\bigg|_E = \mathcal{V}_6 \qquad \frac{\partial \mathcal{V}_n}{\partial D_{\alpha\beta}}\bigg|_E = \mathcal{V}_1 g^{\alpha\beta} + \mathcal{V}_2 b^{\alpha\beta}.
\end{aligned}
\qquad (10.4)
$$

Assuming that at least one of the transport coefficients $\mathcal{V}_1, \mathcal{V}_2$ or $\mathcal{V}_6$ does not identically vanish, relations (10.3) yield

$$
g^+|_E = g^-|_E = g_\varsigma|_E. \qquad (10.5)
$$

Thus the specific free enthalpies of the adjacent bulk materials and the specific free enthalpy of the phase boundary have the same values in equilibrium.

Remark 2: Comparison of $(6.10)_6$ and $(7.11)_1$ yields $\mathcal{V}_1 = \mathcal{V}_2 = \mathcal{V}_4 = \mathcal{V}_5 = \mathcal{V}_6 = 0$ and $\mathcal{V}_3 = 1/\mathcal{F}$, in contradiction with the assumptions from which (10.5) is deduced. However, the results (6.10) are based on the simplest possible boundary layer profiles. Relaxing these, by e.g. dispensing with the similarity assumption for the normal slopes, it can be shown that $\mathcal{V}_1 \neq 0$, $\mathcal{V}_2 \neq 0$, but $\mathcal{V}_6 = 0$, so that (10.5) can be maintained in a limiting sense. ▲

Remark 3: The first result of (10.5), $g^+|_E = g^-|_E$ is a well known result of thermostatics of phase equilibria and has been derived in various ways, see e.g. BAEHR (1979, p. 167). Investigating the behaviour of jump quantities at flat non-material phase boundaries leads to the same result, see e.g. MÜLLER (1972, p. 98). However, the second part of (10.5), $g^-|_E = g_\varsigma|_E$, is new and implies important consequences for the understanding of nucleation phenomena. It contradicts the basic assumption made in classical nucleation theory, that surface tension equals the free energy of the surface per unit area (see BECKER, 1961, p. 59). In our notation this assumption would read $\sigma_E = \varrho_\varsigma^E(u_\varsigma^E - T_\varsigma s_\varsigma^E)$ or $g_\varsigma^E = 0$. We conclude that the classical nucleation theory is thermodynamically inconsistent. Note that this conclusion is independent of the possibility that $\varrho_\varsigma = 0$. ▲

Remark 4: Equation (10.5) implies further inferences. To this end it will be assumed that polycrystalline ice is a compressible, visco-elastic solid, whereas water is a compressible

Newtonian fluid. (Omitting the elastic contributions makes the solid also a viscous fluid which is a classical assumption). For these materials the GIBBS relations can be written in a common form; in non-equilibrium, it is

$$ds = \frac{1}{T}\left\{du - \frac{p_E}{\varrho^2}\,d\varrho - \frac{1}{\varrho}\,\mathrm{tr}\{(\hat{\boldsymbol{t}}|_E\,d\boldsymbol{G})\,\boldsymbol{G}^{-1}\}\right\} \tag{10.6}$$

(ALTS, 1971, 1979), where p_E is the hydrostatic pressure, $\hat{\boldsymbol{t}}|_E = \boldsymbol{t}|_E + p_E\boldsymbol{1}$ is the trace-free Cauchy-stress in equilibrium and

$$\boldsymbol{G} := \frac{\boldsymbol{F}}{[\det(\boldsymbol{F})]^{1/3}}, \qquad \det(\boldsymbol{G}) = 1, \tag{10.7}$$

is a generalized shear and is constructed with the deformation gradient $\boldsymbol{F}$ and $\det(\boldsymbol{F}) = \varrho_R/\varrho$ (ϱ_R reference density in an undeformed state). Legendre-transformations of (10.6) for the bulk materials and of (9.20) for the surface material yield

$$\begin{aligned} dg &= -s\,dT + \frac{1}{\varrho}\,dp_E + \frac{1}{\varrho}\,\mathrm{tr}\{(\hat{\boldsymbol{t}}|_E\,d\boldsymbol{G})\,\boldsymbol{G}^{-1}\}, \\ dg_\varsigma &= -s_\varsigma\,dT_\varsigma - \frac{1}{\varrho_\varsigma}\,d\sigma. \end{aligned} \tag{10.8}$$

The free enthalpy $g = g(T, p_E, \boldsymbol{G})$ for ice depends on temperature, pressure and shear in non-equilibrium, $g = g(T, p_E)$ for water is a function of temperature and pressure since $\hat{\boldsymbol{t}}|_E = \boldsymbol{O}$, and $g_\varsigma = g_\varsigma(T_\varsigma, \sigma)$ depends on temperature and surface tension.

In equilibrium $T_+ = T_- = T_\varsigma$, as follows from (10.1) and $(10.3)_6$. Moreover, because the fluid-like boundary layer cannot transmit equilibrium shear stresses, the adjacent solid bulk material must equally be free of shear stresses, explicitly $\hat{\boldsymbol{t}}_+|_E = \boldsymbol{0}$ or $\hat{\boldsymbol{t}}_-|_E = \boldsymbol{0}$, depending on at which side the ice is situated. Consequently, in equilibrium relations (10.8) reduce to

$$\begin{aligned} dg_E^\pm &= -s_E^\pm\,dT_\varsigma + \frac{1}{\varrho_E^\pm}\,dp_E^\pm, \\ dg_\varsigma^E &= -s_\varsigma^E\,dT_\varsigma + \frac{1}{\varrho_\varsigma^E}\,d\sigma_E. \end{aligned} \tag{10.9}$$

These relations will frequently be used in what follows. ▲

Apart from (10.5), the mechanical equilibrium conditions must be satisfied. These follow from the momentum equations $(3.7)_{2,3}$ by using $\hat{\boldsymbol{t}}_\pm|_E = p_E\boldsymbol{1}$, $(7.10)_5$ and (9.19):

$$\begin{aligned} (p^+ - p^-)|_E &= (2\sigma K_M + \varrho_\varsigma \boldsymbol{e}\cdot\boldsymbol{g})|_E, \\ \sigma_{,\alpha}|_E &= -(\varrho_\varsigma \boldsymbol{\tau}_\alpha\cdot\boldsymbol{g})|_E. \end{aligned} \tag{10.10}$$

Differentiating $(10.10)_1$ with respect to ξ^α, using $(10.10)_2$ and writing $p^\pm_{,\alpha}|_E = (\varrho^\pm\,\boldsymbol{\tau}_\alpha\cdot\boldsymbol{g})|_E$ as obtained from the barometric formulae yields

$$2\sigma K_{M,\alpha}|_E = \left\{\left[[\![\varrho]\!] + \varrho_\varsigma\{2K_M + \frac{\partial\varrho_\varsigma}{\partial\sigma}(\boldsymbol{e}\cdot\boldsymbol{g})\}\right]\boldsymbol{\tau}_\alpha + \varrho_\varsigma b_\alpha^\beta\boldsymbol{\tau}_\beta\right\}\Big|_E\cdot\boldsymbol{g}\,, \tag{10.11}$$

in which the gravity vector g is constant. It follows that the curvature possesses a gradient tangential to the phase boundary; the latter is therefore non-spherical. However, as discussed by ALTS & HUTTER (1986), the right hand side of (10.11) is negligibly small whenever the dimensionless curvature $|k_M| \geq 10^{-10}$, which is well satisfied for temperate glacier ice. Henceforth gravitational contributions to (10.11) shall be ignored.

A second set of conclusions can be inferred from (10.5) and $(10.10)_1$, which now read

$$
\begin{aligned}
g^{\pm}(T_\varsigma, p_E^{\pm}) &= g_\varsigma(T_\varsigma, \sigma_E), \\
p_E^{+} - p_E^{-} &= = 2\sigma_E K_M .
\end{aligned}
\qquad (10.12)
$$

Because in equilibrium the phase boundary is fixed in space, K_M is fixed and (10.12) are three equations for the determination of the four fields $p_E^{\pm}, \sigma_E$ and T_ς. Assuming T_ς to be prescribed and supposing that (10.12) can be inverted, it follows that

$$
\begin{aligned}
p_E^{\pm} &= p^{\pm}(T_\varsigma, K_M), \\
\sigma_E &= \sigma(T_\varsigma, K_M).
\end{aligned}
\qquad (10.13)
$$

For $K_M \neq 0$, the equilibrium pressures of ice and water differ from each other. With (10.13) the equilibrium densities, entropies and internal energies of the bulk materials and the phase boundary become functions of the temperature (and curvature) alone, viz.

$$
\begin{aligned}
\varrho_E^{\pm}(T_\varsigma, K_M) &= \varrho^{\pm}(T_\varsigma, p^{\pm}(T_\varsigma, K_M)), \\
\varrho_\varsigma^{E}(T_\varsigma, K_M) &= \varrho_\varsigma(T_\varsigma, \sigma(T_\varsigma, K_M)),
\end{aligned}
\qquad (10.14)
$$

and so on. It should be emphasized that this curvature dependence enters in order to satisfy the equilibrium equations (10.10)-(10.12). The constitutive relations (9.13) remain independent of curvature. Change of curvature at constant temperature shifts the phase equilibrium as does change of temperature at constant curvature. The inferences, however, differ from each other and will be discussed in due course.

10.2 Change of curvature at fixed temperature

Differentiating the equilibrium conditions (10.12) with respect to K_M and using (10.13) and (10.9) yields

$$
\begin{aligned}
\frac{\partial p_E^{\pm}}{\partial K_M} &= -\frac{\varrho_E^{\pm}}{\varrho_\varsigma^{E}} \frac{\partial \sigma_E}{\partial K_M}, \\
\frac{\partial p_E^{+}}{\partial K_M} - \frac{\partial p_E^{-}}{\partial K_M} &= 2\frac{\partial \sigma_E}{\partial K_M} K_M + 2\sigma_E ,
\end{aligned}
\qquad (10.15)
$$

from which by eliminating the pressure derivatives one obtains

$$
\frac{1}{\sigma_E} \frac{\partial \sigma_E}{\partial K_M} = \frac{-1}{K_M + [\![p_E]\!] / 2\varrho_\varsigma^{E}} . \qquad (10.16)
$$

This is a differential equation for the determination of the curvature dependence of the surface tension. When neglecting gravity (10.11) implies that the mean curvature is constant along the equilibrium phase boundary. The equilibrium shapes of all isotropic phase boundaries that separate isotropic bulk materials are then either spheres or circular cylinders.

Inserting now the result (5.5) into (10.16), and using the dimensionless curvature $k_M = K_M d_\varsigma$, we integrate the resulting expression to obtain

$$\frac{\sigma_E}{\sigma_E^0} = \exp\left\{ - \int_0^{k_M} \frac{\varrho_\varsigma F_1(k, k_G)}{1 + \varrho_\varsigma k F_1(k, k_G)} dk \right\}, \tag{10.17}$$

where $\sigma_E^0 = \sigma_E(T_\varsigma, k_M = 0)$ is the surface tension for the flat phase boundary and where $F_1(k_M)$ equals either $F_1(k_M, k_G = k_M^2)$ for the sphere or $F_1(k_M, k_G = 0)$ for the circular cylinder. Equation (10.17) exhibits a distinct and very interesting curvature dependence of the surface tension.

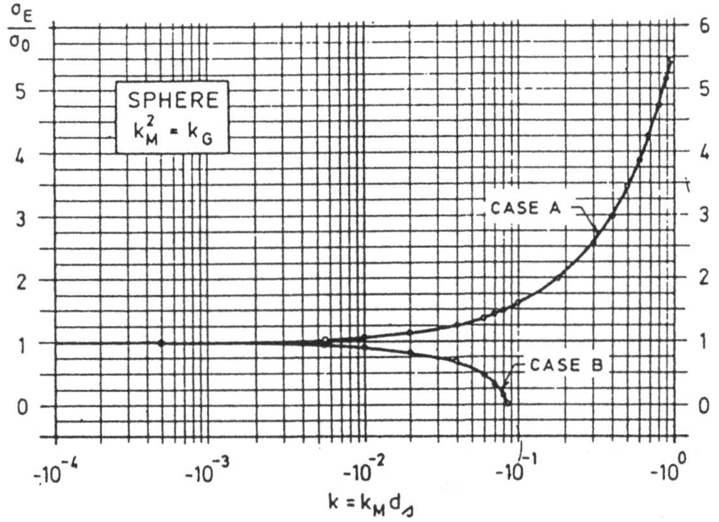

Fig. 14: Normalized surface tension σ_E/σ_E^0 as a function of curvature according to (8.23) for spherical geometry. Case A: spherical water inclusion on ice. Case B: spherical ice inclusion in water.

For spherical geometry the result of numerical integration of (10.17) with the values for $F_1(k_M)$ is drawn in Figure 14. Due to the different signs of $F_1(k_M)$ the surface tension for a water inclusion in ice increases with increasing curvature $|k_M|$ (case A); for an ice inclusion in water (case B), it decreases with increasing curvature and vanishes when $|k_M| = 0.09$. *Vanishing surface tension, however, corresponds physically to disappearance of the phase boundary.* We conclude, that ice inclusions in water of nucleation dimensions $|k_M| \geq 0.09$ do not exist. This supports our earlier conclusion in section 5 about the unlikelyhood of ice nucleation in pure water.

As far as we know, this is a new result on the curvature dependence of the surface tension with important consequences for the physical understanding and the mechanism of nucleation. The reason for this novelty is that experiments on surface tension are performed at too small dimensionless curvatures as to cover the regime where the surface tension splits up into the two branches. It also describes a rather general phenomenon of isotropic phase boundaries. (Anisotropic phase boundaries possess in equilibrium a tensor of surface tension and correspondingly more curvature measures!).

Whereas embedding of the denser phase in the less dense phase (water in ice or water in vapour) is possible down to nucleation dimensions, inclusion of the less dense in the denser phase (ice in water or vapour in water) is not possible at very small inclusion size. This may explain the phenomena of undercooling and overheating observed in water.

Let us repeat the assumptions on which (10.17) is based upon. Besides isotropy, the boundary-layer theory required (i) that the thickness d_ς of the layer be independent of the curvature and (ii) that the density profile is a cubic function of the boundary layer co-ordinate. Because the layer is extremely thin (of the order of a few molecular distances) and the cubic profile guarantees a smooth connection of the bulk phases, these are fairly realistic assumptions. Minor changes are possible, they certainly will not change the qualitative behaviour, as discussed in Figure 14 and will result in only small quantitative changes. In particular, the splitting of the $\sigma_E(k_M)$-relation into two curves must be preserved under any reasonable boundary-layer assumptions. This is so because the dynamic compatibility condition requires two families of solutions for the eccentricity parameter $\alpha = a_\varsigma/d_\varsigma$. And the compatibility condition, in turn, resulted from the necessity to define material lines for the motion within the surface, a concept without which a surface tension could not have been introduced. So the chain of argumentations closes. Basic to (10.17) is thus dynamic consistency of the boundary layer theory of singular surfaces. By purely static reasoning a result like (10.17) for the curvature dependence of the static surface tension could never have been achieved.

With the result (10.17) the adjacent bulk values of the pressure $p_E^\pm$, internal energy $u_E^\pm$ and entropy $s_E^\pm$ can be computed as functions of curvature k_M and temperature. To this end, however, the equations of state for the bulk materials must be known. The results will thus depend upon the materials in question. The equilibrium conditions (10.15) and the boundary-layer result (5.6) form four equations for the six fields $p_E^\pm, \sigma_E, u_E^\pm, \varrho_\varsigma^E$. To integrate them the equations of state, $\varrho^\pm(T^\pm, p^\pm)$ and $T^+ = T^- = T_\varsigma$ serve as the complementing constitutive relations. For low pressures, at which ice Ih (hexagonal ice arising in the geophysical environment) exists, one may without essential loss of accuracy impose the incompressibility assumption and thus write

$$\varrho_E^\pm = \varrho_0^\pm(T_\varsigma). \tag{10.18}$$

(For the water-vapour phase transition at low pressure, water can be considered approximately as incompressible and steam as an ideal gas satisfying $p = \varrho RT$, R = specific gas constant for H_2O. We shall, however, not go into the details for this phase transition). With

(10.18), the boundary-layer result (6.11) yields the equilibrium surface density in the form

$$\varrho_\varsigma^E = \varrho_\varsigma(T_\varsigma, k_M) = [\![\varrho_0(T_\varsigma)]\!] \frac{d_\varsigma}{2} F_1(k_M) \tag{10.19}$$

(in non-equilibrium ϱ_ς would be a function of both curvatures). With (10.17)-(10.19) relations $(10.15)_1$ can be integrated to yield

$$p_E^\pm(T_\varsigma, k_M) = p_E^0(T_\varsigma) + \frac{\varrho_0^\pm(T_\varsigma)}{[\![\varrho_0(T_\varsigma)]\!]} \frac{2}{d_\varsigma} \int_0^{k_M} \frac{\sigma_E(T_\varsigma, k)}{1 + kF_1(k)} dk. \tag{10.20}$$

$p_E^0(T_\varsigma)$ is the melting pressure for the flat phase boundary. Inserting (10.20) into $(10.12)_3$ yields

$$\int_0^{k_M} \frac{\sigma_E(T_\varsigma, k)}{1 + kF_1(k)} dk = \sigma_E(T_\varsigma, k_M)\, k_M. \tag{10.21}$$

This is a relation equivalent to the differential equation (10.16). Substituting this into (10.20), we obtain the simpler form

$$p_E^\pm(T_\varsigma, k_M) = p_E^0(T_\varsigma) + \frac{\varrho_0^\pm(T_\varsigma)}{[\![\varrho_0(T_\varsigma)]\!]} \frac{2}{d_\varsigma} \sigma_E(T_\varsigma, k_M)\, k_M\,. \tag{10.22}$$

For later use we record here expressions for the specific internal energy and the entropy of the incompressible bulk materials adjacent to the phase boundary. These can be shown to be

$$\begin{aligned} u_E^\pm(T_\varsigma, p_E^\pm) &= u_0^\pm(T_\varsigma, p_E^0) + \frac{T_\varsigma}{(\varrho^\pm)^2} \frac{d\varrho_0^\pm}{dT_\varsigma} (p_E^\pm - p_E^0), \\ s_E^\pm(T_\varsigma, p_E^\pm) &= s_0^\pm(T_\varsigma, p_E^0) + \frac{1}{(\varrho^\pm)^2} \frac{d\varrho_0^\pm}{dT_\varsigma} (p_E^\pm - p_E^0), \end{aligned} \tag{10.23}$$

where $u_0^\pm$ and $s_0^\pm$ are the adjacent equilibrium values at a flat phase boundary, given by

$$\begin{aligned} u_0^\pm(T_\varsigma, p_E^0) &= u_{tr}^\pm + \int_{T_{tr}}^{T_\varsigma} c_p^\pm(T', p_{tr})\, dT' + \frac{T_\varsigma}{(\varrho^\pm)^2} \frac{d\varrho_0^\pm}{dT_\varsigma} (p_E^0(T_\varsigma) - p_{tr}) - p_{tr} \left\{ \frac{1}{\varrho^\pm(T_\varsigma)} - \frac{1}{\varrho_{tr}^\pm} \right\}, \\ s_0^\pm(T_\varsigma, p_E^0) &= s_{tr}^\pm + \int_{T_{tr}}^{T_\varsigma} \frac{c_p^\pm(T', p_{tr})}{T'} dT' + \frac{1}{(\varrho^\pm)^2} \frac{d\varrho_0^\pm}{dT_\varsigma} (p_E^0(T_\varsigma) - p_{tr}), \end{aligned} \tag{10.24}$$

where $T_{tr} = 273.16$ K and $p_{tr} = 0.006$ bar are the triple point temperature and pressure for H_2O, and $u_{tr}^\pm, s_{tr}^\pm$ denote the internal energy and entropy of bulk ice or water at the triple point. (By bulk ice we mean ice specimens for which size and boundary effects will not affect the measured material properties). Moreover, $c_p^{\pm}(T, p_{tr})$ is the specific heat capacity of bulk ice or water at the same pressure. Details may be found in ALTS (1984). Setting

$$u_E^{\pm}(T_\varsigma, p_E^{\pm}(T_\varsigma, k_M)) = u_E^{\pm}(T_\varsigma, k_M), \qquad u_0^{\pm}(T_\varsigma, p_E^0(T_\varsigma)) = u_0^{\pm}(T_\varsigma),$$

etc., and using (10.22), the final form of (10.23) becomes

$$\begin{aligned} u_E^{\pm}(T_\varsigma, k_M) &= u_0^{\pm}(T_\varsigma) - T_\varsigma \frac{\beta_0^{\pm}(T_\varsigma)}{[\![\varrho_0(T_\varsigma)]\!]} \frac{2}{d_\varsigma} \sigma_E(T_\varsigma, k_M)\, k_M, \\ s_E^{\pm}(T_\varsigma, k_M) &= s_0^{\pm}(T_\varsigma) - \frac{\beta_0^{\pm}(T_\varsigma)}{[\![\varrho_0(T_\varsigma)]\!]} \frac{2}{d_\varsigma} \sigma_E(T_\varsigma, k_M)\, k_M \end{aligned} \tag{10.25}$$

where $\beta_0^{\pm}(T_\varsigma)$ is the thermal volume expansion coefficient for bulk ice or water

$$\beta_0^{\pm}(T_\varsigma) := \frac{-1}{\varrho^{\pm}(T_\varsigma)} \frac{d\varrho_0^{\pm}}{dT_\varsigma}. \tag{10.26}$$

Various other relations can be derived, but we only list the specific free enthalpies of the bulk materials in equilibrium. These are obtained from (8.5) by inserting (10.15) and (10.22):

$$g_E^{\pm}(T_\varsigma, k_M) = g_0^{\pm}(T_\varsigma) + \frac{1}{[\![\varrho_0(T_\varsigma)]\!]} \frac{2}{d_\varsigma} \sigma_E(T_\varsigma, k_M)\, k_M. \tag{10.27}$$

10.3 Change of temperature at fixed curvature

Let us now turn to the consequences of the equilibrium conditions (10.12), when the temperature is shifted at fixed curvature. Differentiating (10.12) with respect to T_ς and using (10.9) yields

$$\begin{aligned} -s_E^{\pm} - \frac{1}{\varrho_E^{\pm}} \left.\frac{\partial \sigma_E}{\partial T_\varsigma}\right|_{K_M} &= -s_\varsigma^E + \frac{1}{\varrho_\varsigma^E} \left.\frac{\partial p_\varsigma^E}{\partial T_\varsigma}\right|_{K_M}, \\ \left.\frac{\partial p_E^{+}}{\partial T_\varsigma}\right|_{K_M} - \left.\frac{\partial p_E^{-}}{\partial T_\varsigma}\right|_{K_M} &= 2K_M \left.\frac{\partial \sigma_E}{\partial T_\varsigma}\right|_{K_M}, \end{aligned} \tag{10.28}$$

from which, on eliminating the right hand sides of $(10.28)_{1,2}$ we deduce

$$\left[\!\!\left[\frac{1}{\varrho_E}\right]\!\!\right] \left.\frac{\partial p_E^{\pm}}{\partial T_\varsigma}\right|_{K_M} = [\![s_E]\!] - 2K_M \frac{1}{\varrho_E^{\mp}} \left.\frac{\partial \sigma_E}{\partial T_\varsigma}\right|_{K_M}. \tag{10.29}$$

In the limits $K_M \to 0$, $s_E^{\pm} \to 0$, $\varrho_E^{\pm} \to \varrho_0^{\pm}$, and $p_E^{\pm} \to p_E^0$, the relations (10.29) reduce to the well known CLAUSIUS-CLAPEYRON relation

$$\left[\!\!\left[\frac{1}{\varrho_0}\right]\!\!\right] \frac{dp_E^0}{dT_\varsigma} = [\![s_0]\!] \tag{10.30}$$

for the melting pressure at flat phase boundaries. For this reason (10.29) will be called the *generalized* CLAUSIUS-CLAPEYRON relations. With water on the + side of the interface we may identify $s_0^+ = s_W^0$, $s_0^- = s_I^0$, $\varrho_0^+ = s_W^0$, $\varrho_0^- = s_I^0$, and then obtain, since $s_W^0 - s_I^0 = [\![s_E^0]\!] > 0$, $1/\varrho_W^0 - 1/\varrho_I^0 = [\![1/\varrho_E^0]\!] < 0$, $dp_E^0/dT_\varsigma < 0$. Consequently, an increase of the pressure reduces the melting temperature, as expected for pressure melting of ice.

The generalized CLAUSIUS-CLAPEYRON equations depend in a complex fashion on the materials, the kind of phase equilibrium (solid-fluid, solid-vapour, fluid-vapour) and on curvature. A considerable simplification can be achieved for phase equilibrium between ice and water; indeed if one assumes both bulk materials to be approximately incompressible such that

$$\varrho^\pm(T_\varsigma, p^\pm) \approx \varrho^\pm(T_\varsigma) = \varrho_0^\pm(T_\varsigma), \tag{10.31}$$

then the bulk densities are independent of curvature. By inserting (10.30) and $(10.25)_2$ into (10.29) we obtain

$$\left[\!\!\left[\frac{1}{\varrho_0}\right]\!\!\right]\left\{\left.\frac{\partial p_E^\pm}{\partial T_\varsigma}\right|_{K_M} - \frac{dp_E^0}{dT_\varsigma}\right\} = -\frac{2K_M}{[\![\varrho_0]\!]}\left\{[\![\beta_0]\!]\sigma_E + \frac{[\![\varrho_0]\!]}{\varrho_0^\mp}\left.\frac{\partial \sigma_E}{\partial T_\varsigma}\right|_{K_M}\right\}. \tag{10.32}$$

These relations describe the departures of the equilibrium pressures in the adjacent bulk materials at curved phase boundaries from the usual pressure melting curve. These departures can be considerable when curvatures are large and cannot be neglected for inclusions of nucleation dimensions.

Eqs. (10.29) and (10.30) or (10.32) may be used together with (10.17) to evaluate surface entropy s_ς^E and surface energy u_ς^E in terms of p_E^0, T_ς, K_M. This is done for incompressible bulk materials by ALTS & HUTTER (1986). The results of the lengthy manipulations are

$$s_E^\pm - s_\varsigma^E = \frac{1}{\varrho_0^\pm}\frac{dp_E^0}{dT_\varsigma} + \frac{2k_M\sigma_E}{d_\varsigma[\![\varrho_0]\!]}\left\{\frac{1}{\varrho_0^\pm}\frac{d\varrho_0^\pm}{dT_\varsigma} - \frac{1}{[\![\varrho_0]\!]}\frac{d[\![\varrho_0]\!]}{dT_\varsigma}\right\} + \frac{2}{d_\varsigma[\![\varrho_0]\!]}\left.\frac{\partial\sigma_E}{\partial T_\varsigma}\right|_{K_M}\frac{1+k_M F_1(k_M)}{F_1(k_M)}, \tag{10.33}$$

$$\begin{aligned} u_E^\pm - u_\varsigma^E &= -\frac{1}{\varrho_0^\pm}\left\{p_E^0 - T_\varsigma\frac{dp_E^0}{dT_\varsigma}\right\} - \frac{2}{d_\varsigma[\![\varrho_0]\!]}\left\{\sigma_E - T_\varsigma\left.\frac{\partial\sigma_E}{\partial T_\varsigma}\right|_{K_M}\right\}\frac{1+k_M F_1(k_M)}{F_1(k_M)} \\ &+ \frac{2k_M\sigma_E}{d_\varsigma[\![\varrho_0]\!]}T_\varsigma\left\{\frac{1}{\varrho_0^\pm}\frac{d\varrho_0^\pm}{dT_\varsigma} - \frac{1}{[\![\varrho_0]\!]}\frac{d[\![\varrho_0]\!]}{dT_\varsigma}\right\}. \end{aligned} \tag{10.34}$$

Taking the differences of these yields expressions for the jumps $[\![u_E]\!]$ and $[\![s_E]\!]$ as follows

$$\begin{aligned} [\![u_E]\!] &= -\left[\!\!\left[\frac{1}{\varrho_0}\right]\!\!\right]\left\{p_E^0 - T_\varsigma\frac{dp_E^0}{dT_\varsigma}\right\} - \frac{2k_M\sigma_E}{d_\varsigma[\![\varrho_0]\!]}T_\varsigma[\![\beta_0]\!], \\ [\![s_E]\!] &= \left[\!\!\left[\frac{1}{\varrho_0}\right]\!\!\right]\frac{dp_E^0}{dT_\varsigma} - \frac{2k_M\sigma_E}{d_\varsigma[\![\varrho_0]\!]}[\![\beta_0]\!], \end{aligned} \tag{10.35}$$

where (10.26) has also been used, and σ_E is thought to be expressed in terms of σ_E^0 and k_M as shown in (10.17). Now, the melting pressure $p_E^0(T_\varsigma)$ and the surface tension $\sigma_E^0(T_\varsigma)$ for the flat phase boundary and the densities $\varrho_0^\pm(T_\varsigma)$ for the bulk phases are known functions of the temperature. Moreover, $F_1(k_M) = \{F_1(k_M, k_G = k_M^2)$ for the sphere, $F_1(k_M) = F_1(k_M, k_G = 0)$ for the cylinder $\}$, is a known function of the dimensionless curvature k_M. Hence, the entropy differences in (10.33), the differences of internal energy in (10.34) and the jump quantities (10.35) are known functions of temperature and curvature, whenever the thickness parameter d_ς is chosen.

We want to emphasize that, of the boundary-layer results discussed so far, only the "safe" relation, namely $\varrho_\varsigma = [\![\varrho]\!]\,(d_\varsigma/2)\,F_1$, has been used. All results in this section are otherwise consequences of the equilibrium conditions and the assumption of incompressibility of the bulk materials ice and water, assumptions which are reasonably well satisfied for the low pressures that arise in ordinary ice.

All jumps, surface fields and normal slopes which govern the thermostatic equilibrium are traced back to known functions of the curvature and measurable quantities at the flat phase interface. They are therefore known functions of the temperature T_ς and the dimensionless curvature k_M. Numerical values can be determined as soon as the boundary-layer thickness d_ς is chosen. This is fortunate as d_ς is the only adjustable parameter left in this theory.

A final remark in connection with the results (10.35): alternative expressions for these are given in $(6.10)_{3,4}$; the two pairs of relations can then be shown to conflict. Scrutiny for the cause shows that the similarity assumptions for internal energy and entropy made in (5.15)-(5.16) must be abandoned. ALTS & HUTTER (1986) show that alternative cubic profiles for internal energy and entropy can be derived such that all above equilibrium relations can be satisfied, thus removing the inconsistency.

11. Linear transport equations on and across phase-boundary-layers

Relation $(10.2)_2$, expressing the positve semi-definiteness of the Hessian matrix of the entropy production density imposes further restrictions on the transport coefficients arising in the linear constitutive equations $(7.10)_{3,5}$, and (7.11). The evaluation of these restrictions is the content of this section.

The thickness of the phase-boundary layer between ice and water is on the order of a few molecular distances, say $d_\varsigma = 100$ Å. This is much smaller than the dimensions of the smallest available thermometer. By temperature measurements it is therefore impossible to distinguish between the temperatures $T_\pm$ of the adjacent bulk materials and the temperature T_ς of the dividing surface of the bulk phases. This is especially so, since on the surface of the thermometer new boundary layers must develop, when it is inserted into the material whose temperature is to be measured. We therefore assume

(i) The temperature is continuous across the phase boundary and equal to the surface temperature

$$T_+ = T_- = T_\varsigma. \tag{11.1}$$

To set $T_+ = T_- = T_\varsigma$ is the only reasonable possibility if the two bulk materials have comparable densities. In evaporation processes it may be more appropriate to choose $T_\varsigma =$

T_+, where the material on the +-side is denser, and to suppose that the vapour temperature T_- differs from T_+. This would yield yet a different formulation.

The processes of freezing and melting or evaporation and condensation are slow and near equilibrium, where the entropy production density π_ς vanishes. A further possible idealization is therefore achieved by requiring vanishing entropy production density. Consequently, we *assume*

(ii) The phase change process is reversible, implying

$$\pi_\varsigma = 0, \text{ for all thermodynamic processes.} \tag{11.2}$$

These two requirements are consistent with the entropy principle for phase-boundary layers and allow a considerable simplification of the transport equations $(7.10)_{3,5}$ and (7.11) without reducing them to triviality. We discuss their consequences first in the hope of obtaining insight into allowable simplifications of the transport equations for real phase boundaries.

Applying the continuity assumption (11.1) to the transport equations (7.11) requires $\theta_+ = T_+ - T_\varsigma = 0$, $\mathcal{K} = [\![1/T]\!] = 0$ and

$$\mathcal{K}_j = 0, \qquad (j = 1, 2, \ldots, 5). \tag{11.3}$$

Moreover, it is no longer meaningful to define transport coefficients associated with θ_+; thus

$$\mathcal{V}_6 = \mathcal{G}_6 = \mathcal{D}_6 = \mathcal{K}_6 = 0. \tag{11.4}$$

For $T_+ = T_- = T_\varsigma$, the equilibrium pressures and surface tension are $p^\pm_E = p^\pm_E(T_\varsigma, K_M)$ and $\sigma = \sigma_E(T_\varsigma, K_M)$, respectively, and the specific free enthalpies $g^\pm(T_\pm, p^\pm_E)$, $g^\pm(T_\pm, \sigma)$ in nonequilibrium are continuous, see (10.5). Thus

$$g^\pm = g^\pm_E = g^E_\varsigma = g_\varsigma. \tag{11.5}$$

Inserting (11.3) - (11.5) into the entropy production density (9.16) and using (9.18) and (6.9) yields for reversibility the relation

$$T_\varsigma \pi_\varsigma = S^{\alpha\beta}_V D_{\alpha\beta} - \frac{1}{T_\varsigma} q^\alpha_\varsigma T_{\varsigma,\alpha} + \mathcal{G}_\alpha (W^\alpha_+ - \mathcal{V}^\alpha) + \mathcal{V}_\alpha \hat{t}^\alpha_{n+} + \mathcal{G}_n (W^+_n - \mathcal{V}_n) + \mathcal{V}_n \hat{t}^+_n = 0, \tag{11.6}$$

valid for every thermodynamic process. Relation (11.6) imposes further restrictions on the transport equations $(7.10)_{3,5}$ and (7.11). Note that it does not contain the jump $[\![q_n]\!]$ of the normal component of the heat flux. It follows that $[\![q_n]\!]$ is not restricted by the entropy principle when continuity of temperature and reversibility of phase-transition processes are assumed. Instead, the balance equation $(3.7)_4$ for surface internal energy now serves as a constraint, from which $[\![q_n]\!]$ can be determined. This will be done later.

As a matter of simplification, let us now use the "safe" boundary-layer result $\mathcal{V}^\alpha = W_+/\mathcal{F}$, as given in (6.10) and (6.11). Comparison with (7.11) yields

$$\mathcal{D}_1 = \mathcal{D}_2 = 0, \quad \mathcal{D}_3 = \frac{1}{\mathcal{F}}, \quad \mathcal{D}_4 = \mathcal{D}_5 = \mathcal{D}_6 = 0. \tag{11.7}$$

Inserting $(7.10)_3$, (7.11), $(9.15)_2$, (11.3), (11.4) and (11.7) into (11.6) yields a lengthy expression which is identically zero for arbitrary thermodynamic processes, if and only if

$$
\begin{aligned}
&\kappa = \kappa_2 = 0, \\
&\mu_1 = \mu_3 K_M, \quad \mu_2 = -\mu_3, \quad \mu_4 = \mu_5 = 0, \quad \mu_6 \neq 0, \\
&\mathcal{C}_1 = \mathcal{C}_2 = \mathcal{C}_3 = \mathcal{C}_4 = 0, \quad \mathcal{C}_5 = -\frac{\mathcal{D}_3}{1-\mathcal{D}_3}, \quad \mathcal{C}_6 = 0, \\
&\mathcal{V}_1 = \mathcal{V}_2 = 0, \quad \mathcal{V}_3 \neq 0, \quad \mathcal{V}_4 = \mathcal{V}_5 = 0; \\
&\mathcal{G}_1 = \mathcal{G}_2 = \mathcal{G}_3 = 0, \quad \mathcal{G}_4 = -\frac{\mathcal{G}_3}{1-\mathcal{G}_3}, \quad \mathcal{G}_5 = 0.
\end{aligned}
\tag{11.8}
$$

The following linear transport equations are therefore necessary and sufficient conditions for the occurrence of reversible phase transitions:

$$
\begin{aligned}
&q_\varsigma^\alpha = 0, \quad \phi_\varsigma^\alpha = 0, \\
&S^{\alpha\beta} = \sigma_E g^{\alpha\beta} + \mu_3(I_D b^{\alpha\beta} - II_D g^{\alpha\beta}) + \mu_6 g^{\mu\nu}(b_\mu^{(\alpha} D_\nu^{\beta)} - D_\mu^{(\alpha} b_\nu^{\beta)})
\end{aligned}
\tag{11.9}
$$

and

$$
\begin{aligned}
\mathcal{V}_n &= \mathcal{V}_3 W_n^+, \qquad & \mathcal{G}_n &= -\frac{\mathcal{V}_3}{1-\mathcal{V}_3}\hat{t}_n^+, \\
\mathcal{V}^\alpha &= \mathcal{D}_3 W_+^\alpha, \qquad & \mathcal{G}^\alpha &= -\frac{\mathcal{D}_3}{1-\mathcal{D}_3}\hat{t}_{n+}^\alpha.
\end{aligned}
\tag{11.10}
$$

Notice that the viscosities μ_3 and μ_6 are the only phenomenological quantities arising in (11.9). Comparing, finally, $(11.10)_1$ with the boundary layer result $(6.10)_6$ we conclude that

$$
\mathcal{V}_3 = \frac{1}{\mathcal{F}}. \tag{11.11}
$$

For this to hold it was tacitly assumed that the similarity argument of the boundary-layer theory applies to the normal velocity jump. With it and with $(11.7)_3$ we deduce that

$$
\begin{aligned}
\mathcal{V}_n &= \frac{1}{\mathcal{F}} W_n^+, \qquad & \mathcal{G}_n &= \frac{1}{1-\mathcal{F}}\hat{t}_n^+, \\
\mathcal{V}^\alpha &= \frac{1}{\mathcal{F}} W_+^\alpha, \qquad & \mathcal{G}^\alpha &= \frac{1}{1-\mathcal{F}}\hat{t}_{n+}^\alpha.
\end{aligned}
\tag{11.12}
$$

A more symmetrical form of these relations can be derived by inserting $(6.10)_1$, (6.2) and the incompressibility assumptions (10.18) for bulk ice and water:

$$\begin{aligned}(F_1\varrho_0^+ + \frac{1}{3}F_2[\![\varrho_0]\!])W_n^+ &= (F_1\varrho_0^- + \frac{1}{3}F_2[\![\varrho_0]\!])W_n^-,\\ (F_1\varrho_0^- + \frac{1}{3}F_2[\![\varrho_0]\!])\hat{t}_{+n} &= (F_1\varrho_0^+ + \frac{1}{3}F_2[\![\varrho_0]\!])\hat{t}_n^-,\\ (F_1\varrho_0^+ + \frac{1}{3}F_2[\![\varrho_0]\!])W_+^\alpha &= (F_1\varrho_0^- + \frac{1}{3}F_2[\![\varrho_0]\!])W_-^\alpha,\\ (F_1\varrho_0^- + \frac{1}{3}F_2[\![\varrho_0]\!])\hat{t}_{n+}^\alpha &= (F_1\varrho_0^+ + \frac{1}{3}F_2[\![\varrho_0]\!])\hat{t}_{n-}^\alpha.\end{aligned} \tag{11.13}$$

Relations (11.12) or (11.13) for reversible transport across the phase boundary are free of new transport coefficients. Hence, given the relative velocities $W_n^+ = v_n^+ - w_n$ and $W_+^\alpha = v_+^\alpha - w^\alpha$, and the stress deviators $\hat{t}_n^+ = e \cdot t_+ e + p_E^+$ and $\hat{t}_{n+}^\alpha = g^{\alpha\beta}\tau_\beta \cdot t_+ e$ on one side of the phase boundary, as well as its curvatures and the adjacent bulk densities $\varrho_0^\pm(T_\varsigma)$, the relative velocities and stress deviators on the other side can be computed. (11.12) or (11.13) are therefore *generalized boundary conditions for tangential and normal velocities and shear and normal stresses at reversible phase boundaries.*

According to $(11.9)_1$, reversible phase transitions require the surface heat flux and consequently the heat conductivities κ and κ_2 to vanish. It is not difficult to show by using the boundary layer results (5.18) that for reversible phase transitions and in this boundary layer formulation, κ and κ_2 are given by the jump of the bulk conductivities, viz.

$$\kappa = -[\![\kappa]\!]\frac{d_\varsigma}{2}(J_1 + \frac{k_M}{4}J_2), \qquad \kappa_2 = -[\![\kappa]\!]\frac{d_\varsigma^2}{8}J_2. \tag{11.14}$$

ALTS & HUTTER (1986) have estimated orders of magnitude for ice-water phase transitions. In fact, with J_1 and J_2 being order unity, and with

$$\begin{aligned}\kappa_{\text{water}} &= 0.57\ \text{Wm}^{-1}\text{K}^{-1}\,, & T &= 273.15°K,\\ \kappa_{\text{ice}} &= \left(\frac{488.19}{T} + 0.4682\right)\ \text{Wm}^{-1}\text{K}^{-1}\,, & 273.15°K &\le T \le 108°K\end{aligned} \tag{11.15}$$

(numerical values from HOBBS, 1973) and for a phase boundary thickness $d_\varsigma = 100$ Å, one may deduce

$$\kappa = 0(10^{-8})WK^{-1}, \quad \kappa_2 = 0(10^{-17})WmK^{-1}. \tag{11.16}$$

Strictly, this contradicts the results (11.9), but the values (11.16) are so small in comparison with the heat conductivities of the bulk material that the heat flux along the phase interface is negligible. Consequently, the assumption of reversibility of phase transitions is a well motivated approximation to realistic situations.

A further, physically important point should be raised in connection with the representations (11.14). With the numerical data provided by ALTS & HUTTER (1986), we obtain the following estimates:

(i) For spherical water inclusions in ice (case A) one has

$$J_1 + \frac{k_M}{4} J_2 \leq 0, \quad J_2 \leq 0, \quad 0 \leq k_M \leq 1$$

for all dimensionless curvatures k_M. Therefore, (11.14) implies that both κ and κ_2 are greater than zero, as expected.

(ii) For spherical ice inclusions in water (case B), however,

$$J_1 + \frac{k_M}{4} J_2 \geq 0, \quad J_2 \geq 0, \quad 0 \leq |k_M| \leq 0.086$$

for a limited range of dimensionless curvature for which both κ and κ_2 are greater than zero. A stability argument identifies this as the range of stable ice inclusions in water.

Boundary conditions similar to (11.12) or (11.13) can be deduced for the deviators of the adjacent membrane stresses. ALTS & HUTTER (1986) show that

$$\begin{aligned}[\![t^{\alpha\beta} + p_E g^{\alpha\beta}]\!] &= -\frac{2}{d_\varsigma^2}[\mu_3[(I_D\mathcal{M}_1 + d_\varsigma II_D\mathcal{M}_2)d_\varsigma b^{\alpha\beta} \\ &+ (I_D\mathcal{M}_3 + d_\varsigma I_D\mathcal{M}_4)g^{\alpha\beta}] \qquad (11.17) \\ &+ \mu_6\mathcal{M}_5 d_\varsigma g^{\mu\nu}(b_\mu^{(\alpha} D_\nu^{\beta)} - D_\mu^{(\alpha} b_\nu^{\beta)}),\end{aligned}$$

where $\mathcal{M}_j(k_M, k_G)$ $(j = 1, 2, \ldots, 5)$ are known functions of the curvatures.

Reversibility of phase transitions does not require the surface viscosities μ_3, μ_6 to vanish. (This distinguishes phase boundaries from bulk materials, in which reversibility would require the bulk viscosities to be zero). Once these are known, the tensor of surface tension (11.9) and the jump (11.17) for the deviators of the adjacent membrane stresses can be determined. Now, measurements of the surface viscosity of the interface between ice and water indicate that it is extremely slippery justifying the assumptions

$$\mu_3 = \mu_6 = 0. \qquad (11.18)$$

This reduces $(11.9)_3$ to

$$S^{\alpha\beta} = \sigma_E g^{\alpha\beta}, \quad [\![t^{\alpha\beta} + p_E g^{\alpha\beta}]\!] = 0. \qquad (11.19)$$

Hence, the tensor of surface tension of the inviscid phase interface is given by the scalar surface tension in equilibrium, and the stress deviator of the adjacent membrane stresses is continuous across the phase boundary.

The assumption of continuity of temperature across the phase boundary, $T_+ = T_- = T_\varsigma$ makes the balance equation $(3.7)_4$ *for internal energy a constraint for the determination of the normal jump* $[\![q_n]\!]$ *of the heat flux. The exploitation of this condition requires lengthy* manipulations and will not be given here. The result is

$$\mathcal{Q}_1 = T_\varsigma \left(\frac{\partial \sigma}{\partial T_\varsigma}\right)^E_{\varrho_\varsigma}, \quad \mathcal{Q}_2 = \mathcal{Q}_4 = \mathcal{Q}_5 = \mathcal{Q}_6 = 0. \qquad (11.20)$$

$$
\begin{aligned}
\mathcal{Q}_3 &= -\varrho_0^+\left[h_E^+ - h_\varsigma^E - T_\varsigma\frac{1}{\varrho_\varsigma^E}\left(\frac{\partial\sigma}{\partial T_\varsigma}\right)_{\varrho_\varsigma}^E\right] \\
&\quad +\varrho_0^-\left[h_E^- - h_\varsigma^E - T_\varsigma\frac{1}{\varrho_\varsigma^E}\left(\frac{\partial\sigma}{\partial T_\varsigma}\right)_{\varrho_\varsigma}^E\right]\left(1-\frac{1}{\mathcal{F}}\right),
\end{aligned}
\tag{11.21}
$$

where

$$
h_E^\pm = u_E^\pm + \frac{p_E^\pm}{\varrho_0^\pm}, \qquad h_\varsigma^E = u_\varsigma^E - \frac{\sigma_E}{\varrho_\varsigma^E}
\tag{11.22}
$$

are the specific enthalpies of the adjacent bulk materials and the surface material in phase equilibrium. Finally, inserting (11.20), (11.21) into $(7.11)_3$ and using $(11.12)_1$ yields two equivalent forms for the normal jump of the heat flux,

$$
\begin{aligned}
\mathcal{Q} &= \mathcal{Q}_1 I_D + \mathcal{Q}_3 W_n^+ \\
&= T_\varsigma\left(\frac{\partial\sigma}{\partial T_\varsigma}\right)_{\varrho_\varsigma}^E I_D - \left[h_E^+ - h_\varsigma^E - T_\varsigma\frac{1}{\varrho_\varsigma^E}\left(\frac{\partial\sigma}{\partial T_\varsigma}\right)_{\varrho_\varsigma}^E\right]\varrho_0^+ W_n^+ \\
&\quad + \left[h_E^- - h_\varsigma^E - T_\varsigma\frac{1}{\varrho_\varsigma^E}\left(\frac{\partial\sigma}{\partial T_\varsigma}\right)_{\varrho_\varsigma}^E\right]\varrho_0^- W_n^+
\end{aligned}
\tag{11.23}
$$

in which $I_D = D_\alpha^\alpha$ is defined in (7.9).

The classical result would be obtained for an "incompressible" phase boundary with $d\varrho_\varsigma/dt = 0$, $I_D = 0$ and and $\mathcal{R} = \varrho_0^+ W_n^+ - \varrho_0^- W_n^- = 0$; namely

$$
\mathcal{Q} = -[\![h_E]\!]\varrho_0^- W_n^+,
\tag{11.24}
$$

where $[\![h_E]\!]$ is the latent heat of freezing or melting. However, the assumption of incompressibility for phase boundaries between ice and water is unacceptable, as the boundary-layer theory has already shown.

We thus have the remarkable result that all transport coefficients for reversible phase transitions across inviscid phase boundaries are known functions of the geometry and temperature. The boundary-layer theory for a reversible phase transition between ice and water contains thus as a phenomenological parameter one single constant, namely the thickness $d_\varsigma(T_\varsigma)$ at some reference temperature.

12. Summary of results

The continuum description of phase boundaries in this article regarded phase change surfaces as two dimensional, curved and non-material. Restrictions imposed by the rules of material frame indifference, material symmetry and the entropy principle permitted together with a few *ad hoc* assumptions the deduction of a fairly simple continuum model that is likely to be applicable to explicit calculations. The adopted entropy principle enabled us to show that for a heat conducting phase boundary (i)$_D$ the surface entropy flux equals surface heat flux divided by absolute surface temperature,

(ii)$_D$ the absolute surface temperature is a universal function not depending on the material undergoing the phase change process,

(iii)$_D$ the inverse of the absolute temperature is the divisor of the Pfaffian form

$$ds_\varsigma = \frac{1}{T_\varsigma}\left(du_\varsigma + \frac{\sigma}{\varrho_\varsigma^2}\, d\varrho_\varsigma\right)$$

relating the total differential of surface entropy to those of internal energy, surface tension and surface density,

(iv)$_D$ surface internal energy and entropy can be computed, once the specific surface heat capacity at constant density, c_ϱ^ς is known as a function of temperature at one particular value of the surface density and the surface tension is known as a function of surface temperature and density.

GIBBS' differential for the surface entropy, as derived here for phase boundaries, is somewhat different from the form that is assumed in the classical theories of surface tension. This has important consequences for freezing and melting processes and is manifested in the existence of two CLAUSIUS-CLAPEYRON relations (see below).

Additional restrictions to the constitutive relations emerged from the exploitation of the thermostatic conditions (8.4) and partly the application of the boundary layer approximation of the phase change surface.

(i)$_S$ The temperatures of the phase interface and the adjacent bulk materials are the same in equilibrium, i.e. $T_\varsigma|_E = T^\pm|_E$.

(ii)$_S$ The specific free enthalpies have in equilibrium the same values:

$$g_\varsigma|_E = g_\varsigma^\pm|_E$$

(iii)$_S$ When the influence of the body force (gravity) is ignored, equilibrium shapes of isotropic phase boundaries are either spheres or circular cylinders.

(iv)$_S$ The surface tension, the pressures, densities, entropies, internal energies of the bulk materials and the phase boundary, all in equilibrium, are functions of the temperature and the mean curvature.

(v)$_S$ Pressure melting at the convex and concave side of the phase boundary is not the same because the generalized CLAUSIUS-CLAPEYRON relation contains a term depending on surface tension that differs in the two cases (see e.g. (8.35) and (8.37)).

(vi)$_S$ By the boundary-layer theory and the assumption of incompressiblity of bulk ice and water it is possible to reduce all surface fields and all jump conditions to expressions in which only explicitly known curvature dependent contributions and known bulk contributions at flat interfaces appear. This allows the calculation of orders of magnitude for all fields and yields the proof, that phase transitions across the interface are, to first approximation, reversible processes. It determines all transport coefficients in terms of one single phenomenological constant, the *boundary-layer thickness at some reference temperature* and makes the formulation free of adjustable constants and the theory simple and controllable by experiments.

Acknowledgements: I thank Mrs. R. Danner for typing the text, and Bob Svendsen and Klaus Jöhnk for their TEXpertise in TEXing a manuscript with the many tedious formulas. Finally, I profitted from the discussions with the lecturers and the audience at CISM.

14. References

Alts, T.: Zur Thermodynamik der Vorgänge in festen und fluiden Mischungen, Ph.D. Dissertation, Rheinisch-Westfälische Technische Hochschule Aachen, 1971.

Alts, T.: Thermodynamik elastischer Körper mit thermo-kinematischen Zwangsbedingungen-fadenverstärkte Materialien. Habilitation-Dissertation, Technische Universität Berlin, Universitätsbibliothek der TU Berlin, Abteilung Publikationen, 1979.

Alts, T.: The principle of determinism and thermodynamics of simple fluids with thermo-kinematic constraints, J. Non-Equil. Thermo., 10 (1984), 145-162.

Alts, T. and P. Strehlow: Frozen deformation, frozen stresses and structure-induced Na-distributions in sodium silicate glasses, J. Thermal Stresses, 7 (1984), 317 -359.

Alts, T. and K. Hutter: Towards a theory of temperate glaciers: Dynamics and thermodynamics of phase boundaries between ice and water, Mitt. No. 82 der Versuchsanstalt für Wasserbau, Hydrologie und Glaziologie, (D. Vischer, Hrsg.), ETH Zürich, 1986.

Alts, T. and K. Hutter: Continuum description of the dynamics and thermodynamics of phase boundaries between ice and water, part I: Surface balance laws and their interpretation in terms of three-dimensional balance laws averaged over the phase change boundary layer, J. Non-Equil. Thermo., 13 (1988a), 221-257.

Alts, T. and K. Hutter: Continuum description of the dynamics and thermodynamics of phase boundaries between ice and water, part II: Thermodynamics, J. Non-Equil. Thermo., 13 (1988b), 259-280.

Alts, T. and K. Hutter: Continuum description of the dynamics and thermodynamics of phase boundaries between ice and water, part III: Thermostatics and its consequences, J. Non-Equil. Thermo., 13 (1988c), 301-329.

Alts, T. and K. Hutter: Continuum description of the dynamics and thermodynamics of phase boundaries between ice and water, part IV: On thermostatic stability and well-posedness, J. Non-Equil. Thermo., 14 (1989), 1-22.

Baehr, H.D.: Thermodynamik, Springer-Verlag, Heidelberg (3. Auflage)1979.

Becker, R.: Theorie der Wärme, Springer-Verlag, Heidelberg, 1961.

Bowden, F.P. and Tabor, D.: The Friction and Lubrication of Solids, Clarendon Press, Oxford, 1954.

Buff, F.P.: Curved fluid interfaces, I: The generalized Gibbs-Kelvin equation, J. Chem. Phys., 25 (1956), 146-153.

Buff, F.P. and Saltsburg, H.: Curved fluid interfaces, II: The Generalized Neumann Formula, J. Chem. Phys., 26 (1956a), 23-31.

Buff, F.P. and Saltsburg, H.: Curved fluid interfaces, III: The dependence of the free energy on parameters of external force, J. Chem. Phys., 26 (1956b), 1526-1533.

Carathéodory, C., Math. Annalen, 67 (1909), 355, in: Gesmmelte math. Schriften, Bd. II, S. 129, C.H. Beck'sche Verlagsbuchhandlung, München, 1955.

Deemer, A.D. and Slattery, J.C.: Balance equations and structural models for phase interfaces, Int. J. Multiphase Flow, 4 (1978), 171-192.

Faraday, M.: Notes from Sept. 8, 10, 26, 1842 and Oct. 3, 1845, in: Faraday's Diary, ed. by T. Martin, vol. IV, G. Bell & Sons, Ltd., London, 1933.

Fletcher, N.H.: Surface structure of water and ice, Philosophical Magazine, Eighth Ser., 7 (1961), 255-269.

Fletcher, N.H.: Surface structure of water and ice, II: A revised model, Philosophical Magazine, Eighth Ser., 18 (1966), 1287-1300.

Gibbs, J.W.: The collected works of J. Willard Gibbs, V. 1, p. 219, Yale University Press, New Haven, 1929.

Gibbs, J.W.: Collected Works, V. 1, pp. 55-353, Yale University Press, New Haven, 1948.

Glen, J.W.: The creep of polycrystalline ice, Proc. Roy. Soc. London, A 228(1955), 519-538.

Golecki, I., Jaccard, D.: The surface of ice near 0 C studied by 100 keV proton channeling, Physics Letters, 63A(1977), 374-376.

Grauel, A.: Thermodynamics of an interfacial fluid membrane, Physica, 103A (1980), 468-520.

Gubler, H.: Strength of bonds between ice grains after short contact times, J. Glaciology, 28 (1982), 457-473.

Gurtin, M.E. and Murdoch, A. I.: A continuum theory of elastic material surfaces, Arch. Rat. Mech. Anal., 57 (1975), 291-323.

Hobbs, P.V.: Ice Physics, Clarendron Press, Oxford, 1974.

Hutter, K. and Trösch, J.: Über die hydromechanischen und thermodynamischen Grundlagen der Seezirkulation, Mitt. No. 20 der Versuchsanstalt für Wasserbau, Hydrologie und Glaziologie an der ETH Zürich, 1976.

Lindsay, K.A. and Straughan, B.: A thermodynamic viscous interface theory and associated stability problems, Arch. Rat. Mech. Anal., 71 (1979), 307-326.

Liu, I-Shiu: Method of Lagrange multipliers for exploitation of the entropy principle, Arch. Rat. Mech. Anal., 46, (1972), 131-148.

Moeckel, G.P.: Thermodynamics of an interface, Arch. Rat. Mech. Anal., 57 (1974), 255-280.

Müller, I.: Thermodynamik - Grundlagen der Materialtheorie, Bertelsmann Universitätsverlag, Düsseldorf, 1972.

Nakaya, U. and Matsumoto, A.: Simple experiment showing the existence of "liquid water" film on the ice surface, J. Colloid Sci., 9 (1954), 41-49.

Nye, J.F. and Frank, F.C.: Hydrology of intergranular veins in a temperate glacier, in: Association Internationale d'Hydrologie Scientific. Symp. on the Hydrology of Glaciers, Cambridge, 1969, 157 - 161.

Scriven, L.E.: Dynamics of a fluid interface, Chem. Engng. Sci., 12 (1960), 98-108.

Slattery, J.C.: General balance equations for a phase interface, I & EC Fundamentals, 6 (1967), 108-115.

Steinemann, S.: Experimentelle Untersuchungen zur Plastizität von Eis, Beiträge zur Geologie der Schweiz. Hydrologie, No. 10, 1958.

Thomson, J.: Note on Professor Faraday's recent experiments on regelation, Proc. Roy. Soc. London, A11 (1861), 198-204.

Tyndall, J.: On some physical properties of ice, Proc. Roy. Soc. London, 9 (1858a), 76-80.

Tyndall, J.: Über einige physikalische Eigenschaften des Eises, Ann. d. Physik u. Chemie, 103 (1858b), 157 - 162.

Weyl, W.A.: Surface structure of water and some of its physical and chemical manifestations, J. Colloid Sci., 6 (1951), 389-405.

MOVING COMMON LINES, THIN FILMS, AND DYNAMIC CONTACT ANGLES

J. C. Slattery
Texas A&M University, College Station, TX, USA

Abstract

When a common line or three-phase line of contact moves over a rigid solid, an unbounded force must be generated at the common line, if the usual no-slip boundary condition of fluid mechanics is valid.

Since an unbounded force is unrealistic, our description of the physics in this statement must be incorrect. There are various possibilities. Real solids are not rigid. Perhaps the no-slip boundary condition fails within the immediate neighborhood of the common line. In some situations there is undoubtedly mass transfer in the neighborhood of the apparent common line, and there is displacement over an existing film of fluid without a common line ever having been formed. In other situations, a sequence of stationary common lines may be formed in the thin film of fluid within the immediate neighborhood of the apparent common line, giving the appearance of a moving common line to an observer on the macroscale.

Section 1 is a review of what is known about moving common lines on relatively rigid solids. In Sec. 2, I illustrate with at least one computation that continuum mechanics can be applied successfully to very thin films, such as those within the immediate neighborhood of the common line. In Sec. 3 I discuss contact angles, with a demonstration computation for the dynamic contact angle as a function of the apparent speed of displacement of the common line.

1 Moving common lines

Let us begin by asking what happens as a common line or three-phase line of contact moves.

1.1 qualitative description

In our discussion of the motion of multiphase bodies, we have not as yet mentioned their common lines.

A **common line** (contact line or three-phase line of contact) is the curve formed by the intersection of two dividing surfaces. When a drop of water sits on a china plate, the water-air dividing surface intersects the solid in a common line. We are primarily concerned here with the motion of common lines. For example, when the plate is tipped, the drop begins to flow resulting in the displacement of the common line.

If we dip a glass capillary tube in a pan of water, the water-air dividing surface rises in the tube to some equilibrium level. When it has come to rest, the dividing surface intersects the wall of the tube in an easily observed common line. This experiment suggests some questions. Is the air displaced by the water at the wall of the tube or is the water separated from the tube wall by a thin film of air? How does the common line move with respect to the adjoining air, water, and glass?

In attempting to answer these questions, let us begin with a qualitative picture of the phenomena involved. Some simple experiments are helpful.

1.1.1 Initial formation of common line

Dussan V. and Davis ([24], [94, Figure 1.2.9-1]) released a small drop of water in a tank of silicone oil. It appeared to be spherical as it fell and remained spherical for 5-10 seconds after it came to rest on the bottom. Suddenly the water drop *popped* onto the Plexiglas base with a markedly different configuration. Apparently what happened is that, when the drop initially came to rest, it was separated from the Plexiglas by a thin film of silicone oil. When the spherical drop appeared to be resting on the bottom of the tank, it was actually squeezing the silicone oil film, forcing it to become thinner. After 5-10 seconds the film was so thin that it became unstable. It ruptured, rolled back, and the water drop came into contact with the Plexiglas base.

Dussan V. and Davis ([24], [94, Figure 1.2.9-2]) obtained a more detailed view of this popping phenomena using a glycerol drop (whose viscosity is about 1500 times that for water). Following its rupture, the thin film of silicone oil rolled back: a curve dividing two distinct types of reflections appeared after the initial rupture of the film of silicone oil and swept across the lower surface of the drop.

After the thin film of silicone oil ruptured and rolled back allowing the glycerol drop to come into direct contact with the Plexiglas surface, some silicone oil was almost certainly left behind, adsorbed in the glycerol-Plexiglas interface. As time passed and the multiphase system approached a new equilibrium, some of this adsorbed silicone oil was then desorbed into the glycerol bulk phase. (For a more detailed view of the multicomponent dividing surface, see [94, Chapter 5]).

In both of these experiments, there is the strong implication that, following its rupture, the thin film of bulk silicone oil was displaced without violating the requirement that the tangential components of velocity be continuous across a phase interface. The silicone oil film was simply rolled to one side in much the same manner as we might handle a rug.

1.1.2 Precursor film

Sometimes a very thin film or precursor film is observed to precede the advance of a macroscopic film.

Bascom *et al.* [4] studied the spontaneous spreading behavior of hydrocarbon liquids on both horizontal and vertical surfaces using interference microscopy and ellipsometry, which enabled them to study the thickness of the spreading films in detail. All of the hydrocarbons showed zero contact angles on the metal surfaces employed. In all cases, breath patterns formed by breathing over the spreading liquids showed that the outer edge of the film was considerably beyond the edge of the macroscopic film as detected by the first-order interference band: a very thin precursor film preceded the advance of the macroscopic film. The presence of this precursor film was also demonstrated by placing ahead of the macroscopic film minute drops of another liquid having a higher surface tension against air. Placed on the precursor film, these drops would immediately retract from the spreading liquid; placed on a clean metal surface, they would spread uniformly in all directions. For the first few hours, they were not able to detect the precursor film using ellipsometry. After a relatively long period of time (18 hours in the case of squalane), they were able to determine that the precursor film was as much as several millimeters long although less than 50 Å thick. Considering the relatively low volatility of the liquids employed, they attributed the formation and movement of the precursor film to capillary flow in microscratches on the surface and to surface diffusion [94, Sec. 5.9.8] rather than to evaporation from the macroscopic film and subsequent condensation upon the plate. Teletzke *et al.* [95], who have re-examined these experiments, offer new evidence supporting the importance of surface diffusion.

Radigan *et al.* [83] used scanning electron microscopy to detect during the spreading of drops of glass on Fernico metal at 1000°C a precursor film whose height was of the order of 1 μm.

In studies described by Williams [99, 100], the leading edge of this precursor film has a periodic structure. Depending upon perhaps both the liquid and the solid, either it assumes a scalloped periodic structure or it moves by a series of random advances with some approximate periodicity.

Williams ([99], [94, Figure 1.2.9-3]) photographed the advancing front of absolute ethyl alcohol spreading over aluminum that had been evaporated onto a glass substrate. A precursor film of alcohol 1,000 - 2,000 Å thick moved ahead of the drop as much as 1 mm. In this case, the leading edge of the precursor film maintained an early sinusoidal form as it advanced. Williams [99] found similar results with a variety of liquids spreading on aluminum: isopropyl alcohol, n-heptane, toluene, n-octane, and dimethyl silicone oil (molecular weight 340).

Williams ([99], [94, Figure 1.2.9-4]) also photographed the leading edge of an advancing precursor film of a non-volatile, hydrocarbon-substituted silicone oil (General Electric SF-1147, mean molecular weight 2,000) spreading over a surface of evaporated gold. Although the configuration of the leading edge of the precursor film is more random than that of the alcohol precursor film discussed above, it still displays an approximate periodicity. As in the case of alcohol, Williams [99] found that the precursor film advanced far ahead of the bulk liquid, and, after several hours, covered nearly all of the solid surface.

Williams ([100], [94, Figures 1.2.9-5 and 1.2.9-6]) spread the same silicone oil used above as a stripe on a surface of SiO_2. As time passed, the leading edge of the precursor film grew progressively more convoluted.

Teletzke *et al.* [95] suggest that these precursor films are probably the result of what they refer to as *primary* and *secondary spreading* mechanisms.

By primary spreading, they refer to the manner in which the first one or two molecular layers are deposited on the solid. For a nonvolatile liquid, primary spreading is by surface diffusion [4, 17, 8, 87]. For a volatile liquid, primary spreading may be by the condensation of an adsorbed film [39].

By secondary spreading, they refer to the motion of films whose thickness varies from several molecules to roughly a micron. These films spread as the result of either a positive disjoining pressure or a surface tension gradient. Surface tension gradients can be the result of either temperature gradients [70] or composition gradients, perhaps resulting from the evaporation of a more volatile component [4].

For more on wetting and spreading including precursor films, see Teletzke *et al.* [95] and an excellent review by de Gennes [20].

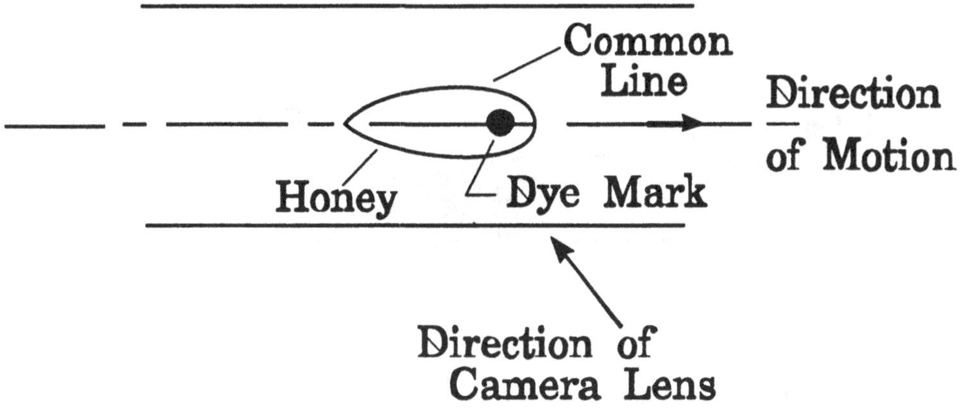

Figure 1.1-1: Plane view of drop of honey on a Plexiglas surface.

1.1.3 In the displacement of one fluid by another, will a common line move?

Not necessarily.

Dip a clean piece of metal partially into a can of oil-based paint and slowly withdraw it. The apparent common line does not move, although the film of paint begins to thin and drain. Before the film of paint can thin appreciably by drainage, loss of solvent through the paint-air interface causes the film to solidify.

More generally, thin films of liquid left behind on a solid as a liquid phase retreats can be stabilized by electrostatic forces, by reaction with the solid phase, or in the case of paint by the loss of solvent and resultant solidification.

1.1.4 If a common line moves, will it move smoothly?

Generally not.

In discussing the precursor film that sometimes precedes an advancing macroscopic film, we noted Williams' [99, 100] observations. The configuration of the leading edge of the precursor film is often markedly different from that of the apparent leading edge of the macroscopic film. It may have a periodic or even random structure. Here I would like to consider the apparent leading edge of the macroscopic film.

Take a glass from the back of your cupboard, preferably one that has not been used for a long time, and add a little water to it, being careful not to wet the sides. Now hold the glass up to the light, tip it to one side for a moment, and return it

Figure 1.1-2: Cross-sectional view of a drop of glycerol that is partially dyed

to the vertical. The macroscopic film of water left behind on the side retreats very rapidly and irregularly, sometimes leaving behind drops that have been cut off and isolated. Now take a clean glass and repeat the same experiment. The macroscopic film of water now retreats much more slowly and regularly. But if you look closely, the apparent common line still moves irregularly, although the irregularities are on a much smaller scale than with the dirty glass. Very small drops that have been cut off and isolated in the irregular retreat are still left behind.

Poynting and Thomson [82] forced mercury up a capillary tube and then gradually reduced the pressure. Instead of falling, the mercury at first adjusted itself to the reduced pressure by altering the curvature of the air-mercury interface. When the pressure gradient finally grew too large, the configuration of the meniscus became unstable and the mercury fell a short distance in the tube before stopping, repeating the deformation of the interface, and falling again when a new instability developed.

Yarnold [102] saw this same sticking phenomena as a liquid index moved slowly through a glass capillary tube.

Elliott and Riddiford [28] described a technique for measuring dynamic contact angles. In their experiment, one fluid was displaced by the other in the gap between two narrowly spaced horizontal parallel plates made of or coated with the solid material under investigation. The interface moved as fluid was injected or withdrawn at the center of the circle by means of a syringe controlled by a cam. For proper interpretation of the data, the projection of the interface between the fluids should have been a circle. Elliott and Riddiford [28] reported that, when fluid was withdrawn through the syringe, the interface often did not move concentrically, irregular jerking or sticking being observed instead. Wilson [101] pointed out that the flow which results when fluid is withdrawn through the syringe must always be unstable. The flow is more likely to be stable when fluid is injected through the syringe, even if the less viscous fluid is required to displace the more viscous one.

In some experiments, observed irregularities and episodic movements probably may be attributable to variations in the contact angle along the common line, which in turn may be due to a non-uniform distribution of contaminants. In other cases, the multiphase flow may be hydrodynamically unstable and no amount of cleaning can alter the situation.

1.1.5 In a displacement, one of the phases exhibits a rolling motion.

In an effort to understand how material adjacent to a phase interface moves with a displacement of the common line, Dussan V. and Davis ([24], [94, Sec. 1.2.9]) carried out the following four experiments.

i) Approximately one cm^3 of honey was placed on a horizontal Plexiglas surface. A small dye mark, consisting of honey and food coloring, was inserted by means of a needle in the honey-air interface at the plane of symmetry of the honey. The Plexiglas was tilted, the honey began to flow, and the trajectory of the dye mark was photographed from the direction shown in Figure 1.1-1. As the honey flowed down the plane, the dye mark approached the common line. Finally, it made contact with the Plexiglas.

ii) Two drops of glycerol, one transparent and the other dyed with food coloring, were placed side by side on a solid bee's wax surface. The surface was tilted, causing the two drops to merge, and it was then returned to its original horizontal position. In cross section, the resulting single drop probably looked like that shown in Figure 1.1-2. The left end was then lowered. After a finite length of time, the entire common line was composed of clear glycerol. The dyed glycerol adjacent to the right end portion of the common line moved to assume a position adjacent to the glycerol-air interface.

iii) A rectangular Plexiglas container was tilted with respect to the horizontal, partially filled with glycerol, and then with silicone oil. The result is sketched in Figure 1.1-3. A small drop of glycerol mixed with food coloring was added to the glycerol-silicone oil interface near the common line. The right end of the container was slowly lowered, the common line moved to the right, and the dye mark approached the common line. It finally became part of the common line and disappeared from sight; it was no longer adjacent to the glycerol-oil interface, but remained adhered to the Plexiglas.

iv) With the same system as in (iii), the common line was forced to move left by slowly raising the right end of the container. First a drop of dyed glycerol was placed on the Plexiglas initially covered with silicone oil. After a few moments, the drop *popped* onto the Plexiglas. The right end of the container was lowered, the clear glycerol moved forward and eventually merged with the dyed glycerol drop. The right end of the container was then gradually raised, the common line moved to the left, and the dyed mark came off the bottom surface.

All four of these experiments illustrate a *rolling* motion exhibited by one of the phases[1] in a displacement. In a **forward rolling motion**, the material in this phase adjacent to the fluid-fluid interface moves down to the common line and is left adjacent to the fluid-solid interface. In a **backward rolling motion**, material in this phase originally adjacent to the fluid-solid interface is lifted off as the common line passes, and it is transported up along the fluid-fluid interface.

Allen and Benson [3] have closely observed the motion in a drop moving down an inclined plane. Although this three-dimensional motion is more complicated than the primarily two-dimensional flows that we have just finished discussing, the motion in the neighborhood of the common line continues to exhibit a rolling character.

1.2 emission of material surfaces [24]

If, as suggested by the experiments described in the last section, one phase exhibits a rolling motion as a common line moves, what is the character of the motion of the other phase?

In Figure 1.2-1, fluid A is being displaced by fluid B as the common line moves to the left over the solid S. I suggest we make the following assumptions.

i) There is a forward rolling motion in phase B in the sense defined in the previous section.

ii) The tangential components of velocity are continuous across all phase interfaces.

iii) The $A - S$ dividing surface is in chemical equilibrium [94, Sec. 5.10.3] with phase A prior to displacement. There is no mass transfer between the bulk phase A and the solid surface.

iv) The mass density of phase A [94, Sec. 1.3.1] must be finite everywhere. A finite mass of phase A must occupy a finite volume.

Let us use referential descriptions for the motion in phase A [94, Sec. 1.1.1]

$$\mathbf{z}^{(A)} = \boldsymbol{\chi}_{\boldsymbol{\kappa}}^{(A)}(\mathbf{z}_{\boldsymbol{\kappa}}^{(A)}, t) \tag{1.2-1}$$

and for the motion in the solid S

[1] Dussan V. and Davis [24] do not report either static or dynamic contact angles.

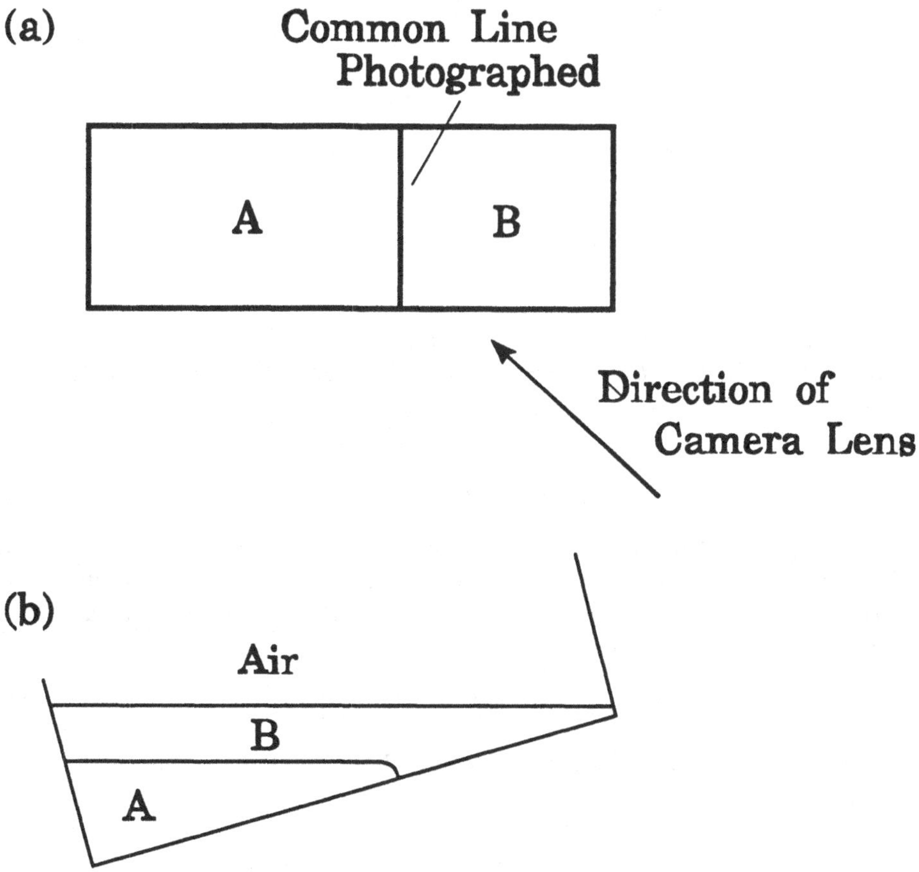

Figure 1.1-3: (a) Plan view of the bottom surface of the container. (b) Side view of the container. In experiments (iii) and (iv), fluid A is glycerol and fluid B is silicone oil.

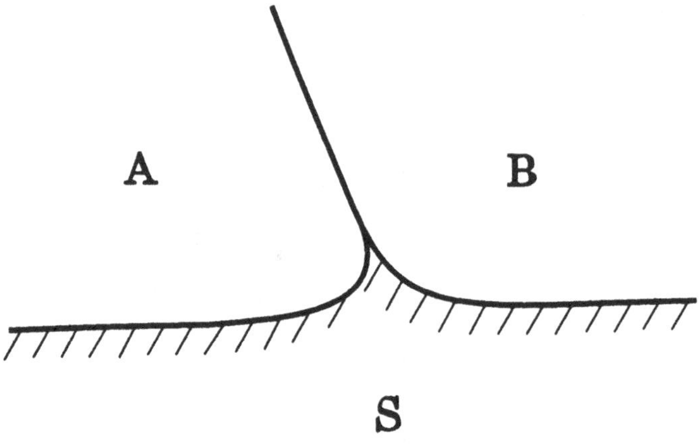

Figure 1.2-1: The common line formed by fluids A and B on the solid S. The solid is shown with a dimple at the common line, although the analysis described in this section applies equally well to both flat and deformed solids.

$$\mathbf{z}^{(S)} = \boldsymbol{\chi}_{\boldsymbol{\kappa}}^{(S)}(\mathbf{z}_{\boldsymbol{\kappa}}^{(S)}, t) \tag{1.2-2}$$

To be specific, let us focus our attention on the position z_0, which in the reference configurations for phases A and S denotes a position on the $A - S$ dividing surface. Our particular concerns are the trajectory of the material particle of phase A that is adjacent to z_0 in the reference configuration

$$\mathbf{z}_0^{(A)}(t) \equiv \text{limit } \mathbf{z}_{\boldsymbol{\kappa}}^{(A)} \to \mathbf{z}_0 : \quad \boldsymbol{\chi}_{\boldsymbol{\kappa}}^{(A)}(\mathbf{z}_{\boldsymbol{\kappa}}^{(A)}, t) \tag{1.2-3}$$

and the trajectory of the material particle of the solid that is adjacent to z_0 in the reference configuration

$$\mathbf{z}_0^{(S)}(t) \equiv \text{limit } \mathbf{z}_{\boldsymbol{\kappa}}^{(S)} \to \mathbf{z}_0 : \quad \boldsymbol{\chi}_{\boldsymbol{\kappa}}^{(S)}(\mathbf{z}_{\boldsymbol{\kappa}}^{(S)}, t) \tag{1.2-4}$$

Since the common line is moving, we know that at some particular time t_1 the solid point $\mathbf{z}_0^{(S)}(t_1)$ will be adjacent to the common line. We know that $\mathbf{z}_0^{(S)}(t)$ for $t > t_1$ must be adjacent to the $B - S$ dividing surface.

Where is $\mathbf{z}_0^{(A)}(t)$ for $t > t_1$? There are only four possible locations.

1) It remains adjacent to the same solid material point

$$\text{for } t > t_1 : \quad \mathbf{z}_0^{(A)}(t) = \mathbf{z}_0^{(S)}(t) \tag{1.2-5}$$

2) It is mapped adjacent to the $A - B$ dividing surface.

3) It remains adjacent to the common line.

4) It is mapped into the interior of phase A.

Alternative (1) is not possible. Let $\mathbf{z}^{(cl)}(s,t)$ denote position on the common line as a function of arc length s measured along the common line from some convenient point. A moving common line implies that there exists a time $t_2 > t_1$ such that for some D:

$$\text{for all } s: \quad \left|\mathbf{z}_0^{(s)}(t_2) - \mathbf{z}^{(cl)}(s,t_2)\right| > D \tag{1.2-6}$$

Equation (1.2-3) implies that for any $\varepsilon > 0$ there exists an $\eta > 0$ such that

$$\left|\mathbf{z}_0^{(A)}(t_2) - \boldsymbol{\chi}_{\boldsymbol{\kappa}}^{(A)}(\mathbf{z}_{\boldsymbol{\kappa}}^{(A)}, t_2)\right| < \varepsilon \tag{1.2-7}$$

for all material points within phase A satisfying the relation

$$\left|\mathbf{z}_{\boldsymbol{\kappa}}^{(A)} - \mathbf{z}_0\right| < \eta \tag{1.2-8}$$

Physically this means that all of the material of phase A located within a distance η of z_0 in the reference configuration is mapped to within a distance ε of $\mathbf{z}_0^{(A)}(t_2)$ at time t_2. Since ε is arbitrary, we are free to choose $\varepsilon < D$. As a result of the motion of the common line, $\mathbf{z}_0^{(S)}(t_2)$ is located adjacent to the $B - S$ dividing surface, $\mathbf{z}_0^{(A)}(t_2)$ is also adjacent to the $B - S$ dividing surface by (1.2-5), and all of phase A within an η neighborhood of z_0 in the reference configuration must also be located on the $B - S$ dividing surface at time t_2. In view of assumption (iii), this implies that a finite quantity of the bulk phase A has a zero volume at time t_2 which contradicts assumption (iv).

Alternative (2) contradicts assumption (i).

Alternative (3) is also impossible, since it violates assumption (ii).

The only possibility is alternative (4): material adjacent to the $A - S$ dividing surface is mapped into the interior of phase A as phase A is displaced by phase B.

If the motion of the common line in Figure 1.2-1 were reversed, we would have a backward rolling motion in phase B. All of the arguments above would apply with relatively little change, but the final conclusion would be that material from the interior of phase A would be mapped into positions adjacent to the $A - S$ dividing phase as phase B is displaced by phase A.

Motions of this character have been observed in three experiments carried out by Dussan V. and Davis [24].

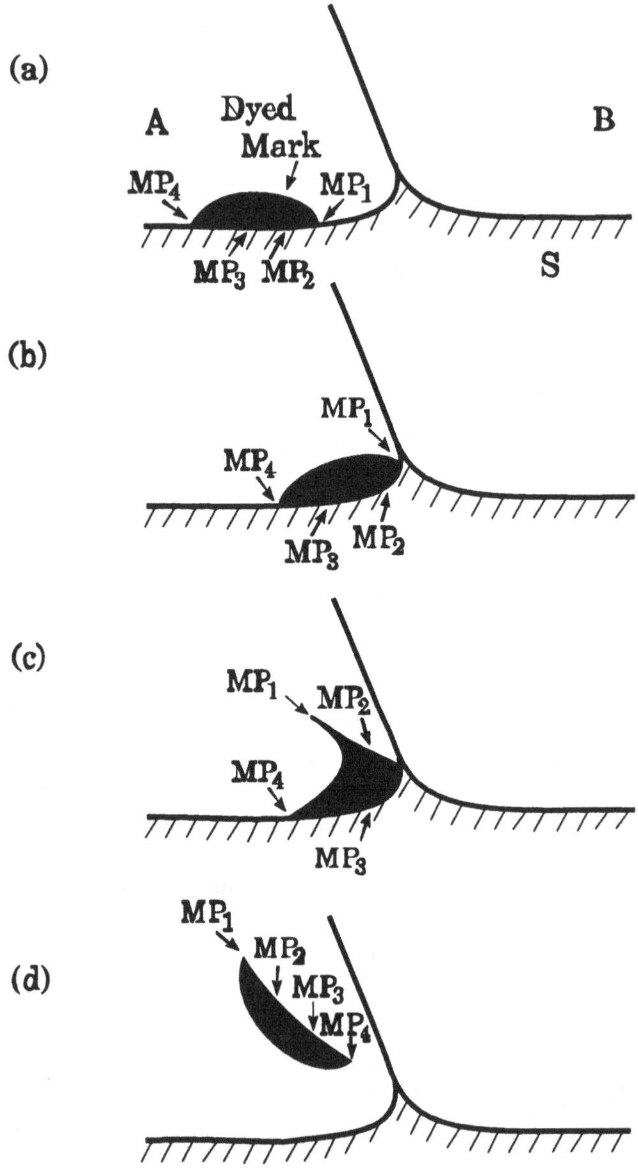

Figure 1.2-2: The trajectory of a dyed mark in the aqueous alcohol solution (A), initially adjacent to the aqueous alcohol solution-Plexiglas (S) phase interface, as the aqueous alcohol solution-silicone oil (B) phase interface moves to the left.

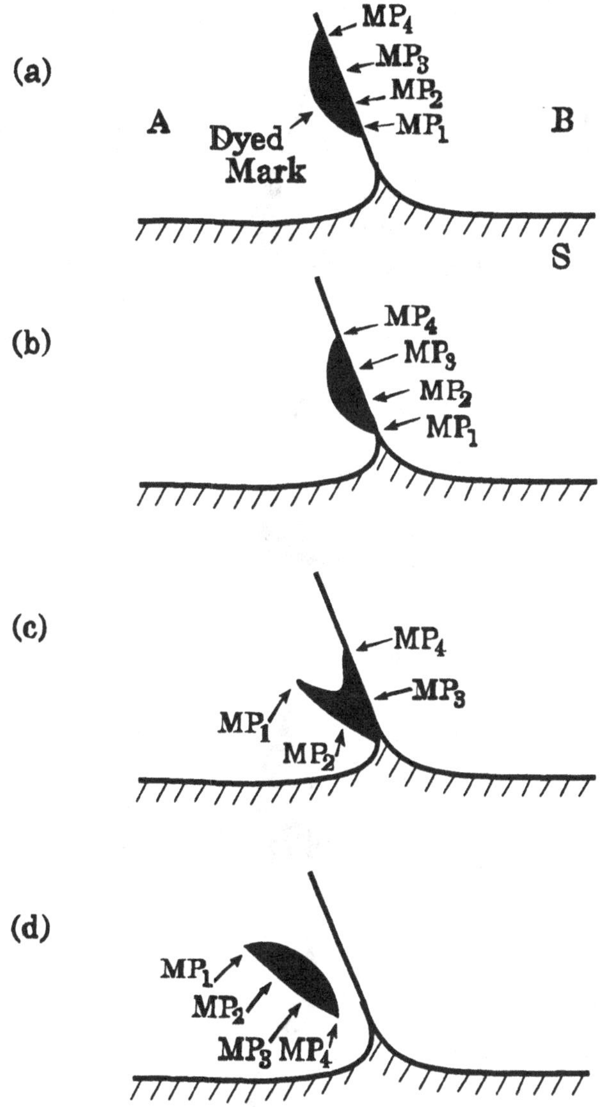

Figure 1.2-3: The trajectory of a dyed mark in the aqueous alcohol solution (A), initially adjacent to the aqueous alcohol-silicone oil (B) interface, as this interface moves to the left.

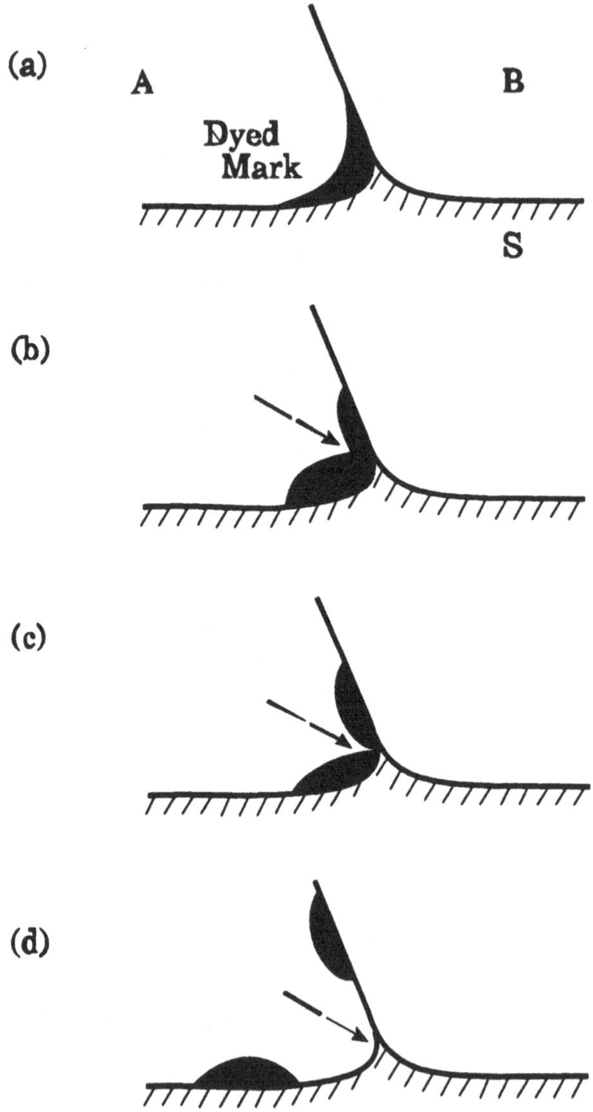

Figure 1.2-4: The trajectory of a dyed mark in the aqueous alcohol solution (A), initially adjacent to the common line, as the aqueous alcohol solution-silicone oil (B) phase interface moves to the right.

a) The rectangular Plexiglas container shown in Figure 1.1-3 was again tilted with respect to the horizontal, partially filled with a mixture of 60% water and 40% methyl alcohol (fluid A), and then silicone oil (fluid B). A drop of the alcohol-water mixture containing food coloring was injected onto the interface between the Plexiglas and the aqueous solution of alcohol. The dye remained undisturbed for a time to allow it to diffuse very close to the wall. The right end of the tank containing the oil was then slowly raised; the common line moved to the left in Figures 1.1-3 and 1.2-1 and approached the dye mark. Their photographs for this experiment are not clear, but I believe that Figure 1.2-2 illustrates what they saw. The dyed mark was lifted off the Plexiglas surface and *ejected* to the interior of the aqueous alcohol solution along an otherwise invisible material surface in the flow field. (A *material* surface is one that is everywhere tangent to the local material velocity.)

b) This same experiment was performed again, but this time the dye mark was adjacent to the aqueous alcohol solution-silicone oil interface. Since their photographs are not clear, I have drawn Figure 1.2-3 from their description. As the right hand end of the container was raised, the common line moved to the left in Figures 1.1-3 and 1.2-1, the dyed mark moved down the fluid-fluid interface to the common line, and it was finally *ejected* to the interior of the aqueous alcohol solution along a material surface. This material surface within the aqueous alcohol solution appeared to coincide with that observed in the previous experiment.

c) The backward version of this same system was also examined. In this case, the common line moves to the right in Figures 1.1-3 and 1.2-1 towards the oil. Figure 1.2-4 illustrates the evolution of a dyed spot placed at the common line within the aqueous alcohol phase. The dyed mark distinctly divides into two parts: one is transported up along the aqueous alcohol solution-silicone oil interface and the other is left behind on the aqueous alcohol solution-Plexiglas interface. In this case, there appears to have been *injection* of aqueous alcohol solution into the immediate neighborhood of the common line along both sides of a material surface within the aqueous alcohol phase.

The preceding discussion and experiments suggest the following generalization concerning the character of the motions involved when one phase displaces another over a solid surface. If one phase exhibits a *forward* rolling motion (see Sec. 1.1), there will be *ejection* from the neighborhood of the common line into the interior of the other phase along both sides of a material surface originating at the common line and dividing the flow field within the phase. If one phase exhibits a *backward* rolling

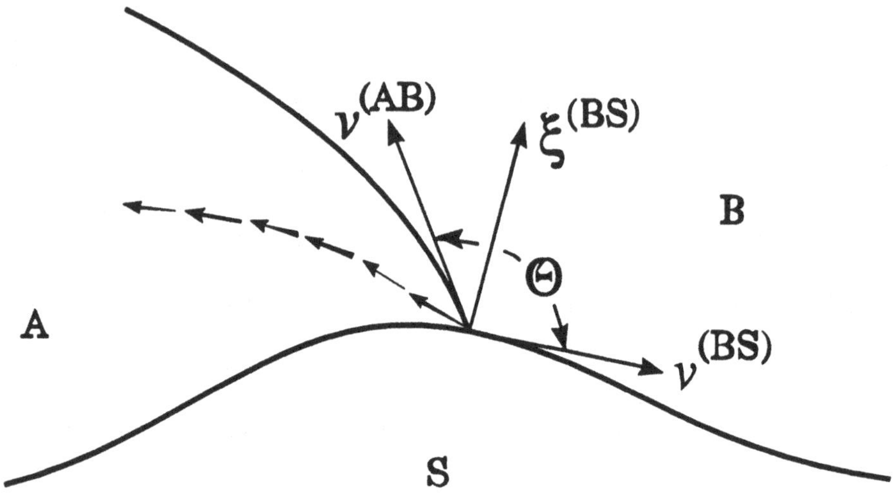

Figure 1.3-1: In the displacement of A by B over S, the solid phase S is assumed not to deform. The material surface ejected from the common line is drawn with the assumption that there is a forward rolling motion within phase B. I denote by $\boldsymbol{\nu}^{(BS)}$ the unit vector normal to the common line and tangent to the $B-S$ dividing surface, by $\boldsymbol{\nu}^{(AB)}$ the unit vector normal to the common line and tangent to the $A-B$ dividing surface, and by $\boldsymbol{\xi}^{(BS)}$ the unit vector that is normal to both the common line and the $B-S$ dividing surface.

motion, there will be *injection* from the interior of the other phase to the immediate neighborhood of the common line along both sides of such a surface.

1.3 velocity is multivalued on a rigid solid

Under certain conditions, the velocity field will be multivalued at a common line moving over a rigid solid. The proof does not require the surface of the solid to be a smooth plane. For this reason, in the displacement of A by B over S in Figure 1.3-1, I have pictured the surface of the solid phase S as an arbitrary curve. In a frame of reference that is fixed with respect to the common line, the velocity distribution in the solid

$$\mathbf{v}^{(S)} = \mathbf{v}^{(S)}(\mathbf{z}, t) \tag{1.3-1}$$

and the velocity distribution in the $A - B$ dividing surface

$$\mathbf{v}^{(\sigma)} = \mathbf{v}^{(\sigma)}(\mathbf{z}, t) \tag{1.3-2}$$

are both functions of position and time. Choosing the origin of our coordinate system to coincide with whatever point is of interest on the common line, we wish to compare

$$\mathbf{v}^{(S)}(\mathbf{0}, t) \equiv \text{limit } \mathbf{z} \to \mathbf{0}, \mathbf{z} \in \text{solid}: \quad \mathbf{v}^{(S)}(\mathbf{z}, t) \tag{1.3-3}$$

$$\mathbf{v}^{(\sigma)}(\mathbf{0}, t) \equiv \text{limit } \mathbf{z} \to \mathbf{0}, \mathbf{z} \in \text{dividing surface}: \quad \mathbf{v}^{(\sigma)}(\mathbf{z}, t) \tag{1.3-4}$$

In this frame of reference, $\mathbf{v}^{(S)}(\mathbf{0}, t)$ is tangent to the solid surface

$$\boldsymbol{\xi}^{(BS)} \cdot \mathbf{v}^{(S)}(\mathbf{0}, t) = 0 \tag{1.3-5}$$

$$\boldsymbol{\nu}^{(BS)} \cdot \mathbf{v}^{(S)}(\mathbf{0}, t) > 0 \tag{1.3-6}$$

and $\mathbf{v}^{(\sigma)}(\mathbf{0}, t)$ does not have a component normal to the $A - B$ dividing surface, which means that it can be written in the form

$$\mathbf{v}^{(\sigma)}(\mathbf{0}, t) = C\boldsymbol{\nu}^{(AB)} + D(\boldsymbol{\xi}^{(BS)} \wedge \boldsymbol{\nu}^{(BS)}) \tag{1.3-7}$$

Equation (1.3-7) implies that

$$\boldsymbol{\xi}^{(BS)} \cdot \mathbf{v}^{(\sigma)}(\mathbf{0}, t) = C(\boldsymbol{\xi}^{(BS)} \cdot \boldsymbol{\nu}^{(AB)}) \tag{1.3-8}$$

and

$$\boldsymbol{\nu}^{(BS)} \cdot \mathbf{v}^{(\sigma)}(\mathbf{0}, t) = C(\boldsymbol{\nu}^{(BS)} \cdot \boldsymbol{\nu}^{(AB)}) \tag{1.3-9}$$

Our objective is to compare these expressions with (1.3-5) and (1.3-6). Four cases can be identified.

Case 1 Let us assume that the contact angle Θ in Figure 1.3-1 is not $0, \pi/2$ or π. This means that

$$\boldsymbol{\xi}^{(BS)} \cdot \boldsymbol{\nu}^{(AB)} \neq 0 \tag{1.3-10}$$

and

$$\boldsymbol{\nu}^{(BS)} \cdot \boldsymbol{\nu}^{(AB)} \neq 0 \tag{1.3-11}$$

If $C \neq 0$, (1.3-5) and (1.3-8) imply

$$\boldsymbol{\xi}^{(BS)} \cdot \mathbf{v}^{(S)}(\mathbf{0}, t) \neq \boldsymbol{\xi}^{(BS)} \cdot \mathbf{v}^{(\sigma)}(\mathbf{0}, t) \tag{1.3-12}$$

If $C = 0$, (1.3-6) and (1.3-9) tell us that

$$\boldsymbol{\nu}^{(BS)} \cdot \mathbf{v}^{(S)}(\mathbf{0}, t) \neq \boldsymbol{\nu}^{(BS)} \cdot \mathbf{v}^{(\sigma)}(\mathbf{0}, t) \tag{1.3-13}$$

Our conclusion is that

$$\mathbf{v}^{(\sigma)}(\mathbf{0}, t) \neq \mathbf{v}^{(S)}(\mathbf{0}, t) \tag{1.3-14}$$

and the velocity vector must have multiple values at the common line.

Case 2 If the contact angle Θ is $\pi/2$

$$\boldsymbol{\nu}^{(BS)} \cdot \boldsymbol{\nu}^{(AB)} = 0 \tag{1.3-15}$$

For all values of C, we find (1.3-13) and (1.3-14). Again the velocity vector must have multiple values at the common lines.

Case 3 If the contact angle Θ is π

$$\boldsymbol{\xi}^{(BS)} \cdot \boldsymbol{\nu}^{(AB)} = 0 \tag{1.3-16}$$

There is a nonzero value of C for which

$$\boldsymbol{\nu}^{(BS)} \cdot \mathbf{v}^{(S)}(\mathbf{0}, t) = \boldsymbol{\nu}^{(BS)} \cdot \mathbf{v}^{(\sigma)}(\mathbf{0}, t) > 0 \tag{1.3-17}$$

and

$$\mathbf{v}^{(\sigma)}(\mathbf{0}, t) = \mathbf{v}^{(S)}(\mathbf{0}, t) \tag{1.3-18}$$

However, if we examine the velocity distribution in the ejected surface within phase A

$$\mathbf{v}^{(e)} = \mathbf{v}^{(e)}(\mathbf{z}, t) \tag{1.3-19}$$

assuming[2] forward rolling motion in phase B, we see that

$$\boldsymbol{\nu}^{(BS)} \cdot \mathbf{v}^{(e)}(0, t) \leq 0 \tag{1.3-20}$$

Our conclusion is that the velocity vector will assume multiple values at the common line, if material is ejected directly from the common line. In this case, our conclusion would be based upon the assumptions predicated in Sec. 1.2 particularly assumptions ii and iii.

Case 4 If the contact angle Θ is 0,

$$\boldsymbol{\nu}^{(BS)} \cdot \mathbf{v}^{(\sigma)}(\mathbf{0}, t) < 0 \tag{1.3-21}$$

assuming (see footnote 2) that there is a forward rolling motion in phase B. This implies (1.3-13) and (1.3-14): the velocity vector is multiple-valued at the common line.

To summarize, we have shown here that under certain conditions the velocity vector will be multiple-valued at a fixed common line on a moving rigid solid.

The discussion given in this section was inspired by that of Dussan V. and Davis [24].

[2]If there is a backward rolling motion in Phase A, the arguments presented in cases 3 and 4 are reversed.

1.4 mass balance at common line

What does conservation of mass require at each point on a common line?

After an application of the transport theorem for a body containing intersecting dividing surfaces, the postulate of mass conservation takes the form

$$\begin{aligned}
&\frac{d}{dt}\left(\int_R \rho dV + \int_\Sigma \rho^{(\sigma)} dA\right)\\
&= \int_R \left(\frac{d_{(m)}\rho}{dt} + \rho \operatorname{div} \mathbf{v}\right) dV\\
&+ \int_\Sigma \left\{\frac{d_{(s)}\rho^{(\sigma)}}{dt} + \rho^{(\sigma)} \operatorname{div}_{(\sigma)} \mathbf{v}^{(\sigma)} + \left[\rho(\mathbf{v}\cdot\boldsymbol{\xi} - v^{(\sigma)}_{(\xi)})\right]\right\} dA\\
&+ \int_{C^{(cl)}} \left(\rho^{(\sigma)}[\boldsymbol{\nu}^{(\sigma)}\cdot\boldsymbol{\nu} - u^{(cl)}_{(v)}]\right) ds\\
&= 0
\end{aligned} \tag{1.4-1}$$

In view of the equation of continuity ([94], Sec. 1.3.5)

$$\frac{d_{(m)}\rho}{dt} + \rho \operatorname{div} \mathbf{v} = 0 \tag{1.4-2}$$

and the jump mass balance ([94], Sec. 1.3.5)

$$\frac{d_{(s)}\rho^{(\sigma)}}{dt} + \rho^{(\sigma)} \operatorname{div}_{(\sigma)} \mathbf{v}^{(\sigma)} + \left[\rho(\mathbf{v}\cdot\boldsymbol{\xi} - v^{(\sigma)}_{(\xi)})\right] = 0 \tag{1.4-3}$$

(1.4-1) reduces to

$$\int_{C^{(cl)}} \left(\rho^{(\sigma)}[\mathbf{v}^{(\sigma)}\cdot\boldsymbol{\nu} - u^{(cl)}_{(v)}]\right) ds = 0 \tag{1.4-4}$$

Since (1.4-1) is valid for every material body and every portion of a material body, (1.4-4) must be true for every portion of a common line. We conclude [94, Exercises 1.3.8-1 and 1.3.8-2]

$$\left(\rho^{(\sigma)}[\mathbf{v}^{(\sigma)}\cdot\boldsymbol{\nu} - u^{(cl)}_{(v)}]\right) = 0 \tag{1.4-5}$$

This can be called the **mass balance at the common line**. It expresses the requirement that mass can be conserved at every point on a common line.

In deriving the transport theorem for a body containing intersecting dividing surfaces [94, Sec. 1.3.7], an important assumption has been made: ρ and $\mathbf{v}$ are piecewise continuous with piecewise continuous first derivatives. This must be reconciled with our observation in Sec. 1.3 that velocity may be multivalued at a common line moving over a rigid solid. Alternatives are discussed in Sec. 1.6.

Equation (1.4-5) tells us that, if there is mass transfer at the common line, $\mathbf{v}^{(\sigma)}\cdot\boldsymbol{\nu} - u^{(cl)}_{(v)}$ may assume different values in each of the dividing surfaces at the common line. This is compatible with the conclusion of Sec. 1.3 that velocity may be multivalued at a moving common line on a rigid solid.

If there is no mass transfer across the common line, we would expect the velocity field to be single-valued, since the velocity field would be continuous everywhere as the common line was approached. (I assume here that the tangential component of velocity that is also tangent to the common line will be continuous across the common line.) In a frame of reference such that the common line is fixed in space, the only motion would be along the common line in the limit as the common line was approached. The common line would be stationary relative to the intersecting dividing surfaces, although it might be in motion relative to the material outside its immediate neighborhood.

If two adjoining phases are saturated with one another, we can with reasonable confidence assume that there is no mass transfer across the phase interface separating them. Under what conditions are we free to say that there is no mass transfer across the common line?

a) If we are concerned with a stationary common line on a solid phase interface, it should be possible after a period of time to assume that chemical equilibrium has been established on the phase interfaces in the neighborhood of the common line and that there is no mass transfer across the common line.

b) If the various species present are assumed not to adsorb in any of the phase interfaces and if the adjoining bulk phases have been pre- equilibrated, it may be reasonable to eliminate mass transfer at the common line.

c) When slip is assumed in the neighborhood of a moving common line on a rigid solid, it is usual to assume that velocity is single-valued at the common line and as a result that there is no mass transfer at the common line (see Sec. 1.6 and [94, Sec. 3.1.1]). With this assumption, no portion of an adsorbed film in one phase interface can be rolled directly into an adsorbed film in another phase interface. All mass transfer must take place through one of the adjoining phases.

In stating (1.4-1) and in deriving (1.4-5) we have assumed there is no mass in the common line. I am not aware of any experimental evidence that suggests mass should be associated with the common line. For related references, see the concluding remarks of Sec. 3.1.

1.5 comment on velocity distribution in neighborhood of moving common line on rigid solid

In Sec. 1.3, we found that under certain conditions velocity will be multivalued at a moving common line on a rigid solid. This suggests that, at least in some cases, one or more derivatives of the velocity component could be unbounded within the adjoining phases in the limit as the common line is approached. What are the physical implications?

If the rate of deformation tensor

$$\mathbf{D} \equiv \frac{1}{2}\left[\nabla\mathbf{v} + (\nabla\mathbf{v})^T\right] \tag{1.5-1}$$

is unbounded as the common line is approached, the stress tensor in the adjoining phases may be unbounded as well. For example, we know that for a Newtonian fluid [93, p. 49]

$$\mathbf{T} = (-P + \lambda \operatorname{div} \mathbf{v})\mathbf{I} + 2\mu\,\mathbf{D} \tag{1.5-2}$$

From a physical point of view, we should require at least the forces in the neighborhood of the common line be finite.

In order to better appreciate the issue, we extend the argument of Dussan V. and Davis [24] to consider a three-dimensional flow of an incompressible, Newtonian fluid within one of the phases in the neighborhood of a common line moving over a *rigid* [3] solid. We will assume that there is no mass transfer between the dividing surfaces and the adjoining bulk phases in the limit as the common line is approached and we will adopt the no-slip boundary condition requiring the tangential components of velocity to be continuous across the dividing surface.

Let us choose a frame of reference in which the moving common line shown in Figure 1.5-1 is stationary. In terms of a cylindrical coordinate system centered upon the common line, assume that the flow is fully three-dimensional but independent of time in phase B:

$$\begin{aligned} v_r &= v_r(r,\theta,z) \\ &= f(\theta,z) + F(r,\theta,z) \end{aligned} \tag{1.5-3}$$

$$\begin{aligned} v_\theta &= v_\theta(r,\theta,z) \\ &= g(\theta,z) + G(r,\theta,z) \end{aligned} \tag{1.5-4}$$

[3] With two restrictions, a two-dimensional plane flow in the neighborhood of a moving common line formed by three deformable phases is not possible [94, Exercise 1.3.9-1]: in the limit as the common line is approached, there is no interphase mass transfer [94, Exercise 1.3.9-2] and all of the surface mass densities must be non-negative with at least one different from zero.

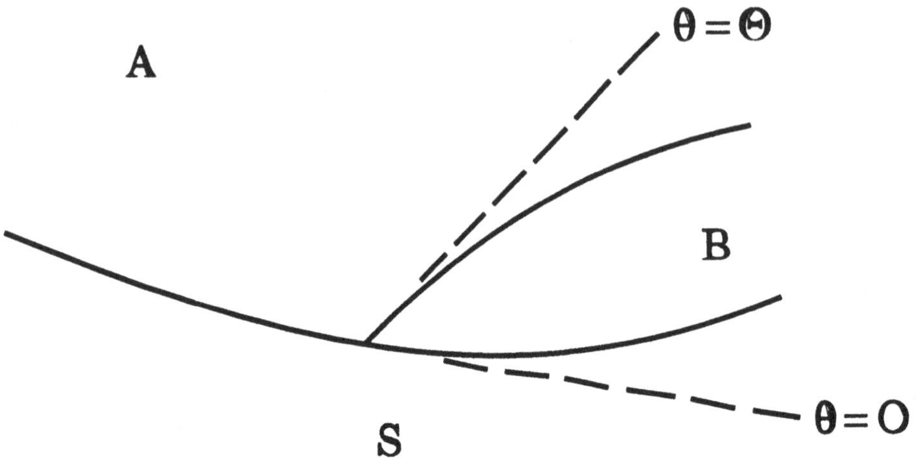

Figure 1.5-1: A frame of reference is chosen in which the moving common line is stationary.

$$v_z = v_z(r, \theta, z) \tag{1.5-5}$$

The form of this velocity distribution is restricted by the equation of continuity for an incompressible fluid

$$\begin{aligned} \text{div } \mathbf{v} &= \frac{\partial v_r}{\partial r} + v_r + \frac{\partial v_\theta}{\partial \theta} + r\frac{\partial v_z}{\partial z} \\ &= r\frac{\partial F}{\partial r} + f + F + \frac{\partial g}{\partial \theta} + \frac{\partial G}{\partial \theta} + r\frac{\partial v_z}{\partial z} \\ &= 0 \end{aligned} \tag{1.5-6}$$

In order to identify $f(\theta, z)$ and $g(\theta, z)$ as the r and θ-components of velocity at the common line, let us require

$$\text{limit } r \to 0, \theta \text{ and } z \text{ fixed}: \quad F(r, \theta, z) = 0 \tag{1.5-7}$$

$$\text{limit } r \to 0, \theta \text{ and } z \text{ fixed}: \quad G(r, \theta, z) = 0 \tag{1.5-8}$$

Following the discussion in Sec. 1.3, we will assume

$$f(0, z) \neq f(\Theta, z) \tag{1.5-9}$$

If this fails in phase B, the comparable relation would necessarily be true in phase A. (The comparable relation would also be true in phase B, if the ejected or emitted material surface described in Sec. 1.2 is in phase B and if we identify Θ with this surface.)

Since our concern here is the possibility of an unbounded force developed within phase B in the neighborhood of the common line attributable to (1.5-9), we need to consider only the r-component of force (per unit width) exerted upon one side of plane $\theta =$ a constant within phase B:

$$\begin{aligned}\mathcal{F}_r &= \int_0^r T_{r\theta}\, dr \\ &= 2\mu \int_0^r D_{r\theta}\, dr \\ &= \mu \int_0^r \left(\frac{1}{r}\frac{\partial v_r}{\partial \theta} + \frac{\partial v_\theta}{\partial r} - \frac{v_\theta}{r} \right) dr \\ &= \mu \int_0^r \left[\frac{1}{r}\left(\frac{\partial f}{\partial \theta} + \frac{\partial F}{\partial \theta} - g - G \right) + \frac{\partial G}{\partial r} \right] dr \end{aligned} \quad (1.5\text{-}10)$$

So long as G is assumed to be absolutely continuous in r for fixed θ,

$$\int_0^r \frac{\partial G}{\partial r} dr = G \quad (1.5\text{-}11)$$

In order that $\mathcal{F}_r$ be finite, we must require

$$\text{limit } r \to 0: \quad \frac{\partial f}{\partial \theta} + \frac{\partial F}{\partial \theta} - g = 0 \quad (1.5\text{-}12)$$

Differentiating (1.5-6) with respect to θ, we have

$$r\frac{\partial^2 F}{\partial r \partial \theta} + \frac{\partial f}{\partial \theta} + \frac{\partial F}{\partial \theta} + \frac{\partial^2 g}{\partial \theta^2} + \frac{\partial^2 G}{\partial \theta^2} + r\frac{\partial^2 v_z}{\partial \theta \partial z} = 0 \quad (1.5\text{-}13)$$

Equation (1.5-13) can be used to eliminate $\partial f/\partial\theta + \partial F/\partial\theta$ from (1.5-12):

$$\text{limit } r \to 0: \quad \frac{\partial^2 g}{\partial \theta^2} + g + r\frac{\partial^2 F}{\partial r \partial \theta} + \frac{\partial^2 G}{\partial \theta^2} + r\frac{\partial^2 v_z}{\partial \theta \partial z} = 0 \quad (1.5\text{-}14)$$

Let us assume that the velocity distribution in the neighborhood of the common line is such that

$$\text{limit } r \to 0: \quad \frac{\partial F}{\partial \theta} = 0 \quad (1.5\text{-}15)$$

$$\text{limit } r \to 0: \quad r\frac{\partial^2 F}{\partial r \partial \theta} = 0 \quad (1.5\text{-}16)$$

$$\text{limit } r \to 0: \quad \frac{\partial^2 G}{\partial \theta^2} = 0 \tag{1.5-17}$$

Because the solid is rigid and because we are assuming that the tangential components of velocity are continuous across the dividing surfaces,

$$\text{limit } r \to 0: \quad r\frac{\partial^2 v_z}{\partial \theta \partial z} = \frac{\partial v_z}{\partial z} = 0 \tag{1.5-18}$$

With the restrictions (1.5-16) through (1.5-18), (1.5-14) reduces to

$$\frac{\partial^2 g}{\partial \theta^2} + g = 0 \tag{1.5-19}$$

Since there is no interphase mass transfer in the limit as the common line is approached,

$$g(0, z) = g(\Theta, z) = 0 \tag{1.5-20}$$

If $\Theta \neq 0$ or π, the only solution is

$$g = 0 \tag{1.5-21}$$

But (1.5-12) and (1.5-15) imply

$$\frac{\partial f}{\partial \theta} - g = 0 \tag{1.5-22}$$

or

$$\frac{\partial f}{\partial \theta} = 0 \tag{1.5-23}$$

which contradicts our assumption (1.5-9) that the r-component of velocity within the $A - B$ dividing surface differs from that within the $B - S$ dividing surface at the common line.

If $\Theta = \pi$ and the emitted or ejected material surface described in Sec. 1.3 is in phase B, the same argument as above applies. It is only necessary to identify $\Theta(\neq \pi)$ with the location of the emitted or ejected surface.

Finally, consider $\Theta = 0$ and the emitted or ejected material surface in phase B. In this case, the solution to (1.5-19) consistent with boundary conditions (1.5-20) is

$$g = B \sin \theta \tag{1.5-24}$$

Equation (1.5-22) consequently requires

$$f = -B \cos \theta + C \tag{1.5-25}$$

The constants B and C are not sufficient to satisfy the constraints of an emitted or ejected surface originating at the common line.

If $\mathcal{F}_r$ is to be finite, one of the other assumptions made in this argument must be incorrect. Notice in particular that this discussion is unaffected by the existence of body forces such as gravity or of mutual forces such as London-van der Waals forces or electrostatic double-layer forces [94, Exercises 2.1.3-1 and 2.1.3-2]. Possibilities are considered in the next section.

1.6 more comments on velocity distribution in neighborhood of moving common line on rigid solid

In Sec. 1.3, we found that under certain conditions velocity may be multivalued at a common line moving over a rigid solid. Although we noted that this is consistent with mass transfer at the common line in Sec. 1.4, it may be inconsistent with the development of the transport theorem for a body containing intersecting dividing surfaces [94, Sec. 1.3.7], and it may lead to unbounded forces at the common line (Sec. 1.5). At least one of the assumptions underlying these inconsistencies must be incorrect. Let us consider the possibilities.

i) A solid is not rigid.

For most solids, the deformation caused by the forces acting at a common line is very small [66, 98]. The deformation of metals and other materials with high tensile strength will be of the same order or smaller than imperfections we will wish to ignore in real surfaces. But this deformation is non-zero. An unbounded force at the common line might be interpreted as a natural response to our refusal to allow the surface to deform.

On the other hand, it can not be denied that a rigid solid is a convenient idealization.

ii) The tangential components of velocity are not continuous at phase interfaces in the limit as a moving common line is approached.

Outside the immediate neighborhood of the common line, there is no debate. Goldstein [34, p. 676] presents a sound case favoring the no-slip boundary condition, except in the limit of a rarefied gas where it is known to fail. Richardson [85] comes to a similar conclusion in considering single phase flow past a wavy wall. He concludes that, even if there were no resistance to relative motion between a fluid and a solid in contact, roughness alone would ensure that the boundary condition observed on a macroscopic scale would be one of no slip. (Still another more restricted argument supporting the continuity of the tangential components of velocity at a dividing surface can be given [94, Exercise 2.1.6-3]).

Hocking [41] finds that, in the displacement of one fluid by another over a wavy surface, a portion of the displaced phase may be left behind, trapped in the valleys. As a result, the displacing fluid moves over a composite surface, particularly within the immediate neighborhood of the moving common line. Hocking [41] concludes that it is appropriate to describe the movement of the displacing phase over this composite surface though a slip boundary condition. On the other hand, his argument could not be used to justify the use of a slip boundary condition on a geometrically smooth surface such as we have considered here.

Huh and Scriven [49] as well as Dussan V. and Davis [24] suggest that slip might be allowed within the immediate neighborhood of the common line. The introduction of slip within the immediate neighborhood of the common line eliminates the appearance of an unbounded force [22, 42, 48, 35, 43, 19].

It is reassuring to note that computations outside the immediate neighborhood of the common line appear to be insensitive to the details of the slip model used [22, 62].

iii) There is interphase mass transfer in the limit as the common line is approached.

Interphase mass transfer could eliminate these inconsistencies. This possibility should be investigated further.

iv) A liquid does not spread over a rigid solid, but rather over a precursor film on the rigid solid.

Teletzke *et al.* [95] suggest that there are three classes of spreading: primary, secondary, and bulk.

By primary spreading, they refer to the manner in which the first one or two molecular layers are deposited on the solid. For a nonvolatile liquid, primary spreading is by surface diffusion [4, 17, 8, 87]. For a volatile liquid, primary spreading may be by the condensation of an adsorbed film [39].

By secondary spreading, they refer to the motion of films whose thickness varies from several molecules to roughly a micron. These films spread as the result of either a positive disjoining pressure or a surface tension gradient. Surface tension gradients can be the result of either temperature gradients [70] or composition gradients, perhaps resulting from the evaporation of a more volatile component [4].

By bulk spreading, they refer to the motion of films thicker than a micron. In these films, the flow is driven by interfacial tension forces, gravity, and possibly a forced convection.

From this point of view, the precursor films discussed in Sec. 1.1 are probably the result of both primary and secondary spreading. There may be no moving common line, but simply a smooth transition from film flow to surface diffusion.

This conception of the spreading process certainly seems to explain some classes of observations. For example, Teletzke *et al.* [95] argue effectively that the experiments of Bascom *et al.* [4] can be understood in this manner.

But it does not explain how common lines move more generally. For example, consider a common line driven by forced convection where a thin film of the liquid exhibits a negative disjoining pressure; a common line driven by forced convection where the time required for displacement is much smaller than the time required the primary and secondary spreading phenomena described above; a common line that recedes as the result of a draining liquid film.

v) The common line does not actually move.

This proposal, perhaps initially surprising, will be investigated further in Sec. 3.4 (see also [94, Sec. 2.1.13]).

No significance should be attached to my developing only proposal v in the remainder of this discussion, other than it is simpler to employ than i, iii or iv and it seems to retain more of the underlying chemistry than ii. I see merit in all of these proposals. The last three in particular are not mutually exclusive. I don't expect to see one mechanism developed to describe all common line movements, since different situations may involve different physical phenomena. It certainly is too early to begin ruling out any ideas.

2 Thin films

The evidence presented in Sec. 1 suggests that, if we are to understand the motion of common lines, we must understand the movement of the thin films within the immediate neighborhood of the common line. In what follows, I would like to illustrate that continuum mechanics can be used successfully to describe these thin films.

2.1 estimating interfacial tension (with D. Li and M. J. Kim)

The vertical thin liquid film shown in Figure 2.1-1 is extended slowly as a function of time by moving the upper solid support. The lower solid support is stationary. The adjoining phase is a gas.

Consider the constant volume system indicated by the dashed lines that includes the liquid film as well as a portion of the adjoining gas phases. We will find it convenient to make the following assumptions.

a) The expansion of the film is sufficiently slow that the effects of inertia, and therefore kinetic energy, may be neglected.

b) There are no external or mutual forces associated with the liquid-gas interfaces.

c) Since the system is chosen to have a constant volume, it has no entrances or exits.

d) The liquid is assumed to be incompressible.

e) There is no mass transfer across the liquid-gas interface.

f) The liquid-gas interfaces are planes. There are at least two ways in which this may be achieved. 1) The solid supports could be such that the dynamic contact angle is 90°. 2) The system could be chosen to include a constant volume of the liquid film but to exclude the three-phase lines of contact on the upper (moving) solid support as well as the adjacent liquid bounded by curved surfaces.

g) The rate at which mechanical energy is dissipated by the action of viscous forces both within the bulk phases and in the dividing surfaces can be neglected.

h) Israelachvili [51] suggests that the distance between the two interfaces can not go to zero due to the finite size of the constituent atoms and that a finite interfacial separation d must be recognized. He recommends that d be viewed as the mean distance between the centers of individual atoms and estimated as

$$d = 0.91649[M/(\rho^{(L)} n_a N)]^{1/3} \tag{2.1-1}$$

Here M is the molecular weight, n_a the number of atoms per molecule, N Avogadro's constant (6.023×10^{23} mol^{-1}), and $\rho^{(L)}$ the mass density of the liquid. In arriving at (2.1-1), the atoms have been assumed to be in a close packing arrangement. For a molecule consisting of a repeating unit, such as an n-alkane with -CH_2- being the repeating unit, we suggest a simple picture in which the molecules are arranged such that the *repeating units* are in a close packing arrangement. Retracing the argument of Israelachvili [51], we have instead

$$d = 0.91649[M/(\rho^{(L)} n_u N)]^{1/3} \tag{2.1-2}$$

where n_u is the number of repeating units in the molecule. This is similar to the suggestion of Padday and Uffindell [79], who replaced M/n_u by the -CH_2- group weight and took the coefficient to be unity.

j) The potential energy of the gas phase attributable to London-van der Waals forces can be neglected with respect to that of the liquid phase.

k) We will neglect the London-van der Waals interactions between the gas and liquid phases [94, Exercise 2.1.3-1] and compute the potential energy per unit volume attributable to both London-van der Waals forces and gravity at any point within the liquid phase as

$$\begin{aligned}\rho^{(L)}\phi^{(L)} &= -4\int_{-(h+d^{(LG)})/2}^{(h+d^{(LG)})/2}\int_0^\infty\int_0^\infty \frac{A^{(LL)}}{\pi^2 r^6}dx\,dy\,dz + \rho^{(L)}gz_1 \\ &= \frac{4A^{(LL)}}{3\pi}\Big[(h+d^{(LG)}+2z_2)^{-3} \\ &\quad + (h+d^{(LG)}-2z_2)^{-3}\Big] + \rho^{(L)}gz_1 \qquad (2.1\text{-}3)\end{aligned}$$

Here

$$r \equiv [x^2 + y^2 + (z - z_2)^2]^{1/2} \qquad (2.1\text{-}4)$$

is the separation distance between a specified point and any other point within the liquid phase, h is the film thickness measured from the center of the first layer of repeating units to the center of the last layer of repeating units, and $d^{(LG)}$ is the liquid-gas interfacial separation, which must be distinguished from the liquid-liquid interfacial separation d. [Equations (2.1-1) and (2.1-2) are intended to be used for interfaces separating condensed phases and can not be employed in estimating $d^{(LG)}$.] This means that the volume of the liquid film at any point in time is $L(h + d^{(LG)})$, where L is the instantaneous length of the film.

We assume that there are no electrostatic effects.

For this system, the mechanical energy balance ([94], Sec. 4.1.4) reduces to

$$\frac{d}{dt}\int_{R_{(sys)}}\rho\phi dV = -\mathcal{W} - \int_{\Sigma_{(sys)}}\gamma\,\text{div}_{(\sigma)}\mathbf{v}^{(\sigma)}dA \qquad (2.1\text{-}5)$$

Neglecting the stresses that the bulk fluid exerts on the moving upper solid support, we find

$$\mathcal{W} = 2\gamma\frac{dL}{dt} \qquad (2.1\text{-}6)$$

We will assume that all new surface is created at the moving upper solid support and that elsewhere the film's surface is immobile or that the tangential components of the surface velocity are zero. This permits us to conclude that

$$\int_{\Sigma_{(sys)}}\gamma\,\text{div}_{(\sigma)}\mathbf{v}^{(\sigma)}dA = 0 \qquad (2.1\text{-}7)$$

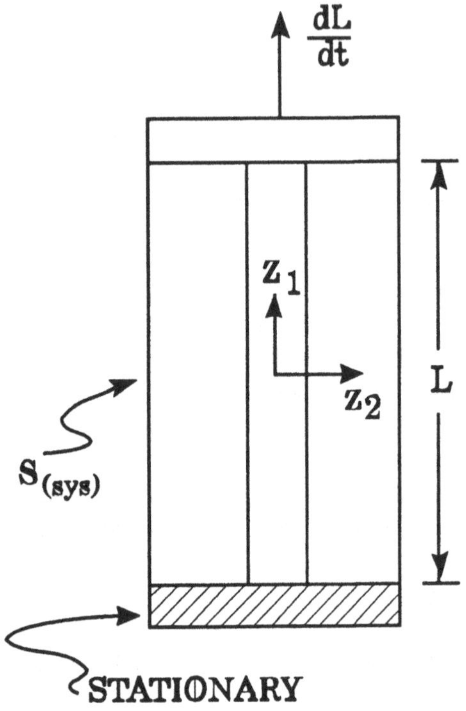

Figure 2.1-1: A vertical thin film is extended slowly as a funtion of time by moving the upper solid support. The lower solid support is stationary.

Considering a unit width of film having a length L, we can determine that, in view of (2.1-6) and (2.1-7), (2.1-5) reduces to

$$\begin{aligned}
&\frac{d}{dt}\int_{-h/2}^{h/2}\int_0^L \frac{4A^{(LL)}}{3\pi}\Big[(h+d^{(LG)}+2z_2)^{-3} \\
&\quad + (h+d^{(LG)}-2z_2)^{-3}\Big]\,dz_1\,dz_2 + \frac{d}{dt}\left[\frac{1}{2}(\rho^{(L)}-\rho^{(G)})ghL^2\right] \\
&= \frac{d}{dt}\left[\frac{8A^{(LL)}h(h+d^{(LG)})L}{3\pi d^{(LG)2}(2h+d^{(LG)})^2}+\frac{1}{2}(\rho^{(L)}-\rho^{(G)})ghL^2\right] \\
&= 2\gamma\frac{dL}{dt}
\end{aligned} \tag{2.1-8}$$

Because the liquid film has a constant volume $L(h+d^{(LG)})$, we can argue that

$$\frac{dh}{dt} = -(h+d^{(LG)})\frac{1}{L}\frac{dL}{dt} \tag{2.1-9}$$

We argue that there is a critical value of $h = h_c$ at which the film ruptures, and we assume that

$$h_c = d \tag{2.1-10}$$

As $h \to h_c$, (2.1-8) and (2.1-9) imply

$$\begin{aligned}
\gamma &= \frac{4A^{(LL)}}{3\pi}\left[\frac{d(d+d^{(LG)})}{d^{(LG)2}}(2d^{(LG)})^2 - \frac{d+d^{(LG)}}{(2d+d^{(LG)})^3}\right] \\
&\quad + \frac{1}{4}(d-d^{(LG)})(\rho^{(L)}-\rho^{(G)})gL
\end{aligned} \tag{2.1-11}$$

Normally gravity could be neglected to obtain

$$\gamma = \frac{4A^{(LL)}}{3\pi}\left[\frac{d(d+d^{(LG)})}{d^{(LG)2}(2d+d^{(LG)})^2} - \frac{d+d^{(LG)}}{(2d+d^{(LG)})^3}\right] \tag{2.1-12}$$

As shown in Table 2.1, (2.1-12) used together with (2.1-2) is in excellent agreement with the experimental measurements of Jasper and Kring [55], when we choose $d^{(LG)} = \alpha d$ and $\alpha = 1.132$. [The experimental study of Richards and Carver [84] indicates that the difference between surface tensions of the n-alkanes against their own vapor and the surface tensions against air is less than 0.5together with (2.1-1) (replacing n_u by n_a), the slope for γ as a function of n is less satisfactory.

Table 2.1 also compares (2.1-12) with

$$\gamma = \frac{A^{(LL)}}{24\pi {D_0}^2} \tag{2.1-13}$$

using $D_0 = 1.65$Å recommended by Israelachvili [52, p. 157]. It represents the data well, although the slope of γ as a function of n is somewhat better predicted by (2.1-12).

Table 2.1-1: Comparisons of (2.1-12) with the experimental measurements of γ for n-alkanes at $20.0 \pm 0.08°C$ reported by Jasper and Kring [55] and with (2.1-13) recommended by Israelachvili [52, p. 157]. In these comparisons, we have employed the values of $A^{(LL)}$ computed by Hough and White [46] and we have assumed $d^{(LG)} = \alpha d$, where α has been found by minimizing the sum of the squared errors.

n	d^a (Å)	d^b (Å)	A^c ($\times 10^{-20}$ J)	$\gamma_{meas}{}^d$ (mN/m)	$\gamma_{calc}{}^e$ (mN/m)	$\gamma_{calc}{}^f$ (mN/m)	$\gamma_{calc}{}^g$ (mN/m)
5	2.054	3.088	3.74	16.05	16.68	18.22	17.36
6	2.028	3.029	4.06	18.40	18.82	19.78	19.33
7	2.012	2.991	4.31	20.14	20.49	21.00	20.85
8	1.999	2.962	4.49	21.62	21.76	21.87	22.01
9	1.990	2.939	4.66	22.85	22.94	22.70	23.05
10	1.982	2.921	4.81	23.83	23.97	23.43	23.98
11	1.976	2.906	4.87	24.66	24.52	23.72	24.43
12	1.971	2.894	5.03	25.35	25.54	24.50	25.36
13	1.966	2.883	5.04	25.99	25.78	24.55	25.54
14	1.962	2.874	5.09	26.56	26.20	24.80	25.90
15	1.959	2.867	5.15	27.07	26.64	25.09	26.28
16	1.956	2.860	5.22	27.47	27.14	25.43	26.72

a) Calculated using (2.1-1).

b) Calculated using (2.1-2).

c) $A \equiv A^{(LL)}$.

d) Measured by Jasper and Kring [55].

e) Calculated using (2.1-12) and (2.1-2). For this case, we found $\alpha = 1.132$.

f) Calculated using (2.1-13) and $D_0 = 1.65$ Å [52, p. 157].

g) Calculated using (2.1-12) and (2.1-1) (replacing n_u by n_a). For this case, we found $\alpha = 1.391$.

3 Dynamic contact angles

In what follows, we will discuss the difference between the intrinsic contact angle Θ_0 defined at a true common line and the contact angle Θ observed by an experimentalist at an apparent common line. This will lead to an illustration of one of the mechanisms that we believe is responsible for the movement of a common line and to a computation of the dynamic contact angle for a draining film.

3.1 Euler's first law at common line

What does Euler's first law require at each point on a common line?

Let us begin by restating Euler's first law [94, Sec. 2.1.3], expressing the stress vector $\mathbf{t}$ in terms of the stress tensor $\mathbf{T}$ [94, Sec. 2.1.4] and the surface stress vector $\mathbf{t}^{(\sigma)}$ in terms of the surface stress tensor $\mathbf{T}^{(\sigma)}$ [94, Sec. 2.1.5]:

$$\frac{d}{dt}\left(\int_R \rho\mathbf{v}dV + \int_\Sigma \rho^{(\sigma)}\mathbf{v}^{(\sigma)}dA\right) = \int_S \mathbf{T}\cdot\mathbf{n}\,dA + \int_C \mathbf{T}^{(\sigma)}\cdot\boldsymbol{\mu}\,ds + \int_R \rho\mathbf{b}\,dV + \int_\Sigma \rho^{(\sigma)}\mathbf{b}^{(\sigma)}dA \tag{3.1-1}$$

The first integral on the right becomes after an application of Green's transformation [93, p. 661]

$$\int_S \mathbf{T}\cdot\mathbf{n}\,dA = \int_R \operatorname{div}\mathbf{T}\,dV + \int_\Sigma \left[\mathbf{T}\cdot\boldsymbol{\xi}\right]dA \tag{3.1-2}$$

Using the surface divergence theorem [94, Exercise A.6.3-1], we can write the second integral on the right of (3.1-1) as

$$\int_C \mathbf{T}^{(\sigma)}\cdot\boldsymbol{\mu}\,ds = \int_\Sigma \operatorname{div}_{(\sigma)}\mathbf{T}^{(\sigma)}dA + \int_{C^{(cl)}}\left(\mathbf{T}^{(\sigma)}\cdot\boldsymbol{\nu}\right)ds \tag{3.1-3}$$

After applying the transport theorem for a body containing intersecting dividing surfaces to the terms on the left of (3.1-1), we can use (3.1-2) and (3.1-3) to write (3.1-1) as

$$\begin{aligned}
&\int_R \left(\rho\frac{d_{(m)}\mathbf{v}}{dt} - \operatorname{div}\mathbf{T} - \rho\mathbf{b}\right)dV \\
&+ \int_\Sigma \left\{\rho^{(\sigma)}\frac{d_{(s)}\mathbf{v}^{(\sigma)}}{dt} - \operatorname{div}_{(\sigma)}\mathbf{T}^{(\sigma)} - \rho^{(\sigma)}\mathbf{b}^{(\sigma)}\right. \\
&\left. + \left[\rho\left(\mathbf{v}-\mathbf{v}^{(\sigma)}\right)(\mathbf{v}\cdot\boldsymbol{\xi} - v_{(\xi)}^{(\sigma)}\right) - \mathbf{T}\cdot\boldsymbol{\xi}\right]\right\}dA \\
&+ \int_{C^{(cl)}}\left(\rho^{(\sigma)}\mathbf{v}^{(\sigma)}[\mathbf{v}^{(\sigma)}\cdot\boldsymbol{\nu} - u_{(\nu)}^{(cl)}] - \mathbf{T}^{(\sigma)}\cdot\boldsymbol{\nu}\right)ds \\
&= 0
\end{aligned} \tag{3.1-4}$$

In view of Cauchy's first law [94, Sec. 2.1.4] and the jump momentum balance [94, Sec. 2.1.6], (3.1-4) reduces to

$$\int_{C^{(cl)}} \left(\rho^{(\sigma)} \mathbf{v}^{(\sigma)} \left[\boldsymbol{\nu}^{(\sigma)} \cdot \boldsymbol{\nu} - u_{(\nu)}^{(cl)} \right] - \mathbf{T}^{(\sigma)} \cdot \boldsymbol{\nu} \right) ds = 0 \tag{3.1-5}$$

Since (3.1-1) is valid for every material body and every portion of a material body, (3.1-5) must be true for every portion of a common line. We conclude [94, Exercises 1.3.8-1 and 1.3.8-2]

$$\left(\rho^{(\sigma)} \mathbf{v}^{(\sigma)} \left[\boldsymbol{\nu}^{(\sigma)} \cdot \boldsymbol{\nu} - u_{(\nu)}^{(cl)} \right] - \mathbf{T}^{(\sigma)} \cdot \boldsymbol{\nu} \right) = 0 \tag{3.1-6}$$

This can be called the **momentum balance at the common line.** It expresses the requirement of Euler's first law at every point on a common line.

If we neglect mass transfer at the common line

$$\text{at } C^{(cl)}: \quad \mathbf{v}^{(\sigma)} \cdot \boldsymbol{\nu} = u_{(\nu)}^{(cl)} \tag{3.1-7}$$

or if we neglect the effect of inertial forces at the common line, the momentum balance at the common line reduces to

$$\left(\mathbf{T}^{(\sigma)} \cdot \boldsymbol{\nu} \right) = 0 \tag{3.1-8}$$

Under static conditions, the surface stress acting at the common line may be expressed in terms of the surface tension and (3.1-8) becomes

$$(\gamma \mathbf{v}) = 0 \tag{3.1-9}$$

In the context of Figure 3.1-1, this says

$$\gamma^{(AB)} \boldsymbol{\nu}^{(AB)} + \gamma^{(AS)} \boldsymbol{\nu}^{(AS)} + \gamma^{(BS)} \boldsymbol{\nu}^{(BS)} = 0 \tag{3.1-10}$$

Here $\gamma^{(AB)}, \gamma^{(AS)}$, and $\gamma^{(BS)}$ are the surface tensions within the $A - B$, $A - S$, and $B - S$ dividing surfaces in the limit as the common line is approached. Equations (3.1-9) and (3.1-10) have been referred to as the **Neumann triangle** [9, p. 288].

In deriving the transport theorem for a body containing intersecting dividing surfaces and therefore in deriving the momentum balance at the common line, we assume that $\mathbf{T}$ is piecewise continuous with piecewise continuous first derivatives (see [94], Sec. 1.3.7). It will be important to keep this in mind in discussing the momentum balance at a common line on a rigid solid in the next section.

Note also that in stating (3.1-1) and in deriving the momentum balance at the common line, I have assumed that there is no momentum or force intrinsically associated with the common line. Gibbs [32, p. 288 and 296] recognized the possibility

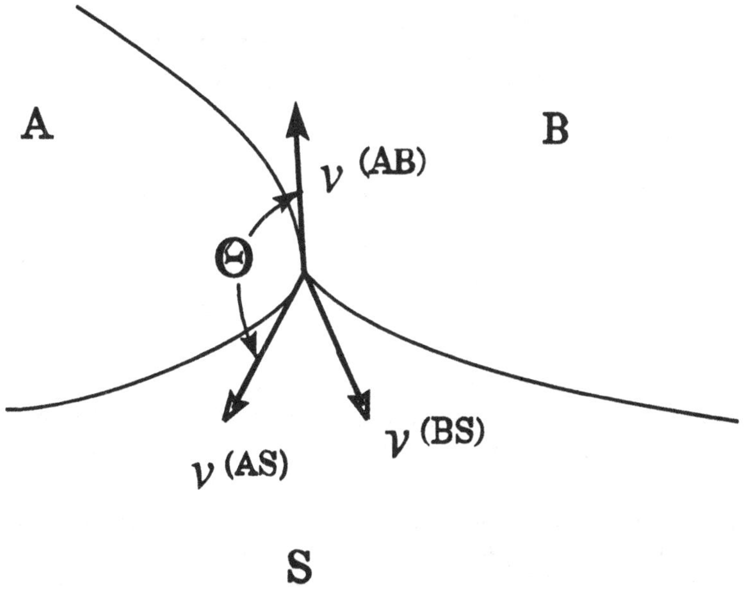

Figure 3.1-1: The common line formed by the intersection of the $A-B$, $A-S$, and $B-S$ dividing surfaces. Here $\boldsymbol{\nu}^{(AB)}, \boldsymbol{\nu}^{(AS)}$, and $\boldsymbol{\nu}^{(BS)}$ are the unit vectors that are normal to the common line and that are both tangent to and directed into the $A-B$, $A-S$, and $B-S$ dividing surfaces respectively; θ is the contact angle measured through phase A.

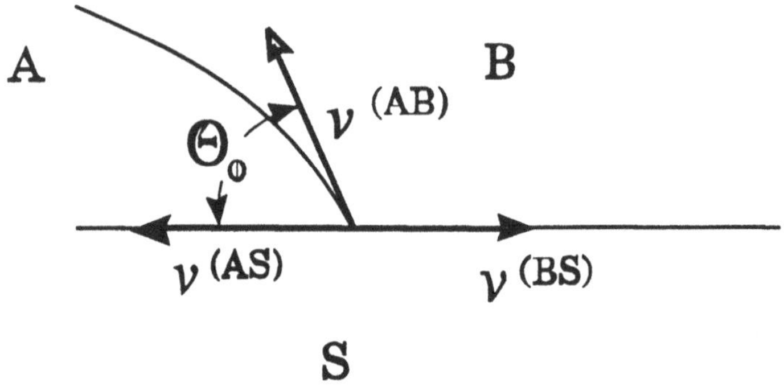

Figure 3.2-1: For a relatively rigid solid, the deformation shown in Figure 3.1-1 normally will not be apparent.

of a force or line tension acting in the common line, but he gave it relatively little attention, suggesting that he considered it to be too small to be of importance. Buff and Saltsburg [10]; see also Buff [9, p. 287] argued that, at least at equilibrium, the effect of the line tension would be too small to be of significance in macroscopically observable phenomena. Yet recent experimental studies suggest that the effects of line tension are observable [81, 64]. The observations of Gaydos and Neumann [31] appear to have a better explanation, as indicated by Slattery [94, Sec. 3.2.7]. On balance, it seems premature to make a judgment on the matter at this writing.

3.2 momentum balance at common line on relatively rigid solid

If the contact angle measured at the common line on a rigid solid through one of the fluid phases were either 0 or π, the solid would not be deformed by the stress exerted on the common line by the fluid-fluid interface, since this stress would act tangent to the solid surface.

If we assume that the contact angle is neither 0 nor π, usual practice is to ignore any deformation in the neighborhood of the common line, arguing that it would be less than the imperfections in a relatively high tensile strength solid surface that we are generally willing to ignore [66, 98]. We will ignore the deformation pictured in Figure 1-1 and idealize the neighborhood of the common line by Figure 3.2-1.

There is an important distinction to be made between choosing to ignore the deformation in a relatively high tensile strength solid and adopting the model of a perfectly rigid solid. It would appear that, with a perfectly rigid solid, **T** would no longer be piecewise continuous with piecewise continuous first derivatives as assumed in deriving the momentum balance at the common line. This assumes that **T** would be bounded in the adjoining fluid phases.

In choosing to ignore the deformation of the solid in the neighborhood of the common line, we eliminate from consideration the effect of the mechanical properties of the solid and the effect of the normal component of the momentum balance at the common line. Only the component of (3.1-6) tangent to the solid surface is significant:

$$\mathbf{P}^{(AS)} \cdot \left(\rho^{(\sigma)} \mathbf{v}^{(\sigma)} [\mathbf{v}^{(\sigma)} \cdot \boldsymbol{\nu} - u^{(cl)}_{(\nu)}] - \mathbf{T}^{(\sigma)} \cdot \boldsymbol{\nu} \right) = 0 \tag{3.2-1}$$

Here $\mathbf{P}^{(AS)}$ is the projection tensor for the $A - S$ dividing surface.

If, as is common, we neglect both mass transfer at the common line and any contribution to the surface stress tensor beyond the interfacial tension, (3.2-1) simplifies to

$$\gamma^{(AB)} \boldsymbol{\nu}^{(AS)} \cdot \boldsymbol{\nu}^{(AB)} + \gamma^{(AS)} - \gamma^{(BS)} = 0 \tag{3.2-2}$$

or

$$\gamma^{(AB)} \cos \Theta_0 + \gamma^{(AS)} - \gamma^{(BS)} = 0 \tag{3.2-3}$$

which is known as **Young's equation** [104]. Here Θ_0 is the contact angle measured through phase A in Figure 3.2-1 and $\gamma^{(AB)}$ is the fluid-fluid interfacial tension. It is important to note that (3.2-3) and Θ_0 refer to the true common line (given a somewhat vague distinction in this continuum picture between a species adsorbed in a solid vapor interface on one side of the common line and a thin film of liquid on the other) rather than an apparent common line seen with 10× magnification. Observe also that, since $\gamma^{(AS)}$ and $\gamma^{(BS)}$ in general include corrections for the actual curvature of the fluid-solid interfaces, they are not thermodynamic interfacial tensions [94, Sec. 5.8.6].[4] One should be cautious in interpreting $\gamma^{(AS)}$ and $\gamma^{(BS)}$ as thermodynamic quantities, unless $\Theta_0 = 0$ or π, in which case the solid would not be deformed.

Although the momentum balance at the common line [94, Sec. 2.1.9], must always be satisfied, a value of Θ_0 can not be found such that Young's equation is satisfied for an arbitrary choice of three phases or therefore for an arbitrary choice of $\gamma^{(AB)}$, $\gamma^{(AS)}$,

[4]Gibbs ([32, p. 326]; see also Dussan V. [23]) suggests that (3.2-3) can be derived by starting with the observation that energy is minimized at equilibrium. The Gibbs free energy is minimized as a function of time as equilibrium is approached for an isolated, isothermal, multiphase body consisting of isobaric fluids [94, Exercise 5.10.3-3]. In arriving at this result, both the Gibbs-Duhem equation and the surface Gibbs-Duhem equation are employed, which again means that all three phases are assumed to be fluids. There is no recognition in these results that the internal energy of a solid is a function of its deformation from its reference configuration. On the basis of the discussion presented in this text, Young's equation is not implied by the observation that the Gibbs free energy is minimized at equilibrium for such a system.

and $\gamma^{(BS)}$. [5] One possible answer to this dilemma is to remember that $\gamma^{(AS)}$ and $\gamma^{(BS)}$ are not thermodynamic quantities and they can not assume arbitrary values.

Since $\gamma^{(AS)}$ and $\gamma^{(BS)}$ are in general not material properties, it is not clear that they can be measured. For a gas-liquid-solid system for example, the common practice would be to measure the contact angles corresponding to two different liquids using perhaps 10× magnification. As pointed out above, (3.2-3) and Θ_0 refer to the true common line, not the apparent common line seen experimentally in this manner.

At the risk of needlessly emphasizing the point, in practice Θ_0 is not measured. Experimentalists report measurements of Θ determined at some distance from the common line, perhaps with 10× magnification. Here the fluid films are sufficiently thick that the effects of mutual forces, such as London-van der Waals forces and electrostatic double-layer forces (see Sec. 3.4 as well as [94], Exercises 2.1.3-1 and 2.1.3-2 and Sec. 3.2.7) as well as structural forces (Israelachvili and McGuiggan 1988), can be neglected.

In summary, if $\gamma^{(BS)} - \gamma^{(AS)}$ can not be measured independently, Young's equation can not be used to determine Θ_0. At this writing, it appears that we will generally have little choice but to assume a value for Θ_0, as we have done in Sec. 3.4 (see also [94, Sec. 3.2.7]).

For another critique of Young's equation, see Dussan V. [23].

---

[5] It is common to express Young's equation in terms of the **spreading coefficient** [2, 71]

$$\begin{aligned} S_{A/B} &\equiv \gamma^{(BS)} - \gamma^{(AS)} - \gamma^{(AB)} \\ &= \gamma^{(AB)}(\cos\Theta_0 - 1) \end{aligned} \tag{3.2-4}$$

If one interprets $\gamma^{(AS)}$ and $\gamma^{(BS)}$ as thermodynamic quantitities, it may not be possible to assign a contact angle Θ_0 such that (3.2-4) can be satisfied for an arbitrary selection of materials. The possibilities are these.

i) If $S_{A/B} > 0$, there is no value of Θ_0 that allows (3.2-4) to be satisfied. The usual interpretation is that fluid A will **spontaneously spread** over the solid, displacing fluid B. The solid prefers to be in contact with phase A, and the thin film of phase A can be expected to exhibit a positive disjoining pressure [94, Exercise 2.1.3-1].

ii) If $S_{A/B} < -2\gamma^{(AB)}$, there again is no value of Θ_0 such that (3.2-4) can be satisfied. The interpretation is that a thin film of fluid A will **spontaneously dewet** the solid, being displaced by fluid B. In this case, the solid prefers to be in contact with phase B, and the thin film of fluid A can be expected to exhibit a negative disjoining pressure [94, Exercise 2.1.3-1].

iii) If $-2\gamma^{(AB)} < S_{A/B} < 0$, there is a value of Θ_0 such that (3.2-4) can be satisfied, and the common line is stationary. In this case, fluid A is said to **partially wet** the solid.

In summary, if $S_{A/B} > 0$ or $S_{A/B} < -2\gamma^{(AB)}$, the usual interpretation is that the common line moves spontaneously over the solid; if $-2\gamma^{(AB)} < S_{A/B} < 0$, the common line is stationary.

For a different point of view, see [94, Sec. 2.1.13].

3.3 factors influencing measured contact angles

Remember that an experimentalist typically measures Θ, the apparent contact angle at some distance from the true common line. (See [94, Secs. 2.1.10 and 3.2.7], for further discussion of this point.) The magnitude of Θ can be influenced by a number of factors.

3.3.1 Contact angle hysteresis

The difference between advancing and receding contact angles is referred to as *contact angle hysteresis*. Either the roughness or heterogeneous composition of the surface is the usual cause [21, 58, 59, 76, 25, 74, 47, 57, 89, 73]. In some cases, the system requires a finite time to reach equilibrium [27], and the rate at which equilibrium is approached may itself depend upon the direction in which the common line recently has moved [103]. Depending upon the system, the approach to equilibrium may be characterized either by the adsorption (desorption) of a species in an interface or by the movement of an advancing (trailing) thin film. When roughness is absent and equilibrium is reached quickly, contact angle hysteresis appears to be eliminated [74].

3.3.2 Speed of displacement of commonline

The contact angle is known to depend upon the speed of displacement of the apparent common line as observed with 10× magnification [1, 86, 27, 28, 88, 29, 18, 90, 91, 38, 83, 44, 11, 12, 63].[6] Several distinct regimes are possible.

If the speed of displacement of the common line is sufficiently small, the advancing contact angle will be independent of it, since the solid within the immediate neighborhood of the common line nearly will have achieved equilibrium with the adjoining phases ([28]; see footnote 6).

If the speed of displacement of the common line is too large for equilibrium to be attained but too small for viscous effects within the adjoining fluid phases to play a significant role, the advancing contact angle will increase with increasing speed of displacement [37, 28, 29, 8, 90, 91, 44, 15, 16]. If we think of the **capillary number**

$$N_{ca} \equiv \frac{\mu u_{(\nu)}^{(cl)}}{\gamma} \tag{3.3-1}$$

as being characteristic of the ratio of viscous forces to surface tension forces, we should be able to neglect the effect of viscous forces upon the contact angle when N_{ca} is sufficiently small [44]. Here μ is the viscosity of the more viscous fluid, $u_{(\nu)}^{(cl)}$ is the speed of displacement of the apparent common line or meniscus as observed experimentally,

[6] Elliot and Riddiford's [28] measurements of the receding contact angle are in error [101].

and γ is the fluid-fluid interfacial tension. Sometimes the advancing contact angle appears to approach an upper limit with increasing speed of displacement [29, 90, 91, 44, 15, 30, 16]. Under these conditions, the common line probably is moving too rapidly for the deviation from equilibrium to have a significant effect upon the contact angle but still too slowly for viscous effects within the adjoining phases to play a significant role.

If the speed of displacement of the common line is sufficiently large for the effects of the viscous forces to be important but too small for inertial effects within the adjoining fluid phases to be considered, the advancing contact angle will once again increase with increasing speed of displacement [44, 17]. If the **Reynolds number**, which characterizes the ratio of inertial forces to viscous forces,

$$\begin{aligned} N_{Re} &\equiv \frac{\rho\, u_{(\nu)}^{(cl)} \ell}{\mu} \\ &= \frac{N_{We}}{N_{ca}} \end{aligned} \tag{3.3-2}$$

is sufficiently small or if the **Weber number**, which characterizes the ratio of inertial forces to surface tension forces,

$$N_{We} \equiv \frac{\rho u_{(\nu)}^{(cl)^2} \ell}{\gamma} \tag{3.3-3}$$

is sufficiently small for a fixed capillary number N_{ca}, we should be able to neglect the effects of the inertial forces [44]. Here ρ is the density of the more dense fluid and ℓ is a length characteristic of the particular geometry being considered.

With increasing speed of displacement, the contact angle approaches 180°. As this upper limit is reached, a continuous, visible film of the fluid originally in contact with the solid is entrained and the common line necessarily disappears [11, 12, 63]. This is referred to as **dynamic wetting**. Inertial forces have become more important than surface forces under these conditions.

3.3.3 Character of solid surface

I mentioned the effect of roughness in connection with contact angle hysteresis. Roughness, scratches or grooves will affect both the advancing and receding contact angles [74, 47, 78, 5, 72, 13]. Chemical heterogeneities undoubtedly also play a significant role [14, 60, 76, 53].

3.3.4 Geometry

A number of techniques have been used to study the effect of the speed of displacement of the common line upon the contact angle.

1. Ablett [1] used a partially immersed rotating cylinder. The depth of immersion was adjusted to give a flat phase interface, which in turn allowed the contact angle to be easily calculated.

2. Rose and Heins [86], Blake *et al.* [7], Hansen and Toong [38], Hoffman [44], and Legait and Sourieau [65] have observed the advancing contact angle in the displacement of one fluid by another through a round cylindrical tube.

3. Coney and Masica [18] studied displacement in a tube with a rectangular cross section.

4. Schonhorn *et al.* [88], Radigan *et al.* [83], Chen [15], and Chen and Wada [16] examined the spreading of sessile drops.

5. Ellison and Tejada [29], Inverarity [50], and Schwartz and Tejada [90, 91] photographed a wire entering a liquid-gas interface.

6. Burley and Kennedy [11, 12] and Kennedy and Burley [63] studied a plane tape entering a liquid-gas interface.

7. Johnson *et al.* [61] and Cain *et al.* [13] measured both advancing and receding contact angles using the Wilhelmy plate.

8. Elliott and Riddiford [26, 28] observed both advancing and receding contact angles in radial flow between two flat plates. Johnson *et al.* [61] found considerably different results in examining similar systems with the Wilhelmy plate. Wilson [101] has discussed the instabilities associated with radial flow between plates.

9. Ngan and Dussan V. [77] measured the advancing dynamic contact angle during upward flow between parallel plates.

Unfortunately, entirely comparable experiments have rarely been carried out in two different geometries.

Gravity plays an important role in at least a portion of the experiments of Ablett [1], Ellison and Tejada [29], Schwartz and Tejada [90, 91], Johnson *et al.* [61], Ngan and Dussan V. [77], and Legait and Sourieau [65]. The principal motions are parallel with gravity.

Inverarity [50] reports the equilibrium contact angle rather than the static advancing contact angle. Since these two contact angles are in general different, there is no basis for comparing his data with those of others.

Johnson *et al.* [61] concluded that effects of adsorption on the solid in the immediate neighborhood of the moving common line might be present in the studies of Elliott and Riddiford ([26, 28]; see footnote 6).

Schonhorn *et al.* [88] observed that the rate of spreading was unchanged when a drop was inverted. This demonstrated that the effect of gravity was negligible in their experiments.

The effects of both inertia and gravity are significant in the studies of Burley and Kennedy [11, 12] and Kennedy and Burley [63].

Coney and Masica [18] considered displacement over a previously wet wall, which sets their work apart from others.

As one exception, Jiang *et al.* [56] have successfully compared Hoffman's [44] data with selected experiments of Schwartz and Tejada [90], who measured the dynamic contact angle formed as a wire entered a liquid-gas interface [94, Sec. 2.1.12].

The equilibrium contact angle observed with 10× magnification appears to be independent of measurement technique or geometry, so long as (see [94, Sec. 3.2.7], for additional restrictions)

$$L/\ell \ll 1 \tag{3.3-4}$$

where ℓ is a length of the macroscopic system that characterizes the radius of curvature of the interface,

$$L \equiv \left(\frac{\gamma}{g\Delta\rho}\right)^{1/2} \tag{3.3-5}$$

is a characteristic length of the meniscus, g the acceleration of gravity, and $\Delta\rho$ the density difference between the two phases. Equation (3.3-4) appears to be valid for most measurements reported to date. Exceptions are the contact angles reported by Gaydos and Neumann [31], which depend upon the diameter of the sessile drop used as we would predict. [We prefer the argument presented by Slattery [94, Sec. 3.2.7], to the explanation of Gaydos and Neumann [31] in terms of line tension [94, Sec. 2.1.9].

We find in Sec. 3.4 that the dynamic contact angle created as a thin film recedes under the influence of a negative disjoining pressure appears to be independent of geometry, so long as (3.3-4) is valid (see Sec. 3.4 for additional restrictions). We expect a similar argument to be developed for advancing contact angles. This expectation appears to be supported by the correlations of Jiang *et al.* [56] and Chen [15] for advancing contact angles [94, Sec. 2.1.12]. It is only in the data of Ngan and Dussan V. [77] and of Legait and Sourieau [65], for which (3.3-4) is not valid, that we see a dependence upon geometry.

3.3.5 Configuration of interface within immediate neighborhood of common line

Chen and Wada [16] and Heslot *et al.* [40] have shown experimentally that the configuration of the interface within the immediate neighborhood of the common line is quite different from that which one would observe with a 10× microscope. For more on this point, see Sections 1.6 and 3.4 as well as Slattery [94, Secs. 2.1.13, 3.1.1, and 3.2.7].

3.4 moving common line, contact angle, and film configuration [67]

The preceding section suggests that continuum mechanics can be used successfully in analyzing the thin films that occur within the immediate neighborhood of a common line and that determine the dynamic contact angle.

A three-phase line of contact (or common line) is the curve formed by the intersection of two dividing interfaces. The movement of a common line over a rigid solid plays an important role in the spreading phenomena which occur during a wide variety of coating and displacement processes: the manufacture of coated paper, the manufacture of photographic film, the application of soil repellents to carpeting, the deposition of photo resists during the manufacture of microelectronic circuits, and the displacement of oil from an oil reservoir.

In Sec. 1.6 (see also [94, Sec. 2.1.13]), I summarized the contradictions in our theoretical understanding of common lines moving across rigid solid surfaces, and I briefly discussed some possibilities open to us for analyzing these problems.

Our premise in what follows is that a common line does not move. As viewed on a macroscale, it appears to move as a succession of stationary common lines are formed on a microscale, driven by a negative disjoining pressure in the receding film. The amount of receding phase left stranded by the formation of this succession of common lines is too small to be easily detected, as in the experiments of Dussan V. and Davis ([24]; [94, Sec. 1.2.9]).

The objective of the analysis that follows is to illustrate this thesis by computing the dynamic contact angle Θ, which might be measured by an experimentalist at some distance from the common line using perhaps 10× magnification. Experimentally observed values of Θ measured through an *advancing* liquid phase against air can be correlated as a function of the static contact angle $\Theta^{(stat)}$ and a capillary number N'_{ca} based upon the speed of displacement of the apparent common line (as seen by an experimentalist using perhaps 10× magnification; see [94, Sec. 2.1.12]). We would expect that a similar correlation could be constructed for Θ measured through a *receding* liquid phase against air, although few measurements of the receding contact

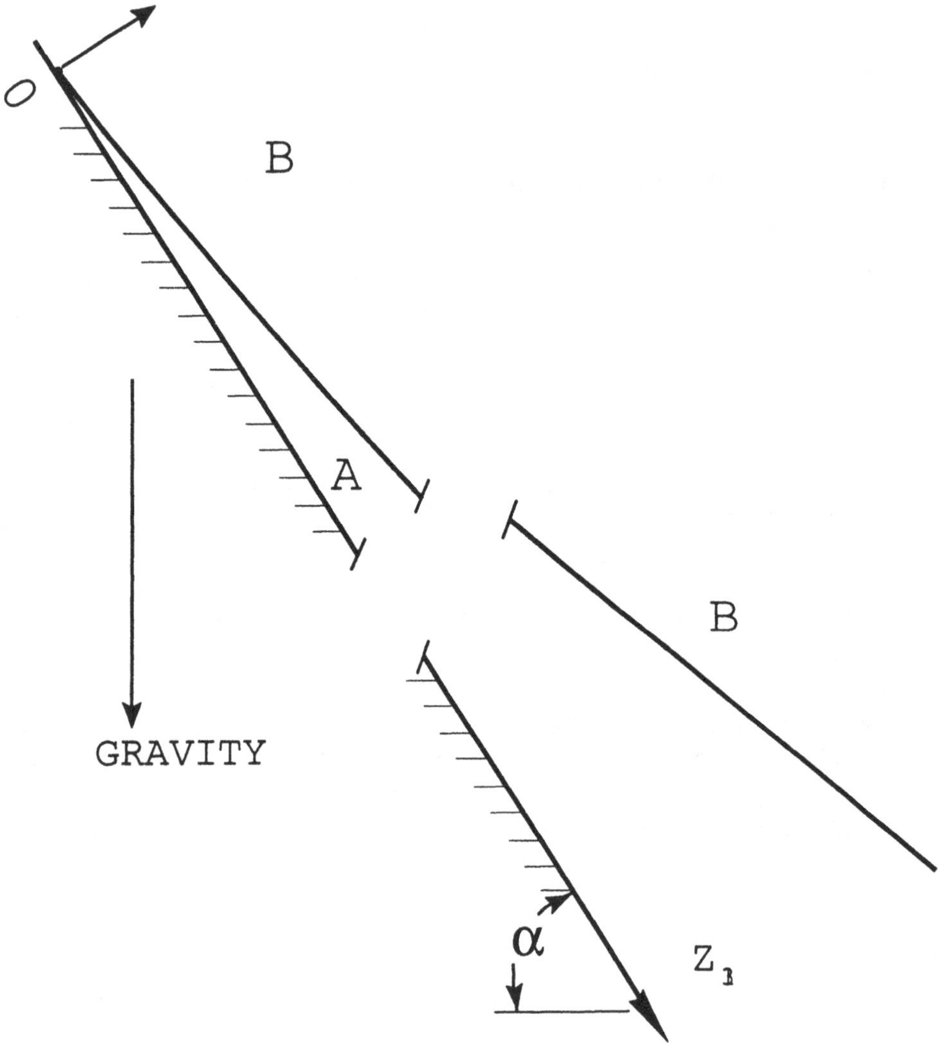

Figure 3.4-1: A draining film of liquid A formed as a plate is withdrawn from a pool of *A* into gas *B* at a constant speed V.

angle are currently available [see for example Johnson *et al.* [61], who used the Wilhelmy plate].

Figure 3.4-1 shows a draining film of liquid A formed as a plate is withdrawn from a pool of A into gas B at a constant speed V. We will assume that the thin film of liquid A on the solid exhibits a negative disjoining pressure against gas B (see [94], Sec. 2.1.13).

We will make a number of assumptions.

i) The solid is rigid, and its surface is smooth and planar. Its orientation with respect to gravity is arbitrary.

ii) Viewed in the rectangular cartesian coordinate system of Figure 3.4-1 in which the origin is fixed on the original common line or one of the common lines formed during the immediate past, the interface bounding the draining film of liquid A takes the form

$$z_3 = h(z_1, t) \tag{3.4-1}$$

and the plate is stationary.

iii) The dependence of h upon z_1 is sufficiently weak that

$$\left(\frac{\partial h}{\partial z_1}\right)^2 \ll 1 \tag{3.4-2}$$

iv) The Reynolds lubrication theory approximation applies in the sense that

$$\left(\frac{h_o}{L_o}\right)^2 \ll 1 \tag{3.4-3}$$

where L_o and h_o are respectively characteristic lengths in the z_1 and z_3 directions that will be defined later.

v) If there is no surfactant present, the interfacial tension is a constant independent of position on the interface, and the interfacial viscosities are zero. We will refer to such an interface as being *mobile.*

vi) If there is surfactant present, the tangential components of velocity at the liquid-gas interface are zero

$$\text{at } z_3 = h: \quad \mathbf{P} \cdot \mathbf{v} = 0 \tag{3.4-4}$$

Here $\mathbf{P}$ is the projection tensor that transforms every vector on an interface into its tangential components. The interfacial tension gradient required to create such an *immobile* interface is very small ([92, 68, 69, 36]; see also [94, Exercise 3.3.2-1]).

vii) The effect of mass transfer is neglected.

viii) The pressure p_o within the surrounding gas B is independent of time and position. Viscous effects within the gas phase are neglected.

ix) The liquid A is an incompressible Newtonian fluid, the viscosity of which is a constant.

x) All inertial effects are neglected.

xi) At the solid surface

$$z_3 = 0 : \quad \mathbf{v} = 0 \tag{3.4-5}$$

xii) All external and mutual body forces [94, Sec. 2.1.3] can be represented as the gradient of a potential energy per unit mass ϕ

$$\mathbf{b} = -\nabla\phi \tag{3.4-6}$$

xiii) We will account for both London-van der Waals forces and gravity within the immediate neighborhood of the common line, neglecting electrostatic double-layer effects.

xiv) Israelachvili [51] suggests that the distance between the two interfaces can not go to zero due to the finite size of the constituent atoms and that a finite interfacial separation d must be recognized. He recommends that d be viewed as the mean distance between the centers of individual atoms and estimated as

$$d = 0.91649[M/(\rho^{(A)} n_a N)]^{1/3} \tag{3.4-7}$$

Here M is the molecular weight, n_a the number of atoms per molecule, N Avogadro's constant ($6.023 \times 10^{23} mol^{-1}$), and $\rho^{(A)}$ the mass density of liquid A. In arriving at (3.4-7), the atoms have been assumed to be in a close packing arrangement. For a molecule consisting of a repeating unit, such as a n-alkane with -CH_2- being the repeating unit, the comparison with experimental data in Sec. 2.1 suggests a simplistic picture in which the molecules are arranged in such a manner that the *repeating units* are in a close packing arrangement. Retracing the argument of Israelachvili [51], we have instead

$$d = 0.91649[M/(\rho^{(A)} n_u N)]^{1/3} \tag{3.4-8}$$

where n_u is the number of repeating units in the molecule. This last is similar to the suggestion of Padday and Uffindell [79], who replaced M/n_u by the $-CH_2-$ group weight and took the coefficient to be unity.

For the interface between phases A and S, we suggest

$$d = (d^{(A)} + d^{(S)})/2 \tag{3.4-9}$$

where, if appropriate, $d^{(A)}$ and $d^{(S)}$ are computed using (3.4-8).

More recently, Israelachvili [52, p. 157] has recommended using $d = 1.65\mathring{A}$ in estimating the surface tension for n-alkanes (see Sec. 2.1).

xv) If the fluid-fluid and fluid-solid interfaces were parallel planes, the difference in the potential energies attributable to London-van der Waals forces and to gravity at the fluid-fluid interface could be described as [94, Exercise 2.1.3-1]

$$\begin{aligned} &\rho^{(A)}\phi^{(A)} - \rho^{(B)}\phi^{(B)} \\ &\quad = \phi_\infty^{(AB)} + \frac{B^{(AB)}}{h'^m} - (\rho^{(A)} - \rho^{(B)})g(z_1 \sin\alpha - h' \cos\alpha) \end{aligned} \tag{3.4-10}$$

in which h' is the thickness of the film and the last term on the right accounts for the effect of gravity. We will follow Wayner [96, 97] in recognizing that the minimum film is a monolayer. Somewhat simplistically, we will estimate the thickness of the monolayer as d, which now might be thought of as the distance between the centers of the last layer of solid atoms and the centers of the first layer of liquid atoms or repeating units. This suggests that, in the continuum model of mechanics used here, we measure the film thickness h from the centers of the first layer of liquid atoms or repeating units and that we replace (3.4-10) by

$$\begin{aligned} &\rho^{(A)}\phi^{(A)} - \rho^{(B)}\phi^{(B)} \\ &\quad = \phi_\infty^{(AB)} + \frac{B^{(AB)}}{(h+d)^m} - (\rho^{(A)} - \rho^{(B)})g(z_1 \sin\alpha - h \cos\alpha) \end{aligned} \tag{3.4-11}$$

In this simplistic model, the common line can be visualized as running through the centers of the liquid atoms or repeating units on the leading edge of the monolayer. As the common line is approached, the London-van der Waals forces remain bounded as the thickness of the liquid film approaches zero. Because the disjoining pressure is assumed to be negative here, $B^{(AB)} > 0$.

Because the dependence of h upon z_1 is weak (assumption iii), we will assume that (3.4-11) applies here.

xvi) Again because the dependence of h upon z_1 is weak (assumption iii), we expect our results to be limited to $\tan\Theta \to 0$, which suggests that

$$\Theta_0 = 0 \tag{3.4-12}$$

In constructing this development, we will find it convenient to work in terms of the dimensionless variables

$$\begin{aligned}
z_1^\star &\equiv \frac{z_1}{L_o} \\
z_3^\star &\equiv \frac{z_3}{h_o} \\
h^\star &\equiv \frac{h}{h_o} \\
d^\star &\equiv \frac{d}{h_o} \\
H^\star &\equiv H L_o \\
t^\star &\equiv \frac{t v_o}{L_o} \\
v_1^\star &\equiv \frac{v_1}{v_o} \\
v_3^\star &\equiv \frac{L_o}{h_o}\frac{v_3}{v_o} \\
\gamma^\star &\equiv \frac{\gamma}{\gamma_o} \\
\phi^{(i)\star} &\equiv \frac{\phi^{(i)}}{{v_0}^2} \\
p^\star &\equiv \frac{p}{\rho^{(A)}{v_0}^2} \\
p_o^\star &\equiv \frac{p_o}{\rho^{(A)}{v_0}^2} \\
\mathcal{P}^\star &\equiv \frac{p + \rho^{(A)}\phi^{(A)}}{\rho^{(A)}{v_0}^2} \\
\mathcal{P}_o^\star &\equiv \frac{p_o + \rho^{(A)}\phi^{(A)}}{\rho^{(A)}{v_0}^2}
\end{aligned} \tag{3.4-13}$$

and the dimensionless Reynolds, Weber, capillary and Bond numbers

$$\begin{aligned}
N_{Re} &\equiv \frac{\rho^{(A)} v_o L_o}{\mu} \\
N_{ca} &\equiv \frac{\mu v_o}{\gamma_o} \\
N_{We} &\equiv \frac{\rho^{(A)} {v_0}^2 L_o}{\gamma_o}
\end{aligned}$$

$$N_{Bo} \equiv \frac{(\rho^{(A)} - \rho^{(B)})gL_0{}^2}{\gamma_o} \tag{3.4-14}$$

Here H is the mean curvature of the liquid-gas interface, γ the surface tension, γ_o the equilibrium surface tension, μ the viscosity of liquid A, g the magnitude of the acceleration of gravity, p_o the pressure in the gas phase, and p the pressure in the liquid phase. We will identify

$$L_o \equiv \left[\frac{\gamma_o}{(\rho^{(A)} - \rho^{(B)})g}\right]^{1/2} \tag{3.4-15}$$

and

$$h_o \equiv L_o \tan \Theta^{(stat)} \tag{3.4-16}$$

which means

$$N_{Bo} = 1 \tag{3.4-17}$$

The characteristic speed v_o will be defined later.

Equation (3.4-1) suggests that we seek a solution in which the velocity distribution takes the form

$$\begin{aligned} v_1^\star &= v_1^\star(z_1^\star, z_3^\star, t^\star) \\ v_3^\star &= v_3^\star(z_1^\star, z_3^\star, t^\star) \end{aligned} \tag{3.4-18}$$

Under these circumstances, the equation of continuity for an incompressible fluid says

$$\frac{\partial v_1^\star}{\partial z_1^\star} + \frac{\partial v_3^\star}{\partial z_3^\star} = 0 \tag{3.4-19}$$

In the limit of assumption iv, Cauchy's first law for an incompressible, Newtonian fluid with a constant viscosity reduces for creeping flow to

$$\frac{\partial \mathcal{P}^\star}{\partial z_1^\star} = \frac{1}{N_{Re}}\left(\frac{L_o}{h_o}\right)^2 \frac{\partial^2 v_1^\star}{\partial z_3^{\star 2}} \tag{3.4-20}$$

$$\frac{\partial \mathcal{P}^\star}{\partial z_3^\star} = \frac{1}{N_{Re}} \frac{\partial^2 v_3^\star}{\partial z_3^{\star 2}} \tag{3.4-21}$$

The z_2-component requires $\mathcal{P}^\star$ to be independent of $z_2^\star$. Equations (3.4-20) and (3.4-21) imply that

$$\frac{\partial \mathcal{P}^\star}{\partial z_3^\star} \ll \frac{\partial \mathcal{P}^\star}{\partial z_1^\star} \tag{3.4-22}$$

and the dependence of $\mathcal{P}^\star$ upon $z_3^\star$ can be neglected. Note that the scaling argument used to neglect inertial effects (assumption x) in arriving at (3.4-20) and (3.4-21) is presumed not to be the one shown here. For this reason, we will regard the magnitude of N_{Re} and the definition of v_o to be as yet unspecified.

The jump mass balance [94, Sec. 1.3.5; the overall jump mass balance of Sec. 5.4.1] is satisfied identically, since we define the position of the dividing surface by choosing $\rho^{(\sigma)} = 0$ [94, Sec. 1.3.6] and since the effect of mass transfer is neglected (assumption vii). The jump mass balance for surfactant [94, Sec. 5.2.1] is not required here, since we assume either that the interfacial tension is constant (assumption v) or that the interfacial tension gradient is so small that its effect and the effect of a concentration gradient developed in the interface can be neglected (assumption vi).

With assumptions vii, viii, and x, the jump momentum balance [94, Sec. 2.4.2] for the interface between phases A and B reduces to

$$\nabla_{(\sigma)}\gamma + 2H\gamma\boldsymbol{\xi} - (\mathbf{T} + p_o\mathbf{I}) \cdot \boldsymbol{\xi} = 0 \tag{3.4-23}$$

Here $\boldsymbol{\xi}$ is the unit normal to the interface pointing out of the liquid film. Under the conditions of assumptions iii and iv, the z_1- and z_3-components of (3.4-23) assume the forms at $z_3^\star = h^\star$ [94, Table 2.4.2-6]

$$\frac{\partial\gamma^\star}{\partial z_1^\star} - 2H^\star\gamma^\star\frac{h_o}{L_o}\frac{\partial h^\star}{\partial z_1^\star} - N_{We}(p^\star - p_o^\star)\frac{h_o}{L_o}\frac{\partial h^\star}{\partial z_1^\star} - N_{ca}\frac{L_o}{h_o}\frac{\partial v_1^\star}{\partial z_3^\star} = 0 \tag{3.4-24}$$

and

$$\begin{aligned} &\frac{h_o}{L_o}\frac{\partial\gamma^\star}{\partial z_1^\star}\frac{\partial h^\star}{\partial z_1^\star} + 2H^\star\gamma^\star + N_{We}(p^\star - p_o^\star) - 2N_{ca}\frac{\partial v_3^\star}{\partial z_3^\star} \\ &\quad + N_{ca}\frac{\partial v_1^\star}{\partial z_3^\star}\frac{\partial h^\star}{\partial z_1^\star} = 0 \end{aligned} \tag{3.4-25}$$

The z_2-component is satisfied identically. Adding $(h_o/L_o)(\partial h^\star/\partial z_1^\star)$ times (3.4-25) to (3.4-24) and recognizing assumptions iii and iv, we have

$$\frac{\partial\gamma^\star}{\partial z_1^\star} - N_{ca}\frac{L_o}{h_o}\frac{\partial v_1^\star}{\partial z_3^\star} = 0 \tag{3.4-26}$$

so that (3.4-24) implies[7]

$$2H^\star\gamma^\star + N_{We}(p^\star - p_o^\star) = 0 \tag{3.4-27}$$

[7]In arriving at (3.4-26) and (3.4-27), it was not necessary to make any statement about the relative magnitudes of N_{ca} and N_{We} or the definition of v_o. But some statement is necessary, in order to establish consistency with (4-25).

Substituting (3.4-26) into (3.4-25), we have

$$2H^\star\gamma^\star + N_{We}(p^\star - p_o^\star) + 2N_{ca}\left(\frac{\partial v_1^\star}{\partial z_3^\star}\frac{\partial h^\star}{\partial z_1^\star} - \frac{\partial v_3^\star}{\partial z_3^\star}\right) = 0$$

For a mobile interface, (3.4-26) requires (see assumptions v and viii)

$$\text{at } z_3^\star = h^\star : \quad \frac{\partial v_1^\star}{\partial z_3^\star} = 0 \tag{3.4-28}$$

For an immobile interface (see assumptions iv and vi)

$$\text{at } z_3^\star = h^\star : \quad v_1^\star = 0 \tag{3.4-29}$$

and we can employ (3.4-26) to calculate the interfacial tension gradient required to create an immobile interface.

Since we neglect the effect of mass transfer on the velocity distribution (assumption vii; see [94, Table 2.4.2-6],

$$\text{at } z_3^\star = h^\star : \quad v_3^\star = \frac{\partial h^\star}{\partial t^\star} + \frac{\partial h^\star}{\partial z_1^\star} v_1^\star \tag{3.4-30}$$

Finally, (3.4-5) requires at the solid surface

$$\text{at } z_3^\star = 0 : \quad v_1^\star = v_3^\star = 0 \tag{3.4-31}$$

Our objective in what follows is to obtain a solution to (3.4-19) and (3.4-20) consistent with (3.4-27), (3.4-30), (3.4-31) and either (3.4-28) or (3.4-29).

3.4.1 Solution

Integrating (3.4-20) twice consistent with (3.4-31) and either (3.4-28) or (3.4-29), we find that

$$v_1^\star = N_{Re} \left(\frac{h_o}{L_o}\right)^2 \frac{\partial \mathcal{P}^\star}{\partial z_1^\star} \left(\frac{1}{2} z_3^{\star 2} - \frac{1}{n} h^\star z_3^\star\right) \tag{3.4-32}$$

where

$$\begin{aligned} n &= 1 \text{ for amobile interface} \\ &= 2 \text{ for an immobile interface} \end{aligned} \tag{3.4-33}$$

Substituting (3.4-32) into (3.4-19) and integrating once consistent with (3.4-31), we have

$$v_3^\star = -\frac{1}{2} N_{Re} \left(\frac{h_o}{L_o}\right)^2 \left[\frac{\partial^2 \mathcal{P}^\star}{\partial z_1^{\star 2}} \left(\frac{1}{3} z_3^{\star 3} - \frac{1}{n} h^\star z_3^{\star 2}\right) - \frac{1}{n} \frac{\partial \mathcal{P}^\star}{\partial z_1^\star} \frac{\partial h^\star}{\partial z_1^\star} z_3^{\star 2}\right] \tag{3.4-34}$$

If follows from (3.4-32) through (3.4-34) and (3.4-36) that

$$|2H^\star \gamma^\star| \gg \left| 2N_{ca} \left(\frac{\partial v_1^\star}{\partial z_3^\star} \frac{\partial h^\star}{\partial z_1^\star} - \frac{\partial v_3^\star}{\partial z_3^\star} \right) \right|$$

in agreement with (3.4-27).

Equations (3.4-30), (3.4-32) and (3.4-34) tell us

$$-\frac{\partial h^\star}{\partial t^\star} = -N_{Re}\frac{3-n}{6n}\left(\frac{h_o}{L_o}\right)^2 \frac{\partial}{\partial z_1^\star}\left(h^{\star 3}\frac{\partial \mathcal{P}^\star}{\partial z_1^\star}\right) \tag{3.4-35}$$

Equation (3.4-27), with the appropriate expression for the dimensionless mean curvature H, says

$$N_{We}(\mathcal{P}^\star - \mathcal{P}_o^\star) - N_{We}\left(\phi^{(A)\star} - \frac{\rho^{(B)}}{\rho^{(A)}}\phi^{(B)\star}\right) = -\frac{h_o}{L_o}\frac{\partial^2 h^\star}{\partial z_1^{\star 2}} \tag{3.4-36}$$

In terms of dimensionless variables, (3.4-11) becomes

$$\begin{aligned}\phi^{(A)\star} - \frac{\rho^{(B)}}{\rho^{(A)}}\phi^{(B)\star} = {} & \frac{1}{N_{We}}\left[\phi_\infty^{(AB)\star}\right.\\ & \left. + \frac{h_o}{L_o}\frac{B^\star}{(h^\star + d^\star)^m} - z_1^\star \sin\alpha + \frac{h_o}{L_o}h^\star \cos\alpha\right]\end{aligned} \tag{3.4-37}$$

where we have recognized (3.4-17), we have defined

$$\phi_\infty^{(AB)\star} \equiv \frac{L_o}{\gamma_o}\phi_\infty^{(AB)} \qquad B^\star \equiv \frac{{L_o}^2 B^{(AB)}}{\gamma_o {h_o}^{m+1}} \tag{3.4-38}$$

and we have set $\gamma^\star = 1$ either by assumption v or by assumption vi. In view of (3.4-37), we can differentiate (3.4-36) to find

$$\begin{aligned}\frac{\partial \mathcal{P}^\star}{\partial z_1^\star} = {} & -\frac{1}{N_{We}}\frac{h_o}{L_o}\left[\frac{\partial^3 h^\star}{\partial z_1^{\star 3}} + \frac{mB^\star}{(h^\star + d^\star)^{m+1}}\frac{\partial h^\star}{\partial z_1^\star}\right.\\ & \left. + \frac{h_o}{L_o}\sin\alpha - \frac{\partial h^\star}{\partial z_1^\star}\cos\alpha\right]\end{aligned} \tag{3.4-39}$$

Equation (3.4-35) together with (3.4-39) defines the configuration of the film as a function of time and position.

Since [94, Exercise 2.1.3-1]

$$B^\star \ll 1 \tag{3.4-40}$$

our objective is to develop a solution that is correct in the limit $B^\star \to 0$ or a perturbation solution that is correct to the zeroth order in $B^\star$. Outside the immediate neighborhood of the common line, (3.4-39) reduces in this limit to

$$\frac{\partial \mathcal{P}^\star}{\partial z_1^\star} = -\frac{1}{N_{We}}\frac{h_o}{L_o}\left(\frac{\partial^3 h^\star}{\partial z_1^{\star 3}} + \frac{L_o}{h_o}\sin\alpha - \frac{\partial h^\star}{\partial z_1^\star}\cos\alpha\right) \tag{3.4-41}$$

Equation (3.4-35) together with (3.4-41) must be solved consistent with

$$\text{as } z_1^\star \to 0: \quad \frac{dh^\star}{dz_1^\star} \to k \tag{3.4-42}$$

as well as three other boundary conditions and an initial condition for the particular macroscopic flow with which one is concerned. Here

$$k \equiv \frac{\tan \Theta}{\tan \Theta^{(stat)}} \tag{3.4-43}$$

where Θ is the receding dynamic contact angle corresponding to the speed of the common line. Note that $k = 1$ corresponds to the static problem. We will refer to this as the *outer problem*, in the sense that it is outside the immediate neighborhood of the common line. Equation (3.4-42) is imposed not at the common line, but as the common line is approached in the outer solution.

Within the immediate neighborhood of the common line, the effects of the London-van der Waals forces must be preserved. We will restrict our consideration here to the case of unretarded London-van der Waals forces (for films less than about 120Å thick; see [94, Exercise 2.1.3-1]) for which

$$\begin{aligned} B^{(AB)} &= \frac{1}{6\pi}(A^{(AA)} - A^{(AS)}) \\ m &= 3 \end{aligned} \tag{3.4-44}$$

Here $A^{(AA)}$ is the Hamaker constant for the interaction of the film fluid with itself; $A^{(AS)}$ is the Hamaker constant for the interaction of the film fluid with the solid. This suggests that we introduce expanded variables within the immediate neighborhood of the common line:

$$\begin{aligned} h^{\star\star} &\equiv \frac{h^\star}{B^{\star 1/2}} \\ d^{\star\star} &\equiv \frac{d^\star}{B^{\star 1/2}} \\ z_1^{\star\star} &\equiv \frac{z_1^\star}{B^{\star 1/2}} \\ t^{\star\star} &\equiv \frac{t^\star}{B^{\star 1/2}} \end{aligned} \tag{3.4-45}$$

In terms of these variables, (3.4-35) and (3.4-39) require

$$-\frac{\partial h^{\star\star}}{\partial t^{\star\star}} = N_{ca}\frac{3-n}{6n}\left(\frac{h_o}{L_o}\right)^3 \left\{ h^{\star\star 3}\frac{\partial^4 h^{\star\star}}{\partial z_1^{\star\star 4}} + 3h^{\star\star 2}\frac{\partial h^{\star\star}}{\partial z_1^{\star\star}}\frac{\partial^3 h^{\star\star}}{\partial z_1^{\star\star 3}}\right.$$

$$+ 3h^{**3}\left[\frac{1}{(h^{**}+d^{**})^4}\frac{\partial^2 h^{**}}{\partial z_1^{**2}} - \frac{1-3d^{**}/h^{**}}{(h^{**}+d^{**})^5}\left(\frac{\partial h^{**}}{\partial z_1^{**}}\right)^2\right]$$
$$+ 3B^\star\frac{L_o}{h_o}\sin\alpha\ h^{**2}\frac{\partial h^{**}}{\partial z_1^{**}} - 3B^\star\ \cos\alpha\ h^{**2}\left(\frac{\partial h^{**}}{\partial z_1^{**}}\right)^2$$
$$\left. -B^\star\ \cos\alpha\ h^{**3}\frac{\partial^2 h^{**}}{\partial z_1^{**2}}\right\} \tag{3.4-46}$$

Since $B^\star \to 0$, the effect of gravity can be neglected in the inner region.

For the sake of simplicity, let us define the characteristic speed as

$$v_o \equiv \frac{3-n}{6n}\frac{\gamma_o}{\mu}\left(\frac{h_o}{L_o}\right)^3 \tag{3.4-47}$$

permitting us to write (3.4-46) as

$$-\frac{\partial h^{**}}{\partial t^{**}} = h^{**3}\frac{\partial^4 h^{**}}{\partial z_1^{**4}} + 3h^{**2}\frac{\partial h^{**}}{\partial z_1^{**}}\frac{\partial^3 h^{**}}{\partial z_1^{**3}}$$
$$+3h^{**3}\left[\frac{1}{(h^{**}+d^{**})^4}\frac{\partial^2 h^{*}}{\partial z_1^{**2}} - \frac{1-3d^{**}/h^{**}}{(h^{**}+d^{**})^5}\left(\frac{\partial h^{**}}{\partial z_1^{**}}\right)^2\right] \tag{3.4-48}$$

Here we have neglected the effects of gravity as indicated following (3.4-46).

Note that [94, Exercise 3.3.3-1]

$$d^{**} = 1 \tag{3.4-49}$$

or

$$\tan\Theta^{(stat)} = \left(\frac{A}{6\pi\gamma_o d^2}\right)^{1/2} \tag{3.4-50}$$

where $\Theta^{(stat)}$ is the static contact angle. This last also follows from the result of Slattery [94, Sec. 3.2.7] in the limit $\Theta^{(stat)} \to 0$. In view of (3.4-49), equation (3.4-48) simplifies to

$$-\frac{\partial h^{**}}{\partial t^{**}} = h^{**3}\frac{\partial^4 h^{**}}{\partial z_1^{**4}} + 3h^{**2}\frac{\partial h^{**}}{\partial z_1^{**}}\frac{\partial^3 h^{**}}{\partial z_1^{**3}}$$
$$+3h^{**3}\left[\frac{1}{(h^{**}+1)^4}\frac{\partial^2 h^{**}}{\partial z_1^{**2}} - \frac{1-3/h^{**}}{(h^{**}+1)^5}\left(\frac{\partial h^{**}}{\partial z_1^{**}}\right)^2\right] \tag{3.4-51}$$

Equation (3.4-51) must be solved consistent with

$$\text{at } z_1^{**} = 0 : \quad h^{**} = 0 \tag{3.4-52}$$

and, in view of (3.4-12),

$$\text{at } z_1^{\star\star} = 0 : \quad \frac{dh^{\star\star}}{dz_1^{\star\star}} = 0 \tag{3.4-53}$$

In addition, the inner and outer solutions must be consistent in some intermediate region

$$\text{as } B^\star \to 0 \text{ and } z_1^{\star\star} \to \infty : \quad \frac{\partial^2 h^{\star\star}}{\partial z_1^{\star\star 2}} \to B^{\star 1/2}\left(\frac{\partial^2 h^\star}{\partial z_1^{\star 2}}\right)_\infty \tag{3.4-54}$$

$$\text{as } B^\star \to 0 \text{ and } z_1^{\star\star} \to \infty : \quad \frac{\partial h^{\star\star}}{\partial z_1^{\star\star}} \to \left(\frac{\partial h^\star}{\partial z_1^\star}\right)_\infty = k \tag{3.4-55}$$

Here the subscript $\ldots_\infty$ indicates a quantity in the outer solution seen as the common line is approached.

Finally, we require an initial configuration of the interface. At first thought, it would appear most natural to take as our initial configuration the static configuration. This would correspond to considering the problem of start-up from rest with a step-change in the speed of the wall from 0 to V and a corresponding step-change in $\partial h^{\star\star}/\partial z_1^{\star\star}$ from 1 to k as $B^\star \to 0$ and $z_1^{\star\star} \to \infty$. The difficulty is that this would be inconsistent with our neglect of inertial effects in assumption x. Instead, we will say that

$$\text{at } t^{\star\star} = 0 : h^{\star\star} = (k^2 z_1^{\star\star 2} + 1)^{1/2} - 1 \tag{3.4-56}$$

which is consistent with (3.4-52) through (3.4-55). We will not define the experimental technique required to achieve this initial condition. Rather, we shall assume that for $t^{\star\star} \gg 1$ the effect of the initial condition is negligible.

From the simultaneous solutions of the inner and outer problems, one can compute the dimensionless **speed of displacement of the apparent common line** (determined by the tangent to the interface as $B^\star \to 0$ and $z_1^{\star\star} \to \infty$ in the inner solution or $z_1^\star \to z_{1\infty}^\star$ in the outer solution)

$$u^{(cl)\star} \equiv \frac{dz_1^{(cl)\star}}{dt^\star} = \frac{dz_1^{(cl)\star\star}}{dt^{\star\star}} \tag{3.4-57}$$

and determine the specific form of the relationship

$$u^{(cl)\star} = g\left[k, B^{\star 1/2}\left(\frac{\partial^2 h^\star}{\partial z_1^{\star 2}}\right)_\infty, t^{\star\star}\right] \tag{3.4-58}$$

In general the dynamic contact angle Θ can be expected to be dependent upon the measurement technique or, equivalently, the geometry of the macroscopic system as a result of its dependence upon the curvature in the outer solution as the common line is approached. [This is analogous with the observations of Ngan and Dussan V. [77] and of Legait and Sourieau [65] for advancing common lines.]

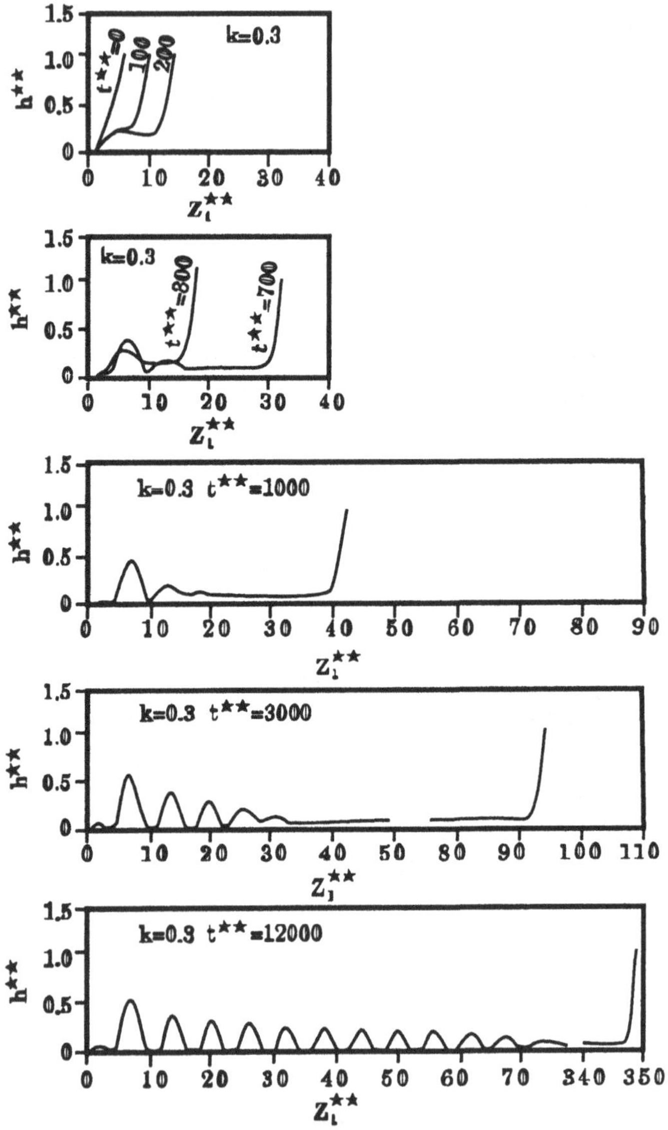

Figure 3.4-2: Dimensionless film thickness h^{**} as a function of z_1^{**} and t^{**} for $k = 0.3$.

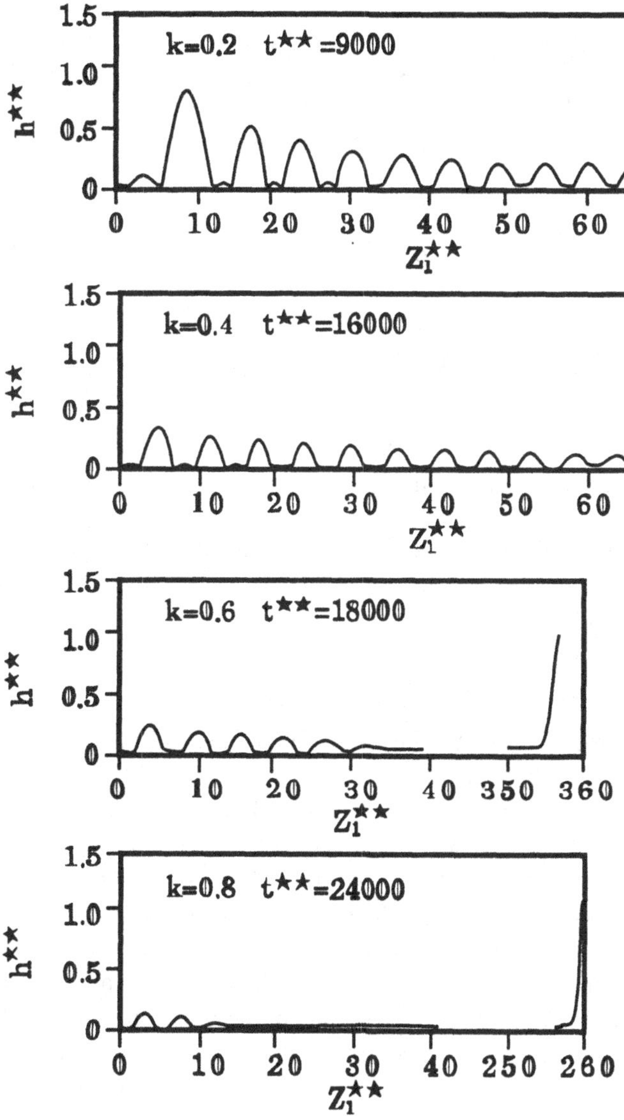

Figure 3.4-3: Dimensionless film thickness h^{**} as a function of z_1^{**} for different values of k.

Physically, the maximum value of the curvature in the outer solution as the common line is approached can be expected to be proportional to $1/\ell$, where ℓ is a characteristic dimension of the macroscopic system. In the limiting case such that $B^\star \to 0$,

$$\left(\frac{\partial^2 h^\star}{\partial z_1^{\star 2}}\right)_\infty = \frac{L_o^{\ 2}}{h_o}\left(\frac{\partial^2 h}{\partial z_1^{\ 2}}\right)_\infty \sim \frac{h_o}{\ell}\left(\frac{L_o}{h_o}\right)^2 \ll 1 \tag{3.4-59}$$

and (3.4-3) are all satisfied, as is true for most experiments used to measure Θ, (3.4-54) reduces to

$$\text{as } B^\star \to 0 \text{ and } z_1^{\star\star} \to \infty : \quad \frac{\partial^2 h^{\star\star}}{\partial z_1^{\star\star 2}} \to 0 \tag{3.4-60}$$

and (3.4-58) takes the form

$$u^{(cl)\star} = g(k, t^{\star\star}) \tag{3.4-61}$$

In this limit, the relationship between $u^{(cl)\star}$ and k no longer depends upon the technique used for its measurement or the geometry of the macroscopic problem [94, Sec. 2.1.12].

In what follows, we consider only the solution of this limiting case. Equation (3.4-51) is solved consistent with (3.4-52), (3.4-53), (3.4-55), (3.4-56), and (3.4-60) using the Crank-Nicolson method [75].

3.4.2 Results and discussion

Figure 3.4-2 shows the dimensionless film thickness $h^{\star\star}$ as a function of $z_1^{\star\star}$ and $t^{\star\star}$ for $k = 0.3$. As $t^{\star\star}$ increases, the bulk of phase A recedes, and a very thin liquid film is left behind. The slope of the surface of the bulk of phase A is the receding contact angle Θ, which remains almost unchanged with $t^{\star\star}$. We will refer to the junction of the thin film and bulk of phase A (identified by the intersection of the tangent to this portion of the surface with the $z_1^{\star\star}$ axis) as the **apparent common line** and to its position as $z_1^{(cl)\star\star}$. As $t^{\star\star}$ increases, waves form in the thin film driven by the negative disjoining pressure with coalescence occurring at the troughs of these waves, and a small residue of phase A is left behind. This coalescence process can not occur at a common line, since this would imply a moving common line and unbounded forces.

Figure 3.4-3 shows $h^{\star\star}$ as a function of $z_1^{\star\star}$ and $t^{\star\star}$ for different values of k. The average thickness of the thin film and therefore the residue of phase A left behind increases as k decreases or as the receding contact angle Θ decreases.

Notice that the thickness of the residual film in Figures 3.4-2 and 3.4-3 is less than molecular dimensions. We believe that this should be interpreted as a continuum description of what in reality is an incomplete monomolecular film.

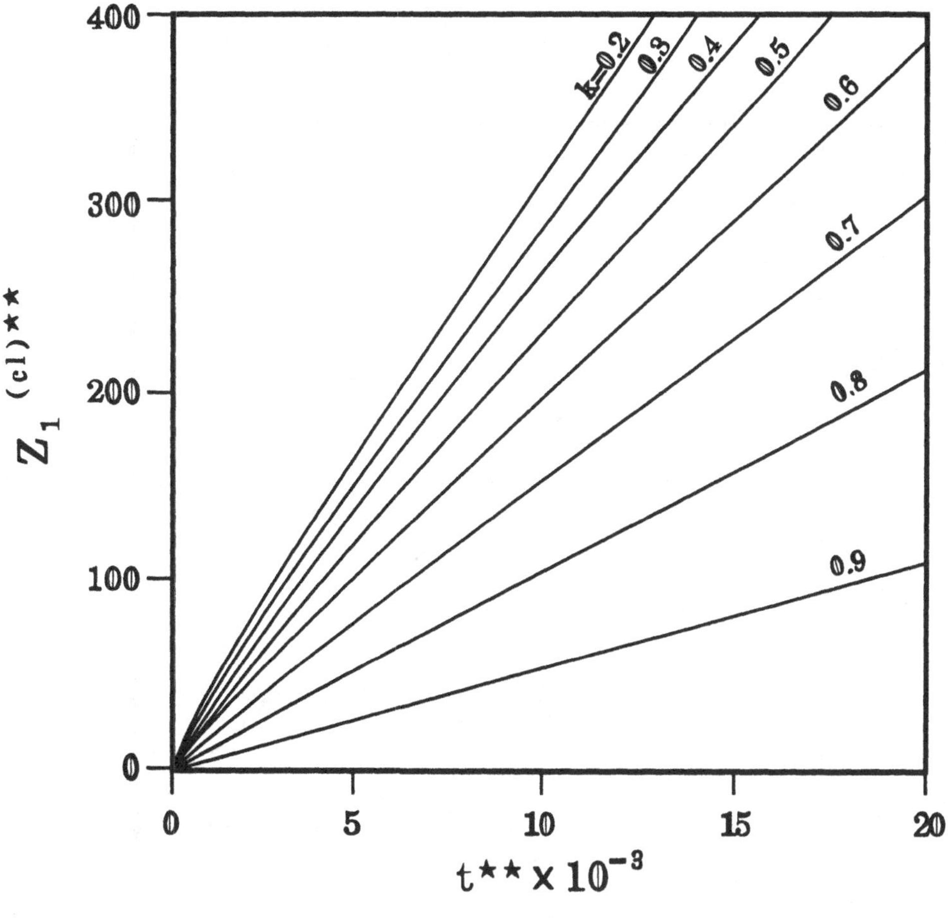

Figure 3.4-4: Position $z_1^{(cl)**}$ of apparent common line as a function of time t^{**}.

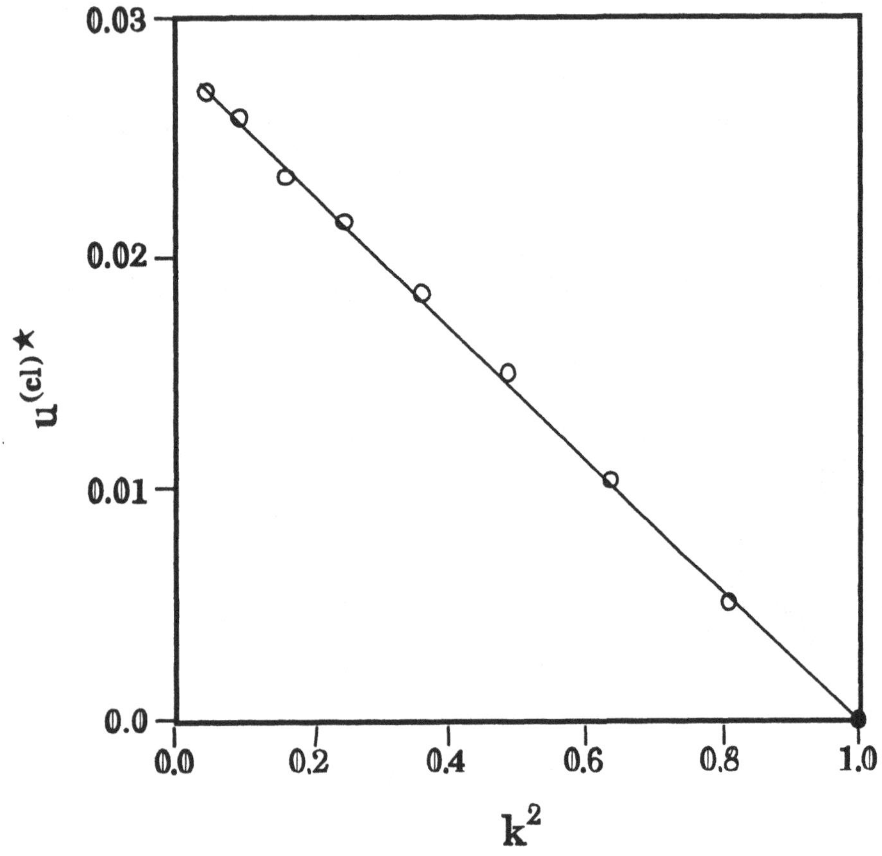

Figure 3.4-5: Dependence of $u^{(cl)\star}$ on k^2.

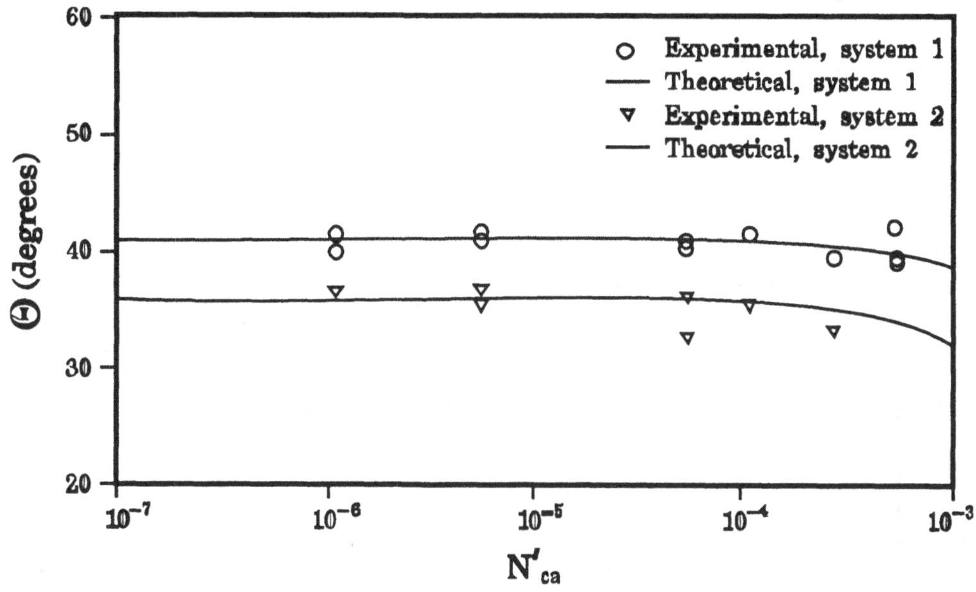

Figure 3.4-6: Comparisons of (3.4-67) with the experimental data of Johnson *et al.* [61]: system 1, hexadecane on teflon; system 2, hexadecane on siliconed glass.

Figure 3.4-4 presents the position $z_1^{(cl)\star\star}$ of the apparent common line as a function of $t^{\star\star}$ for different values of k. After a short initial period attributable to the initial configuration (3.4-56), $u^{(cl)\star}$ becomes a constant. The relationship between $u^{(cl)\star}$ and k is shown in Figure 3.4-5. Alternatively, we can express this relationship as

$$u^{(cl)\star} = 0.028(1 - k^2) \tag{3.4-62}$$

or

$$u^{(cl)\star} = 0.028 \tan^{-2} \Theta^{(stat)}(\tan^2 \Theta^{(stat)} - \tan^2 \Theta) \tag{3.4-63}$$

We will define a new capillary number based on the speed of the apparent common line as

$$N'_{ca} \equiv \frac{\mu u^{(cl)}}{\gamma_0} \tag{3.4-64}$$

According to (3.4-13), the dimensionless speed of the apparent common line can be

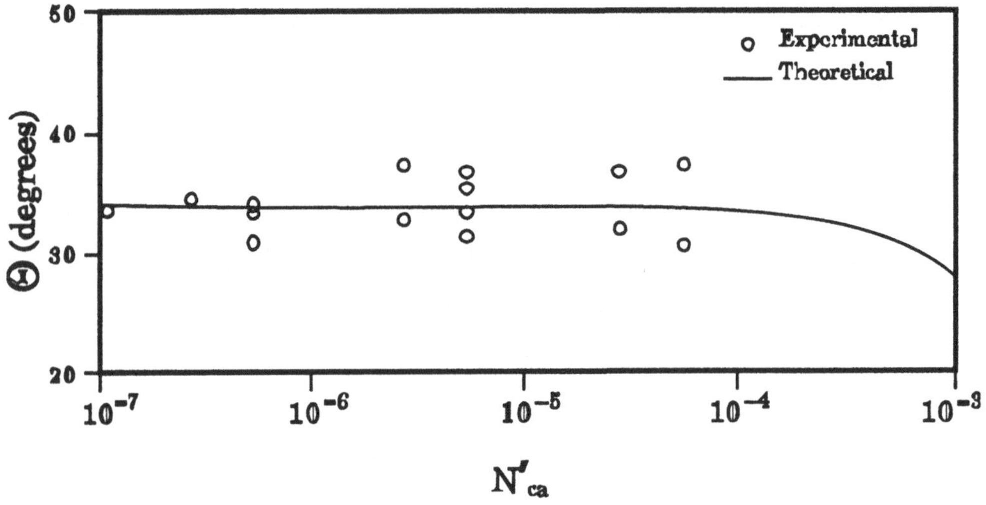

Figure 3.4-7: Comparisons of (3.4-67) with the experimental data of Johnson *et al.* [61]: water on monolayer of trimethyloctadecylammonium chloride on glass.

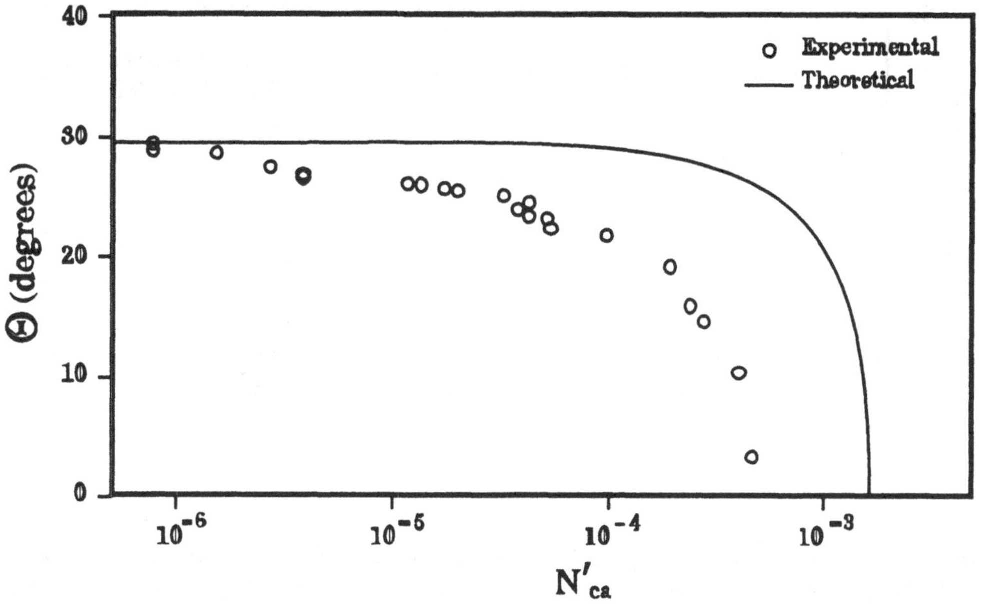

Figure 3.4-8: Comparisons of (3.4-67) with the experimental data of Hopf and Stechemesser [45]: distilled water on monolayer of octadecylamine deposited on quartz by the Langmuir-Blodgett technique.

also written as

$$u^{(cl)\star} = \frac{u^{(cl)}}{v_0} \tag{3.4-65}$$

With (3.4-15), (3.4-16), (3.4-47), (3.4-63) and (3.4-65), equation (3.4-64) becomes

$$N'_{ca} = \frac{3-n}{6n}\tan^3\Theta^{(stat)}u^{(cl)\star} \tag{3.4-66}$$

or

$$N'_{ca} = 0.028\frac{3-n}{6n}\tan\Theta^{(stat)}(\tan^2\Theta^{(stat)} - \tan^2\Theta) \tag{3.4-67}$$

Equation (3.4-67) gives the relationship between the static contact angle $\Theta^{(stat)}$, the receding contact angle Θ, and the capillary number N'_{ca}. When N'_{ca} increases, Θ decreases. In the limit as $\Theta^{(stat)} \to 0$, the apparent common line will not recede.

There have been very few experimental studies of the receding contact angle Θ as a function of the speed of displacement of the apparent common line. Figures 3.4-6

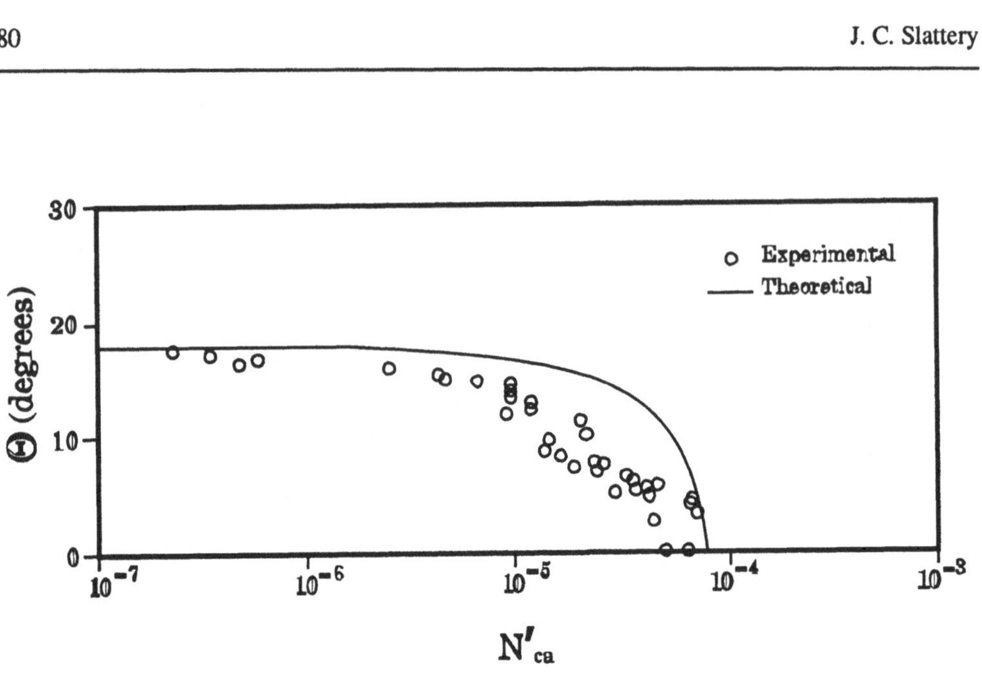

Figure 3.4-9: Comparisons of (3.4-67) with the experimental data of Hopf and Stechemesser [45]: surfactant solution (10^{-5} M dodecylamine + 10^{-2} M KCl at pH = 4.0 adjusted by adding KCl solution) on quartz.

and 3.4-7 compare our prediction (3.4-67) with the experimental data of Johnson *et al.* [61], who used the Wilhelmy plate. Since surfactant was not used in the their experiments, we assumed that the surface is mobile (assumption v) and n = 1. The static contact angles $\Theta^{(stat)}$, which were not measured, were estimated by extrapolating $u^{(cl)} \rightarrow 0$ in each case. Because of assumptions iii, iv, and xvi, we compare (3.4-67) only with their observations for smaller Θ. Unfortunately, their $u^{(cl)}$ were sufficiently small that Θ was nearly a constant.

Figures 3.4-8 through 3.4-10 compare (3.4-67) with the experimental data of Hopf and Stechemesser [45], who measured the receding contact angle as a function of the speed of movement of the common line after rupture of the thin liquid film between a quartz plate and a gas bubble generated at the tip of a capillary. The difference between our prediction and the experimental data is likely attributable to the role of inertial effects in their experiment.

Elliott and Riddiford [28], Black [6], Johnson *et al.* [61], and Petrov and Radoev [80] reported measurements of the receding contact angle as a function of the speed of displacement of the common line for large contact angles. Because of assumptions iii, iv, and xvi, (3.4-67) is not applicable to these cases.

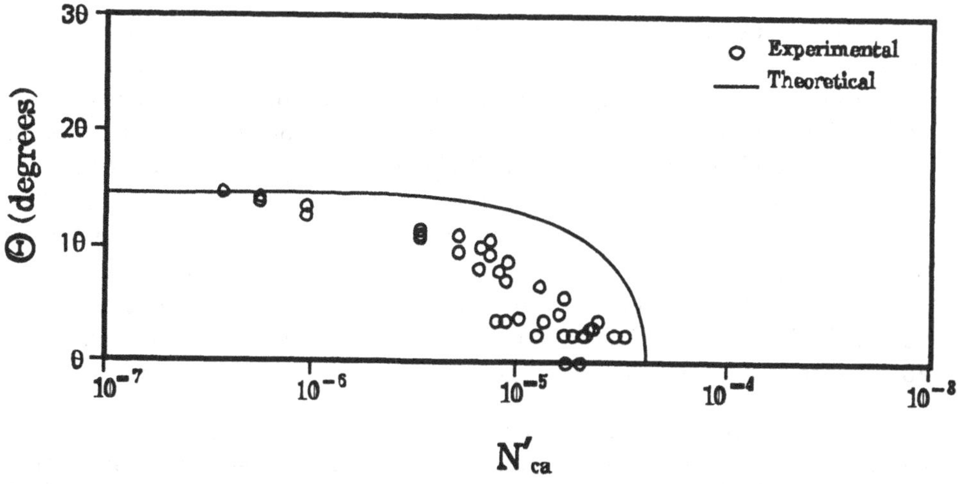

Figure 3.4-10: Comparisons of (3.4-67) with the experimental data of Hopf and Stechemesser [45]: surfactant solution (10^{-5} M dodecylamine + 10^{-2} M KCl at pH = 10.0 adjusted by adding KOH solution) on quartz.

4 Summary

There are several points that I have attempted to make in these lectures.

1. When a common line or three-phase line of contact moves over a rigid solid, an unbounded force must be generated at the common line, if the common no-slip boundary condition of fluid mechanics is valid.

2. Since an unbounded force is unrealistic, our description of the physics in this statement must be incorrect. There are various possibilities. Real solids are not rigid, although deformations attributable to a common line have been estimated to be of the same order of magnitude as the imperfections in real surfaces that we normally ignore. Perhaps the no-slip boundary condition fails within the immediate neighborhood of the common line, although in all other known situations it appears to be valid for relatively dense gases and liquids. In some situations there is undoubtedly mass transfer in the neighborhood of the apparent common line, and there is displacement over an existing film of fluid without a common line ever having been formed. In other situations, a sequence of stationary common lines may be formed in the thin film of fluid within the immediate neighborhood of the apparent common line, giving the appearance of a moving common line to an observer on the macroscale.

3. In Sec. 2, I have illustrated with at least one computation that contimuum mechanics can be applied successfully to very thin films, such as those within the immediate neighborhood of the common line.

4. The contact angle Θ_0 that might be measured at the true common line should in general be expected to be quite different from the contact angle Θ measured by experimentalists with perhaps a $10\times$ microscope.

5. The validity of Young's equation is questionable.

6. Sec. 3 illustrates that a common line which appears to be moving, when viewed on the macroscale, may in fact be a sequence of stationary common lines, when viewed on the microscale. The dynamic receding contact angle is predicted to be a function of the speed of displacement of the apparent common line in agreement with experimental observations.

This material has been taken from Slattery [94], which may be consulted for further details.

References

[1] Ablett, R., *Philos. Mag.* (6) **46**, 244 (1923).

[2] Adamson, A. W., *Physical Chemistry of Surfaces*, fourth edition, John Wiley, New York (1982).

[3] Allen, R. F., and P. R. Benson, *J. Colloid Interface Sci.* **50**, 250 (1975).

[4] Bascom, W. D., R. D. Cottington, and C. R. Singleterry, in *Contact Angles, Wettability and Adhesion, Advances in Chemistry Series* No. 43, p. 355, American Chemical Society, Washington, D.C. (1964).

[5] Bayramli, E., and S. G. Mason, *J. Colloid Interface Sci.* **66**, 200 (1978).

[6] Black, T. D., *VDI-Berichte Nr.* **182**, 117 (1972).

[7] Blake, T. D., D. H. Everett, and J. M. Haynes, in *Wetting*, S. C. I. Monograph No. 25, Society of Chemical Industry, London (1967).

[8] Blake, T. D., and J. M. Haynes, *J. Colloid Interface Sci.* **30**, 421 (1969).

[9] Buff, F. P., *Handbuch der Physik*, vol. 10, edited by S. Flügge, Springer- Verlag, Berlin (1960).

[10] Buff, F. P., and H. Saltsburg, *J. Chem. Phys.* **26**, 23 (1957).

[11] Burley, R., and B. S. Kennedy, *Br. Polym. J.* **8**, 140 (1976a).

[12] Burley, R., and B. S. Kennedy, *Chem. Eng. Sci.* **31**, 901 (1976b).

[13] Cain, J. B., D. W. Francis, R. D. Venter, and A. W. Neumann, *J. Colloid Interface Sci.* **94**, 123 (1983).

[14] Cassie A. B. D., *Fiscuss Faraday Soc.* **3**, 11 (1948).

[15] Chen, J. D., *J. Colloid Interface Sci.* **122**, 60 (1988).

[16] Chen, J. D., and N. Wada, *Phys. Rev. Letters* **62**, 3050 (1989).

[17] Cherry, B. W., and C. M. Holmes, *J. Colloid Interface Sci.* **29**, 174 (1969).

[18] Coney, T. A., and W. J. Masica, *NASA Tech. Note* **TN D-5115** (1969).

[19] Cox, B. G., *J. Fluid Mech.* **168**, 169 (1986).

[20] de Gennes, P. G., *Rev. Mod. Phys.* **57**, 827 (1985).

[21] Dettre, R. H., and R. E. Johnson Jr., in *Contact Angle, Wettability, and Adhesion, Advances in Chemistry Series No. 43*, p. 136, American Chemical Society, Washington, D.C. (1964).

[22] Dussan V., E. B., *J. Fluid Mech.* **77**, 665 (1976).

[23] Dussan V., E. B., *Ann Rev. Fluid Mech.* **11**, 371 (1979).

[24] Dussan V., E. B., and S. H. Davis, *J. Fluid Mech.* **65**, 71 (1974).

[25] Eick, J. D., R. J. Good, and A. W. Neumann, *J. Colloid Interface Sci.* **53**, 235 (1975).

[26] Elliott, G. E. P., and A. C. Riddiford, *Nature London* **195**, 795 (1962).

[27] Elliott, G. E. P., and A. C. Riddiford, *Recent Progr. Surface Sci.* **2**, 111 (1965).

[28] Elliott, G. E. P., and A. C. Riddiford, *J. Colloid Interface Sci.* **23**, 389 (1967).

[29] Ellison, A. H., and S. B. Tejada, *NASA Contract Rep.* **CR 72441** (1968).

[30] Fermigier, M. and P. Jenffer, *Annales de Physique (Colloque n° 2, supplement au n° 3)* **13**, 37 (1988).

[31] Gaydos, J., and A. W. Neumann, *J. Colloid Interface Sci.* **120**, 76 (1987).

[32] Gibbs, J. W., *The Collected Works*, vol. 1, Yale University Press, New Haven, Conn. (1948).

[33] Giordano, R. M., personal communication, April 25, 1979.

[34] Goldstein, S., *Modern Developments in Fluid Dynamics*, Oxford University Press, London (1938).

[35] Greenspan, H. P., *J. Fluid Mech.* **84**, 125 (1978).

[36] Hahn, P. S., J. D. Chen, and J. C. Slattery, *AIChE J.* **31**, 2126 (1985).

[37] Hansen, R. S., and M. Miotto, *J. Am. Chem. Soc.* **79**, 1765 (1957).

[38] Hansen, R. J., and T. Y. Toong, *J. Colloid Interface Sci.* **36**, 410 (1971).

[39] Hardy, W. B., *Phil. Mag. (6)* **38**, 49 (1919).

[40] Heslot, F., N. Fraysse, and A. M. Cazabat, *Nature London* **338**, 640 (1989).

[41] Hocking, L. M., *J. Fluid Mech.* **76**, 801 (1976).

[42] Hocking, L. M., *J. Fluid Mech.* **79**, 209 (1977).

[43] Hocking, L. M., and A. D. Rivers, *J. Fluid Mech.* **121**, 425 (1982).

[44] Hoffman, R. L., *J. Colloid Interface Sci.* **50**, 228 (1975).

[45] Hopf, W., and H. Stechemesser, *Colloids Surf.* **33**, 25 (1988).

[46] Hough, D. B., and L. R. White, *Adv. Colloid Interface Sci.* **14**, 3 (1980).

[47] Huh, C., and S. G. Mason, *J. Colloid Interface Sci.* **60**, 11 (1977a).

[48] Huh, C., and S. G. Mason, *J. Fluid Mech.* **81**, 401 (1977b).

[49] Huh, C., and L. E. Scriven, *J. Colloid Interface Sci.* **35**, 85 (1971).

[50] Inverarity, G., Ph.D. dissertation, Victoria University of Manchester (1969).

[51] Israelachvili, J. N., *J. Chem. Soc. Faraday Trans. II* **69**, 1729 (1973).

[52] Israelachvili, J. N., *Intermolecular and Surface Forces*, p. 157, Academic Press (1985).

[53] Israelachvili, J. N., and M. L. Gee, *Langmuir* **5**, 288 (1989).

[54] Israelachvili, J. N., and P. M. McGuiggan, *Science* **241**, 795 (1988).

[55] Jasper, J. J., and E. V. Kring, *J. Phys. Chem.* **59**, 1019 (1955).

[56] Jiang, T. S., S. G. Oh, and J. C. Slattery, *J. Colloid Interface Sci.* **69**, 74 (1979).

[57] Joanny, J. F., and P. G. de Gennes, *J. Chem. Phys.* **81**, 552 (1984).

[58] Johnson, R. E. Jr., and R. H. Dettre, in *Contact Angle, Wettability, and Adhesion, Advances in Chemistry Series No. 43*, p. 136, American Chemical Society, Washington, D.C. (1964a).

[59] Johnson, R. E. Jr., and R. H. Dettre, *J. Phys. Chem.* **68**, 1744 (1964b).

[60] Johnson, R. E. Jr., and R. H. Dettre, *Surface and Colloid Science*, **2**, 85, edited by E. Matijevic, Wiley-Interscience, New York (1969).

[61] Johnson, R. E. Jr., R. H. Dettre, and D. A. Brandreth, *J. Colloid Interface Sci.* **62**, 205 (1977).

[62] Kafka, F. Y., and E. B. Dussan V., *J. Fluid Mech.* **95**, 539 (1979).

[63] Kennedy, B. S., and R. Burley, *J. Colloid Interface Sci.* **62**, 48 (1977).

[64] Kralchevsky, P. A., A. D. Nikolov, and I. B. Ivanov, *J. Colloid Interface Sci.* **112**, 132 (1986).

[65] Legait, B., and P. Sourieau, *J. Colloid Interface Sci.* **107**, 14 (1985).

[66] Lester, G. R., *J. Colloid Sci.* **16**, 315 (1961).

[67] Li, D., and J. C. Slattery, *J. Colloid Interface Sci.* (1991).

[68] Lin, C. Y., and J. C. Slattery, *AIChE J.* **28**, 147 (1982a).

[69] Lin, C. Y., and J. C. Slattery, *AIChE J.* **28**, 786 (1982b).

[70] Ludviksson, V., and E. N. Lightfoot, *AIChE J.* **17**, 1166 (1971).

[71] Miller, C. A., and P. Neogi, *Interfacial Phenomena*, Marcel Dekker, New York (1985).

[72] Mori, Y. H., T. G. M. van de Ven, and S. G. Mason, *Colloids Surfaces* **4**, 1 (1982).

[73] Morra, M., E. Occhiello, and F. Garbassi, *Langmuir* **5**, 872 (1989).

[74] Morrow, N. R., *J. Can. Pet. Technol.* **14**, 42 (1975).

[75] Myers, G. E., *Analytical Methods in Conduction Heat Transfer*, p. 274, McGraw-Hill, New York (1971).

[76] Neumann, A. W., and R. J. Good, *J. Colloid Interface Sci.* **38**, 341 (1972).

[77] Ngan, C. G., and E. B. Dussan V., *J. Fluid Mech.* **118**, 27 (1982).

[78] Oliver, J. F., C. Huh, and S. G. Mason, *J. Adhes.* **8**, 223 (1977).

[79] Padday, J. F., and N. D. Uffindell, *J. Phys. Chem.* **72**, 1407 (1968).

[80] Petrov, J. G., and B. P. Radoev, *Colloid Polym. Sci.* **259**, 753 (1981).

[81] Platikanov, D., M. Nedyalkov, and V. Nasteva, *J. Colloid Interface Sci.* **75**, 620 (1980).

[82] Poynting, J. H., and J. J. Thomson, *A Text-Book of Physics - Properties of Matter*, p. 142, Charles Griffin, London (1902).

[83] Radigan, W., H. Ghiradella, H. L. Frisch, H. Schonhorn, and T. K. Kwei, *J. Colloid Interface Sci.* **49**, 241 (1974).

[84] Richards, T. W., and E. K. Carver, *J. Am. Chem. Soc.* **43**, 827 (1921).

[85] Richardson, S., *J. Fluid Mech.* **59**, 707 (1973).

[86] Rose, W., and R. W. Heins, *J. Colloid Sci.* **17**, 39 (1962).

[87] Ruckenstein, E., and C. S. Dunn, *J. Colloid Interface Sci.* **59**, 135 (1977).

[88] Schonhorn, H., H. L. Frisch, and T. K. Kwei, *J. Appl. Phys.* **37**, 4967 (1966).

[89] Schwartz, L. W., and S. Garoff, *Langmuir* **1**, 219 (1985).

[90] Schwartz, A. M., and S. B. Tejada, *NASA Contract Rep.* **CR 72728** (1970).

[91] Schwartz, A. M., and S. B. Tejada, *J. Colloid Interface Sci.* **38**, 359 (1972).

[92] Sheludko, A., *Adv. Colloid Interface Sci.* **1**, 391 (1967).

[93] Slattery, J. C., *Momentum, Energy, and Mass Transfer in Continua*, McGraw-Hill, New York (1972); second edition, Robert E. Krieger, Malabar, FL 32950 (1981).

[94] Slattery, J. C., *Interfacial Transport Phenomena*, Springer-Verlag, New York (1990).

[95] Teletzke, G. F., H. T. Davis, and L. E. Scriven, *Chem. Eng. Commun.* **55**, 41 (1987).

[96] Wayner, P. C., *J. Colloid Interface Sci.* **77**, 495 (1980).

[97] Wayner, P. C., *J. Colloid Interface Sci.* **88**, 294 (1982).

[98] Wickham, G. R., and S. D. R. Wilson *J. Colloid Interface Sci.* **51**, 189 (1975).

[99] Williams, R., *Nature London* **266**, 153 (1977).

[100] Williams, R., *personal communication* (1988).

[101] Wilson, S. D. R., *J. Colloid Interface Sci.* **51**, 532 (1975).

[102] Yarnold, G. D., *Proc. Phys. Soc. London* **50**, 540 (1938).

[103] Yarnold, G. D., and B. J. Mason, *Proc. Phys. Soc. London* **B62**, 125 (1949).

[104] Young, T., *Philos. Trans. R. Soc. London* (4 to.) **95**, 65 (1805).

Printed in the United States
By Bookmasters